Risk Analysis and Reduction in the Chemical Process Industry

Risk Analysis and Reduction in the Chemical Process Industry

J. M. Santamaría Ramiro

Department of Chemical Engineering
University of Zaragoza
Zaragoza, Spain

and

P. A. Braña Aísa

Amylum Ibérica, S.A.
Zaragoza, Spain

Translated by J. Hutchinson

BLACKIE ACADEMIC & PROFESSIONAL
An Imprint of Chapman & Hall
London - Weinheim - New York - Tokyo - Melbourne - Madras

363.11
523 ra

Published by Blackie Academic & Professional,
an imprint of Thomson Science, 2–6 Boundary Row, London SE1 8HN, UK

Thomson Science, 2–6 Boundary Row, London SE1 8HN, UK

Thomson Science, 115 Fifth Avenue, New York, NY 10003, USA

Thomson Science, Suite 750, 400 Market Street, Philadelphia, PA 19106, USA

Thomson Science, Pappelallee 3, 69469 Weinheim, Germany

English language edition 1998

© 1998 Thomson Science

Original Spanish language edition – Análisis y Reducción de Riesgos en la
Industria Química – © 1994 J. M. Santamaría Ramiro and P. A. Braña Aísa.

Published by Fundación MAPFRE.

Thomson Science is a division of International Thomson Publishing **I(T)P**

Typeset in 10/12pt Times by AFS Image Setters Ltd., Glasgow

Printed in Great Britain by St Edmundsbury Press Ltd, Bury St Edmunds, Suffolk

ISBN 7514 0374 1

This book has been produced with the intention of contributing to the improvement
of safety in the chemical industry. With this aim in mind, considerable effort has
been made to check the exactness of the contained statements and of the
mathematical expressions used. However, neither the authors nor the publishers nor
its employees guarantee explicitly or implicitly the absence of errors in the text, nor
do they accept responsibility for the consequences of its improper use.

A catalogue record for this book is available from the British Library

Library of Congress Catalog Card Number: 97–61082

⊗ Printed on acid-free text paper, manufactured in accordance with ANSI/NISO
Z39.48-1992 (Permanence of Paper)

Contents

Foreword

As President of the Spanish Chemical Industry Federation (FEIQUE) and on behalf of the chemical sector I wish to express my satisfaction at the publication of a book as interesting and as useful as the one written by Jesus M. Santamaría Ramiro and Pedro A. Braña Aísa.

The chemical industry, which has always occupied a prominent position in defence of the safety of its workers, its products and its installations and the environment, takes great pleasure in welcoming a book which will undoubtedly contribute to the growth of a safety culture from the university lecture rooms in which our young chemists and engineers are shaped.

The book arrives at a very opportune moment, since, whilst any effort in this area has always been justified it is even more so now, when risk analysis is receiving a new impetus as a consequence of the growing number of EC regulations which aim to increase the levels of safety in industrial activities.

It is also satisfying to observe that many of the techniques and procedures gathered together and developed in this book were developed in the chemical industry, and are due to the efforts of its workers. Nowadays, these risk analysis techniques are being applied to other industrial sectors, effectively helping in the protection of human life and of property.

The book covers a wide range of methodologies applicable to very diverse situations. It takes a very practical approach and each chapter is completed with an extensive bibliography, making it even more useful. It will therefore be an appropriate tool for work, as much for those university students who wish to study in depth the techniques involved in risk analysis in the chemical industry, as for the professionals of our industry responsible for the minimization of these risks.

Lastly, I would like to emphasize the collaboration between a university professor and a process engineer from our industry, which has produced such an important book. My most effusive congratulations to the authors.

Juan José Nava Cano
President of FEIQUE

Preface

During the past 30 years, environmental topics have moved up on the list of society's concerns. At the same time the term 'environmental impact' has extended its traditional meaning, principally referring to the contamination/pollution of air, water and soil, to include other aspects. Thus, nowadays, in the forums of environmental debate, subjects ranging from the adequate use of raw materials to landscape alterations caused by a new road are discussed.

Within this extensive environmental concept, a special sensitivity has developed, in view of the possibility of industrial accidents which, because of their magnitude, are capable of causing significant damage to people, property or the environment. This concern, which in the past was principally associated with the nuclear industry, today also includes the chemical industry, even more so after the Flixborough, Seveso and Bhopal accidents.

The quality of life which society now perceives is not only identified with the amount of products and services available, but also with the safety of the industries which produce them. Companies have responded, in general, to the demands of society, which has given rise to the appearance of environmental and safety management programmes which are gradually extending to the majority of the chemical industry. Without a doubt, legislative pressure has contributed to this tendency, but it is only fair to acknowledge there has also been an important change in mentality on the part of industry.

Any programme of action designed to improve the safety level of a particular process should start by carrying out a diagnosis from the beginning, identifying the most problematic areas and evaluating the available alternatives. It is extremely difficult to carry out this task using experience and intuition alone without the help of a tool capable of examining the complex safety problems encountered in a chemical plant. Risk analysis is the discipline which responds to this need, combining a number of techniques to produce a quantitative estimation of the risks involved in a given process.

The reason for the text presented here is the desire to contribute new material, dealing with risk analysis and reduction in a unified way, i.e. trying to integrate the different techniques involved using a global approach. Although this book was originally conceived to serve as a text book for chemical engineering students, we believe that it may also be useful to industrial engineering and chemistry students, as well as to the professionals working in the chemical industry.

Throughout this book we have tried to maintain clarity in its presentation, and, as far as possible, simplicity in dealing with the complex subjects involved. We have included a number of worked examples with the aim of clarifying the concepts presented. In the same way, at the end of each chapter there is a list of questions and problems which are designed to stimulate discussion on the topics presented.

We have also endeavoured to quote at least part of the relevant bibliography so that readers know where to look for in-depth study of the different topics which cannot be dealt with extensively in a text such as this. Lastly, an appendix has been included with details of a few selected industrial accidents, which will give the reader direct reference to real cases.

As expected, given the topics in this book, extensive use has been made of empirical equations taken from the bibliography. Nobody is exempt from using common sense when using expressions of this type. Special care should be taken not to carry out inadequate extrapolations by using the equations outside the conditions for which their validity has been checked. Whenever possible the original reference should be consulted, and in any case, the results obtained should be critically examined to assess their applicability. In situations in which more than one equation can be used, if none can be discarded on an 'a priori' basis, the one which produces a more conservative result from the point of view of safety should be selected. Finally, after having completed a risk analysis course, following this text or any other, one cannot be considered an expert in the subject, but rather as a person more conscious of the problems of safety and of the areas where a significant lack of knowledge still exists.

In expressing appreciation it is difficult not to make omissions, for which we would like to apologise in advance. However we should at least mention our colleagues in the Chemical and Environmental Engineering Department at the University of Zaragoza, whose help and stimulation have always been at hand; the companies Dow Ibérica, Walthon Weir Pacific, Elfab-Hughes and Campo Ebro Industrial; the newspaper El Pais; the National Fire Protection Association; and Editorial Mapfre, who provided or helped to locate illustrations for the book. Special thanks are also due to Mr Francisco Ponz from Walthon Weir Pacific and Mr Jordi Bessa from Dow Ibérica, who made useful comments on Chapters 7 and 9. It is also appropriate to acknowledge here the pre-existing literature, and especially the CCPS's *Guidelines for Chemical Process Quantitative Risk Analysis, Loss Prevention in the Process Industries* (F. P. Lees), and *Chemical Process Safety, Fundamentals with Applications* (D. A. Crowl and J. F. Louvar) on which much of this book is based. Last but not least, our gratitude to our wives, Eva and Pili, for their unwavering support throughout the 18 months it has taken to write this book.

Zaragoza
The Authors

1 Introduction and general concepts

In the summer of 1986 I had an opportunity to spend a few months at BASF Corporation working on a computer simulation project. One day I was visited by Joe Louvar, Director of Chemical Engineering, who asked me if I was interested in working on a few safety related projects. In total ignorance I replied, 'You mean hard hats and safety shoes?'. Joe went on to explain some of the more fundamental aspects of safety, including reactor dynamics, two phase flow during reactor venting, gas dispersion models....

Prof. Daniel A. Crowl (Wayne State University)
Chemical Engineering Education 22, 74 (1988)

1.1   Introduction

During the last 50 years the chemical industry has experienced change on a very large scale. As technological advances have given rise to the appearance of new materials, processes and even new fields of industrial activity, we have witnessed an almost exponential increase in the number and application of chemical products available on the market. Every year hundreds of new products are added to the tens of thousands already commercially available. In spite of the great variety of chemical products on offer, 90% of consumption is concentrated on only 5% of the available products.

Large chemical plants are commonplace nowadays, the capacity of some of them having increased by an order of magnitude in the past 20 years. As may be expected, this growth, as much in the number of plants as in their capacities, has led to an increase in the number of people (both within the plants and the general public), who could be exposed to the consequences of an industrial accident. This, in turn, has raised awareness in industrial safety which now extends to the general public. Administrations at various levels have responded to this growing social sensitivity with a substantial effort to regulate industrial activities in general, and in particular those considered as more hazardous.

When using statistics to express accident probability it is appropriate to begin by stating that the chemical industry possesses a safety record considerably better than the average for the industry as a whole. A commonly used statistic is the FAR (Fatal Accident Rate), which establishes the number of fatal accidents in a particular industry after 10^8 hours of activity (a period which, at the time when the index was defined, corresponded approximately to the working life of a group of 1000 workers). For the chemical industry the FAR index falls between 4 and 5.

By comparison the FAR for an equivalent group of workers in agriculture, mining and construction is 10, 12 and 64 respectively. To place the number of accidents in the chemical industry into context it is necessary to bear in mind that in general about 80% of accidents can be assigned to the non-specific group, such as falls, collisions, contusions, etc., so that only about 20% of fatalities are caused by specific risks associated with the chemical industry. Another way of visualizing the level of accident probability to which chemical industry workers are exposed is the following [1]: a worker who spends all his working life in a chemical plant of 1000 employees will witness about four fatal accidents in that plant. In comparison, about 20 plant workers will die in other types of accidents (primarily on the roads or in their homes), and about 370 will die as result of various illnesses, including 40 as a direct result of smoking tobacco.

The majority of the accidents mentioned above are individual events which involve one or more persons, nearly always within the plant. In spite of the fact, as already shown, that the chemical industry has an accident record better than that of other industrial activities, general public perception is that it is a high-risk industry. Thus a survey carried out in the late 1980s [2] showed that the general view of chemical products was that nearly all of them could be considered as very dangerous, and two out of every three North Americans expected a disastrous accident to happen in the chemical industry, causing thousands of deaths, within the next 50 years.

Without doubt, the principal causes of this belief are the widespread effects and the social impact of major accidents, some of which have reached far beyond the physical limits of the industries involved. Analysing newspaper cuttings of the past decade, it is not difficult to construct a list of accidents that have caused major losses, both human and material. Table 1.1 shows some of these industrial accidents of major impact, related both to the manufacture and transportation of chemical products. To the list of accidents which caused headlines in the press would have to be added another list, longer and of considerably less newspaper coverage, containing the individual, smaller-scale accidents already mentioned, professional illnesses, and harm of varied nature suffered by a number of people working in industrial environments or their areas of influence.

It is difficult to quantify the cost of chemical industry accidents, even in purely economic terms. During 1984, in only five chemical industry accidents, direct losses were estimated at $268 million. Every year hundreds of minor accidents happen, usually without coming to the notice of the general public. To the direct material costs of accidents it is necessary to add costs due to the consequent production disruption and loss of raw materials, costs due to litigation and indemnification for harm caused to people or property and the cost of insurance payments. A very considerable additional cost is that of loss of image and the negative publicity which the company involved suffers, although in practice this is usually only associated with very large accidents.

In view of all this it is not surprising to observe an increased effort by industry in general, and by the chemical industry in particular, dedicated to loss prevention.

Table 1.1   Some notable industrial accidents which have occurred since 1973

Accident	Consequences
Flixborough (UK), 1st June 1974	
In a Nypro plant the rupture of a pipe caused the discharge of between 40 and 80 tons of hot liquid cyclohexane. The resulting cloud caused a powerful and destructive explosion.	28 deaths and hundreds of injured. Complete destruction of the plant.
Seveso (Italy), 9th July 1976	
At the Icmesa (Hoffmann La Roche) plant, a runaway reaction caused the release of between 0.5 and 2 kg of chemicals into the atmosphere. Among these was 0.5 to 2 kg of dioxin (TCCD). The lethal dose of TCDD for a person of average sensitivity is < 0.1 mg.	It was necessary to evacuate more than 1000 people. There were no deaths as a direct result of the accident, but the dioxin released caused skin damage to many people (chloracne), provoked miscarriages, and resulted in ground contamination.
Camping de los Alfaques, San Carlos de la Rápita (Spain), 11th July 1978	
A 30 tonne lorry, overloaded with about 45 m³ of propylene caused a BLEVE explosion when it crashed into the wall of a campsite.	215 deaths.
Cubatao (Brazil), 25th February 1974	
An oil pipeline was damaged. The gasoline which escaped evaporated and ignited causing a fireball.	At least 500 deaths.
Mexico D.F. (Mexico), 19th November 1984	
Various LPG containers exploded in San Juan de Ixhuatepec	452 deaths and more than 4200 injured. The number of persons unaccounted for could be as many as 1000.
Bhopal (India), 17th December 1984	
A leak of toxic gas (methyl isocyanate) ocurred at a Union Carbide plant manufacturing insecticide. The leak spread over an area of approximately 40 km².	2500 deaths due to poisoning and approximately the same number in a critical condition. About 150 000 people required medical treatment. There were long-term effects such as blindness, mental illness, hepatic and kidney injuries, in addition to embryonic malformations.
Guadalajara (Mexico), 23rd April 1992	
A chain of explosions along a 13 km urban sewage network occurred due to leaks of combustible liquid from a pipeline owned by Pemex.	Official information gave an estimate of 200 deaths and 1500 injured, 1200 houses and 450 commercial buildings destroyed. Estimates of economic damage are put at $7000 million.

It is estimated that the North American industry as a whole invested $7700 million in 1985 in measures to increase safety levels of installations and for protection of workers' health. A very significant part of this total investment relates to the chemical industry, and because of it the probability of fatal accidents for a chemical industry worker was only one-quarter of that for the average North American industrial worker in 1985 [3].

The preoccupation with safety in the chemical industry will continue to increase in the near future. Among numerous public manifestations of this tendency in recent years is that of the Chemical Committee of the European Community [4], whose report, referring to the 1990s, states: 'Industrial safety and hygiene will be even more important than now. Attention will have to be paid not only to improving safety within the chemical industry as a whole... but also to the risks that the chemical industry poses for the surrounding population and for the environment in the long run. The European Chemical Industry will have to develop an adequate policy for risk reduction and consequently will have to develop new products, technologies and processes'.

1.2   Risks and hazards

The term **risk** is usually used to indicate the possibility of suffering loss [5], or as a measure of economic loss or damage to people, expressed as the product of the incident likelihood and the magnitude of its consequences [6]. We use the word **hazard** to describe a physical or chemical condition that has the potential for causing damage to people, property or the environment [6].

Once the distinction has been made, it is fitting to point out that in everyday use of the English language the two terms are often interchanged, and their use does not always conform to previous definitions. Thus, a frequent quote is 'a high level of danger exists' when what is really meant is 'the level of risk is high'. In this text we have tried to respect the meaning previously given to both of these terms, except where common usage tended to the contrary.

1.3   Accidents and risk analysis

All of the accidents described in Table 1.1 fall into the category of major accidents. By **accident** we mean any incident which implies an intolerable deviation from the conditions in which a system is designed to operate. More specifically this book is concerned with those accidents whose effects could have adverse consequences on lives, health, property or the environment. In Spain, the Royal Decree 886/1988 on the prevention of major accidents in certain industrial activities (the Spanish version of the so-called **Seveso Directive** of the EC) defines a **major accident** as 'any event, such as an emission, leak, spill, fire or explosion which is a consequence of an uncontrolled development of an industrial activity leading to a situation of serious risk, catastrophe or public calamity, immediate or delayed, for people, the environment or property, whether inside or outside of an installation, and in which at least one of the dangerous substances listed in the Annex of this Royal Decree are involved'.

In spite of the above-mentioned increase in public awareness of potential industrial hazards, the vast majority of people know that any human activity, beneficial as it may be, carries a certain risk. In particular, regarding the chemical industry, it is clear that no matter how many safeguards are introduced, the activity implies risks which can only be completely eliminated by eliminating the industry. Obviously, the chemical industry is necessary in the real world; therefore, the question becomes simply: 'What is the acceptable level of risk in a given installation or process?' or, perhaps more accurately: 'To what degree may a risk be accepted when measured against the benefits derived by assuming it?' This decision, always difficult, is further complicated by a series of factors which frequently occur, e.g. the fact that the risks are not known with sufficient precision, that the people who could be affected (both inside and outside the plant) may not have assumed the risk voluntarily or that they do not have sufficient information on these risks, that the people at risk are not the main beneficiaries of the activity, etc. It also often happens that the alternatives to a given situation are uncertain or impractical, which makes the adoption of solutions even more difficult.

The process of decision-making on the level of acceptable risk is complex because the objectives are multiple and, on occasion, contradictory. It is necessary to bear in mind humanitarian, economic, legal and public relations considerations. Thus the risk of a single catastrophe would be considered less acceptable socially than a group of lesser risks even if the total level of risk for people and property were identical.

It is important to distinguish between the risk which exists objectively, and can be quantified, and the risk perceived by possible passive subjects. Thus, it is well-known that familiarity with a hazardous activity reduces the perceived level of risk. This benefits traditional industries (agriculture, construction) as opposed to newer industries (chemical, nuclear), in which social acceptance is lower in spite of the fact that the probability of accidents is much greater in the first of these groups. Obviously, there is an additional factor present in this perception. However high the annual number of victims in agricultural activities may be, we know that these activities rarely affect people other than those directly involved. On the other hand, it is evident that in industries such as the chemical or nuclear, the potential for damage extends well beyond the limits of the plant considered.

Clearly, a certain level of voluntary risk is accepted by the majority of citizens as part of daily life, even activities whose risks can be considerable, such as smoking or climbing mountains. On the other hand, tolerance towards involuntary risks is minimal, even if these risks are smaller than those assumed voluntarily. While risks derived from activities upon which control can be exercised (e.g. driving a car) are readily accepted, others where the control that passive subjects may exercise is minimum or non-existent tend to be rejected (e.g. close proximity to a nuclear plant, aeroplane accidents, industrial pollution...). Also, natural risks such as earthquakes, lightning or floods seem to be more easily accepted than those derived from human activities, by virtue of the 'unavoidability' of the former.

Finally there is the question of the benefits derived from the risks. There is no need to be reminded that circulation of traffic on the roads means thousands of victims each year in road accidents. The benefits of road transport are so obvious, however, that few people oppose the construction of roads, or the growth in the total number of vehicles, by enacting laws that ban the manufacture of new vehicles. The benefits of the chemical industry are not so obvious to the public. The primary market for the chemical industry is, in general, other industries rather than individual consumers. The average citizen moves in a world of registered trademarks and consumer products and (apart from the exceptions represented by specific products such as petrol and plastics) they do not identify the products used in everyday life with the chemical industry. This makes perception of the benefits of this industry more difficult.

1.3.1   Quantifying the level of risk

If we accept the premise that it is impossible to completely eliminate risk, the basic question is: 'How much safety does 'sufficiently safe' imply?' In many cases the insufficient level of scientific understanding of the processes involved and the lack of available information make it impossible to answer this question with certainty. Despite this, administrations and regulatory bodies have to publish rules for the protection of the public, decide whether to authorize the use of a certain chemical product or the construction of a new plant, limit environmental impact, regulate waste disposal, etc. Uncertainty sometimes gives rise to legislation aimed at protecting against extremely unlikely circumstances (a very low probability, worst-case scenario). This may result in a disproportionate assignment of resources, a decrease in technical innovation and excessive costs. At the same time it is possible that situations which should be legislated for are ignored [2].

To be able to say whether or not a risk is acceptable, an estimation of its magnitude is required, which implies previous analysis. Risk analysis very often means developing a quantitative estimate of the hazards that a given activity may represent to people, property and the environment. The analysis is aimed at providing an estimate of the magnitude of the potential damage and the likelihood of its occurrence. Risk analysis as a discipline combines engineering evaluation of the process with simulation models to assess consequences, and mathematical techniques which allow the estimation of the probable frequency of the accident. The results of risk analysis are used to make decisions (risk management), allowing the comparison of the estimated risk levels with those set as objectives in a particular activity, and helping to set priorities for risk reduction strategies. The principal elements of a risk management programme are set out in Table 1.2.

Risk analysis allows, within the levels of uncertainty associated with the method of analysis used and the available information, quantification of the accident potential existing in a particular installation or process, and, when this is considered to be too high, enables comparison of alternative risk reduction

Table 1.2   Elements of a risk management program

• Hazard identification
• Consequence analysis
• Risk assessment
• Employee training
• Control of plant modifications
• Operating procedures
• Maintenance procedures
• Safety audits
• Accident/incident investigation
• Records and files
• Emergency planning

procedures. Each of the risk reduction alternatives will have a different cost, which should be borne in mind when making the final decision. At this point an example may help illustrate some of these concepts. Let us assume that the result of our risk analysis in a certain plant indicates the possibility of an explosion with an estimated frequency of once every 500 years, with a 30% probability of someone dying in the accident. The analysis does not tell us if this risk is acceptable or not. What it does give us is a figure which can be compared to others, and which we can modify by adding supplementary safety measures (for example, an option A, which reduces the frequency to once every 800 years at a certain cost, and an option B, at a higher cost, which would lower the expected frequency to once every 1000 years). As some level of risk must be accepted, the risk analysis allows us to assign priorities to investments in safety, distributing the available funds in the most effective way.

1.4   Returns on risk analysis

Economic resources in any activity are limited and, as has been pointed out, risk analysis is a valuable tool when making decisions with regards to the destiny of funds available for investment in safety. Given the economic cost that accidents have for industry, it is clear that, even if there were no risk to the life or health of people, a certain level of safety investment would be justified simply by applying the profitability criteria. This corresponds to area 1 in Figure 1.1 where, with a small investment, significant benefits are obtained, and the returns are so high that they are capable of competing advantageously with other possible investments. In area 2, although investment is still advantageous, it probably wouldn't be justifiable as such, solely on economic grounds, when compared to others with a larger rentability. Despite this, the majority of industrial companies also invest in area 2. Here other very important reasons (ethical, image, public relations, etc.), which are difficult or impossible to quantify economically, come into play. These same reasons point to the desirability of maintaining investment in safety, even

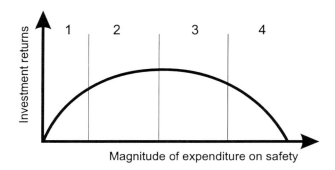

Figure 1.1   Qualitative diagram showing the economic return on investments in safety (adapted from [1]).

though it ceases to be profitable, which takes us into area 3. Increasing further the level of investment and entering area 4 would imply being at a large competitive disadvantage with other industries in the sector. If the risk levels of a particular company are such that the required safety investment is of this magnitude the usual option is to cease the activity.

It is important to point out that, in any decision made on the characteristics of a particular installation, a risk analysis is being carried out, whether consciously or unconsciously. Kletz [1] expresses it in the following way: 'In fixing the height of handrails round a place of work, the law does not ask us to compare the cost of fitting them with the value of the lives of the people who would otherwise fall off. It fixes a height for the handrails (36 inches to 45 inches). A sort of intuitive hazard analysis shows that with handrails of this height the chance of falling over them, though not zero, is so small that we are justified in ignoring it.' Logically, in decisions which are not so obvious, a formal analysis is required, which takes into account the complexity of the different options being considered.

As we have seen, risk analysis applied to an existing plant can help to identify and separate the different areas which appear in Figure 1.1 and to make the corresponding decisions. However, its major potential is for application to plants which have not yet been built. The opportunity for the implementation of intrinsic safety (which will be dealt with in later chapters) in the process is maximum, and with a minimum cost, while the process is still being defined. The sooner possible hazards of an installation are identified, the more opportunities will arise to change the reaction route, the process conditions, or the type of equipment to be used so that the possibility of an accident is reduced, and its possible effects are limited. What cannot be implemented as intrinsic process safety will have to be added as extrinsic safety: controls, alarms, redundant equipment, safety procedures, etc., with the consequent increase in investment. As we advance from process definition and development to the design and construction stage there are still possibilities for extrinsic safety, but the opportunities for selecting conditions which increase intrinsic safety diminish rapidly. It is therefore essential to carry out a basic risk analysis at the very early stages of the process design, which will gradually become more sophisticated as the process is defined with more precision.

1.5   Stages in risk analysis

An analysis of risks orientated to the prevention of accidents generally implies the following stages:

1. Identification of undesirable events which could lead to the materialization of a hazard.

2. Analysis of the mechanisms leading to these undesirable events.

3. Estimation of the undesired consequences and of the frequency with which they could happen.

The different stages in risk analysis correspond to the general questions indicated in Figure 1.2. The first question: 'What could go wrong?' refers to all the circumstances which could reasonably give rise to adverse effects. The nature of the question is purely qualitative, and gives origin to the block 'hazard

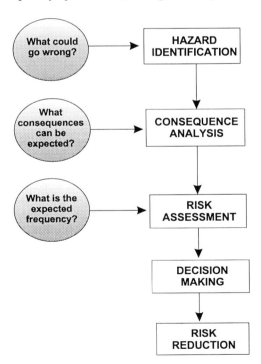

Figure 1.2   Stages in risk analysis.

identification'. In this study phase an exhaustive list, within the limits of the analysis, should be obtained detailing all the deviations which: a) could produce a significant adverse effect, and b) have a reasonable probability of happening. With regard to this section, all deviations that could occur should be retained as a first step, even if their probability is apparently small (though not negligible). For this, engineering judgement should be used which, together with the experience accumulated on the process being studied and in similar ones, should allow the rejection, without a formal mathematical treatment, of the highly improbable deviations.

The identification of circumstances which could give rise to dangerous situations is crucial: an unidentified hazard is one which will not be considered in the following analysis. To avoid omissions in this section, the experience of personnel directly involved in the process is obviously counted on, but the industry has also developed a series of powerful tools: design standards and codes of good practice, test lists, specific data on equipment and component failure, historical record analysis, risk indices, 'what-if' analysis, hazard and operability (HAZOP) analysis, failure modes and effects analysis (FMEA), etc. Some of these methods will be described in detail in the next chapter.

Once the circumstances which could reasonably give rise to adverse effects of a certain magnitude are identified, the following stage is marked by the second question: 'What consequences can be expected?' To answer this, it is necessary to have a model of the system which can be used to estimate the effects originating from the cause identified. A previous stage exists, in which the selection of pertinent models is carried out. Thus, the same incident (e.g. collapse of a tank containing a flammable liquid under pressure) could have different consequences, (unconfined vapour cloud explosion, flash fire, boiling liquid expanding vapour explosion, formation and dispersion of the cloud without ignition, etc.). The various possibilities should be analysed with appropriate models, which in each case will give an estimate of the expected consequences to personnel or the installations. Evasive actions and/or protective measures can be included in the model, thus modifying the anticipated accident effects.

The models for estimating the consequences of different types of accidents are described in Chapters 3 to 5. As an example of the results of a consequence analysis, Figure 1.3 shows the outcome of a hypothetical simulation of the escape of a hazardous substance from an industrial plant. For a given set of circumstances, the simulation predicts the intensity and duration of the escape and the extent of the affected areas. Figure 1.3 is a 'snapshot' of a flammable release, showing the area where concentrations higher than the lower flammability limit exist. In the same way, for a toxic emission, the outline could be calculated for different reference concentrations, such as TLV-C or IDLH values.

The objective of the third stage of the risk analysis is to answer the question: 'What is the expected frequency?' Once the events which could give rise to significant damage have been identified, and the magnitude of their effects has been estimated, the next step requires quantifying the likelihood of the said events,

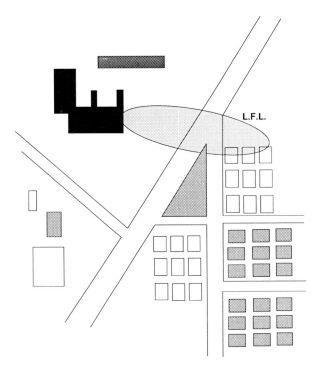

Figure 1.3 A leak of a hazardous substance has occurred in the chemical plant shown in the top left-hand corner. The cloud moves in the direction of the wind, possibly affecting inhabited areas. The consequence analysis models are useful in estimating the extent of the zone where concentration values at or above a certain level may be reached. L.F.L. = lower flammability limit.

e.g. as an estimated frequency (occasions per year) or as the probability of a given event taking place during the expected active life of the installation. The product of the magnitude of the harmful effects and the probability of their occurrence gives us the statistically expected loss (SEL), a very useful tool when making subsequent decisions. Semi-quantitative estimates can be obtained of the likelihood of an incident from historical record analysis. However, there is not always sufficient information in the records available for consultation, due to the difficulties inherent in gathering information on accidents. This is the reason why more structured methods are usually employed, such as Fault Tree Analysis (FTA), or Event Tree Analysis (ETA), where probabilities are assigned to each step in the sequence of events considered. This makes use of the information on failure frequencies of equipment and components which is available in equipment reliability databases.

The different stages in the application of risk analysis to assessing plant designs are shown in Figure 1.4. It is important to recognize that not all of the techniques

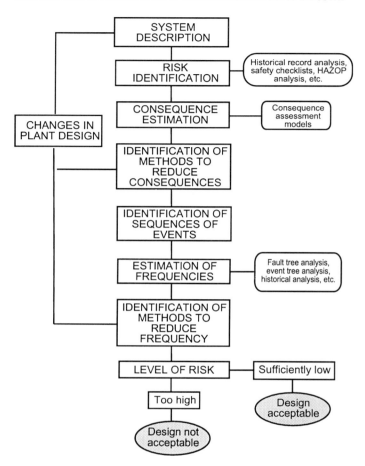

Figure 1.4   Assessment of plant designs using quantitative risk analysis (based on [6]).

required are developed to the same degree. Thus, it may be said that hazard identification techniques as a group have reached maturity and can be used with confidence, i.e. if they are applied correctly, the identification of all the relevant risks should be obtained. It may also be considered that consequence estimation techniques are well-developed, which signifies that, given a scenario, the uncertainties with regard to the effects produced are relatively small, and its magnitude can be approximately estimated. Contrary to this, frequency estimation is comparatively less developed, and will require significant development until its uncertainty decreases to levels comparable to those of the techniques previously mentioned. Nevertheless, the amount of information available in failure and accident databases is steadily increasing, which means that eventually most frequencies and failure probabilities will be estimated with good approximation. It should also be noted that not all of the data entries have the same weight when estimating accident frequencies. On the contrary, often only a few frequencies at the base of the failure tree are critical to the precision of the final result.

Although not considered in Figure 1.4, one of the tasks to be carried out at an early stage of the risk analysis process consists of developing a specific data set for the analysis. Apart from the information contained in the 'system description' block of Figure 1.4, knowledge of external factors is required (topography and use of surrounding land, demographic information, meteorological information, external services, etc.) in addition to data relevant to estimating incident probability (past accident records, equipment reliability data, natural catastrophe records, etc.). Part of the information required can be obtained from unrestricted access databases, although normally the analyst can compile this information with experience and internal data from installations owned by the company.

1.5.1 Making decisions in matters of industrial safety

As already indicated, when facing a particular risk level, it can be accepted as such or efforts can be made to reduce it. In either case, the decision involves carrying out an estimation of the magnitude of the probabilities and consequences involved, as well as the cost of possible corrective measures. Some of the aspects in which risk analysis can be especially useful are shown in Figure 1.5. Risk analysis involves the identification of the possible causes of an accident and the mechanisms leading to it, assessment of its consequences and estimation of the

Figure 1.5 Use of risk analysis in safety decision-making.

probability of its taking place. Once this information is available for the accident scenarios considered, a risk hierarchy can be established, which leads to a list of priorities of risk reduction. In cases where risk reduction action is deemed necessary, in general more than one procedure exists to achieve the desired risk level. Again, risk analysis is the tool which helps decide between the existing options, choosing that which provides the most effective reduction (higher increase in safety for a specific investment).

Last, but no less important, risk analysis permits justification of the decisions taken. It is not enough for a decision to be right, it should, whenever possible, be understood as such by those affected. The results of the analysis provide the means of explaining the level of risk to which the workers of an installation and the surrounding community are exposed, what has been done to increase safety and why one alternative has been chosen over another among those that exist. If the decision is not based on at least a semi-quantitative analysis, but is based on personal opinions and judgements, not only will there be a higher probability of a wrong decision, but, even in the event of it being correct, acceptance by those who could be affected by it will be more difficult.

1.6   Risk analysis in the training of chemical industry professionals

Every day the social importance of environmental and industrial safety is greater and consequently the demands on industry increase. This is leading to a clearer definition of the professional responsibilities of the chemical industry professional, and especially of the chemical engineer, which includes safety, health and welfare of the public [7]. Although in general chemical industry professionals possess the necessary knowledge of industrial safety to develop their work, it seems clear that many of them have not had the opportunity to receive a formal and extensive training in risk analysis. Consequently, pressure is building on universities worldwide to include formal risk analysis contents in chemical and chemical engineering curricula. In countries with a long chemical engineering tradition, such as the USA and Great Britain, the respective professional institutes have for a long time taken the initiative through accreditation programmes, requiring a minimum content of industrial safety in the curriculum of the new graduates, and promoting recycling courses for professionals already at work.

Special mention must be made of institutions existing within the American Institute of Chemical Engineers (AIChE), such as the Safety and Health Division, the Centre for Chemical Process Safety and the Design Institute for Emergency Relief Systems. A special working group has been created with the mission of identifying the key aspects of industrial safety which should be reflected in studies of chemical engineering [8]. A similar approach has been taken by the Institution of Chemical Engineers in Great Britain. Risk analysis should be the cornerstone of any education programme in industrial safety, and a basic tool in the professional activities of chemical engineers.

In spite of the above, and although the trend is towards specific subjects in industrial safety, it is sometimes difficult to include new courses in already crowded curricula. Chemical Engineering departments in Great Britain or North America that offer courses in this field are still in a minority. The rest include the required contents within other courses such as the design project or the chemical engineering laboratories.

Part of the difficulty of including risk analysis in the curricula offered by the different universities stems from the fact that the majority of chemical engineering faculty members did not have the opportunity to study risk analysis in their time as students. Other added problems are that research in this field is still limited to a small number of centres, that risk analysis requires working with probabilities rather than with well-defined situations, and that the studies are of an interdisciplinary nature, frequently including elements from widely different areas, from biology and atmosphere physics to law.

However, it does not appear likely that these difficulties will change the tendency towards a broader education of chemical engineers in industrial safety in general and risk analysis in particular. Not only is the need for this education widely recognized, but also chemical engineers are, by their knowledge of the process, the operating conditions and the techniques and materials involved, the only ones capable of successfully confronting the problem of safety in a chemical plant as a whole.

1.7 References

1. Kletz, T. (1992) *Hazop and Hazan. Identifying and Assessing Process Industry Hazards*, 3rd edn, The Institution of Chemical Engineers, Rugby.
2. Committee on Chemical Engineering Frontiers, National Research Council (1988) *Frontiers in Chemical Engineering. Research Needs and Opportunities*, National Academic Press, Washington.
3. National Safety Council (1985) *Accident Facts*, National Safety Council, Chicago.
4. Comité de Química de las Comunidades Europeas (1990) *Ciencia y Tecnología Químicas: Necesidades Europeas para los años 90*, RSEQ Bulletin, June.
5. Rodellar Lisa, A. (1988) *Seguridad e Higiene en el trabajo*, Marcombo-Boixareu, Barcelona.
6. CCPS (Center for Chemical Process Safety) (1989) *Guidelines for Chemical Process Quantitative Risk Analysis*, American Institute of Chemical Engineers, New York.
7. American Institute of Chemical Engineers (1987) AIChE code of ethics, in *Guide to AIChE, a Handbook for AIChE Members*, AIChE, New York.
8. Crowl, D. A. and Louvar, J. F. (1988) Safety and loss prevention in the undergraduate curriculum. *Chem. Eng. Educ.* **22**(2), 74.

2 Hazard identification techniques

...for although no danger frightens me, still it causes me misgivings to think that powder and lead may deprive me of winning fame and renown by the strength of my arm and the edge of my sword, over all the known earth.

Miguel de Cervantes Saavedra,
The Adventures of Don Quixote,
Chapter XXXVIII

2.1 Introduction

As already stated, the first stage in the study of risk analysis consists of hazard identification. The existing methods for achieving this objective differ in their qualitative or quantitative character as much as in their level of systemization. In any case, the formal hazard identification techniques have been extended and become popular in recent years, becoming standard practice in a large part of the modern chemical industry. In this respect the evolution of industrial mentality has been very notable, from the traditional approximation to the identification of hazards which Kletz [1] ironically observes in the first chapter of his book Hazop and Hazan, consisted in building a plant and observing what happened, up to the computer-assisted identification methods which are used more and more frequently.

The identification of hazards is, in fact, the most important step in the analysis, as any hazard which is not identified cannot be the object of study. On the other hand, once a significant hazard is identified, it is probable that measures will be taken to reduce it, even if the following quantitative evaluation is defective.

Sometimes hazards are obvious and do not need special procedures to be revealed. This would be, for example, the case of a reactor in which hydrocarbons and oxygen are mixed close to the flammability interval. In other cases hazards are not so evident and they require an analysis of a certain depth to unravel the accidents which could take place. In any case, to say that in a given installation an explosion or a toxic leak *could* happen is not sufficient. Instead, a study is required which indicates what are the mechanisms or sequences of events which could lead to a certain type of accident. In this way, there is an opportunity to act upon the sequence of events to reduce the probability of the accident or its consequences. The first event in the chain is known as the **initiating event**. Generally, between the initiating event and the accident a sequence of actions are found which include the responses of the system and the operators, and also other concurrent events. All of these factors are known as elements of the accident: some of the most common are shown in Figure 2.1. The consequences of the accident vary depending on the specific evolution of the sequence of events, i.e. of the elements which give rise to it. Thus, the same initiating event could have different adverse consequences (or not), depending on the combination of intermediate events of propagation or mitigation.

Hazard identification and characterization can and should be carried out throughout the entire life of the installation. However, as shown in the previous chapter, the sooner you begin, the greater are the advantages which could be expected with respect to efficiency of risk reduction and also as to the cost of the

Hazardous circumstances	Initiating events	Propagating events	Mitigating events	Accident consequences
Significant inventories of hazardous materials (flammable and combustible materials, unstable materials, toxic substances, inert gases, materials at very high or very low temperature, etc.) **Highly reactive substances** (reactants, products, by-products, intermediate products) **Reaction rates specially sensitive to impurities or process parameters**	**Machinery and equipment malfunctions.** (pumps, valves, engines, instruments, sensors, etc.) **Containment failures** (piping, vessels, storage tanks, gaskets, etc.) **Operator errors** (operations, maintenance, testing) **Loss of utilities** (water, electricity, compressed air, steam) **External events** (floods, earthquakes, storms, high winds, high velocity impacts, vandalism, sabotage, etc.) **Method/ information errors**	**Process parameter deviations** (pressure, temperature, flow rate, concentration, phase/state change) **Containment failures** (piping, vessels, tanks, gaskets, bellows, input/output, venting, etc.) **Material releases** (combustibles, explosive materials, toxic materials, reactive materials) **Ignition/ explosion** **Operator errors** (commission, omission, diagnosis, decision-making) **External events** (delayed warning, unwarned) **Method/ information errors**	**Safety system responses** (relief valves, back-up utilities and components, redundant systems, etc.) **Mitigation measures** (vents, dykes, flares, sprinklers, etc.) **Control and operator responses** (planned, 'ad hoc') **Contingency operations** (alarms, emergency procedures, personnel safety equipment, evacuation, etc.) **External agents** (early detection, early warning) Adequate flow of information	**Fires** **Explosions** **Impacts** **Dispersion of toxic material** **Dispersion of highly reactive material** **Environmental pollution**

Figure 2.1   Some common elements of accidents (adapted from [3]).

safety measures installed. Hazard identification in the definition phase of the process could allow for reducing these costs, by selecting process routes of greater intrinsic safety, regarding the process itself, the materials and reagents used, the required inventory levels, etc. Hazard identification continues throughout the design and construction stages of the plant, in the start-up, during its operation, when carrying out modifications to the plant, during programmed maintenance shut-downs, and finally in the dismantling of the plant at the end of its useful life. Each phase may require different depths of study and in some simple cases the formal analysis can be omitted, although the safety considerations carried out in previous analysis should always be present.

The methods of hazard identification can be divided into the three categories shown in Figure 2.2. The comparative methods are based on previous accumulated experience in a particular field, and can take the form of a record of previous accidents or design and operation codes and safety checklists. Risk indices, although they do not usually identify specific hazards, are useful to point out the areas of high concentration of risk, which require a deeper analysis or supplementary safety measures. Finally, generalized methods are in principle applicable to any situation, which makes them versatile and very useful analysis tools. Specific mention should also be made of the so-called **expert systems**, which are being proposed as a useful tool in safety analysis (especially for complex, unsteady situations, e.g. start-up [2], etc.). However, the application of these techniques to safety analysis in the chemical process industry can still be considered to be in a developmental stage and, given the scope of this book (i.e. a general introduction to risk analysis), will not be treated further.

2.2 Comparative methods of hazard identification

The comparative methods of hazard identification are used to assess the safety of an installation in the light of the experience acquired in previous operations of the

---

Comparative methods
- Engineering codes and practices
- Safety checklists
- Historical record analysis

Risk indices
- Dow index
- Other indices: Dow-Mond, IFAL, etc.

Generalized methods
- HAZard and OPerability analysis (HAZOP)
- Failure Modes and Effects Analysis (FMEA)
- Fault Tree Analysis (FTA)
- Event Tree Analysis (ETA)
- 'WHAT IF' analysis

Figure 2.2 Classification of methods used in hazard identification.

company or in external organizations. Thus in chemical companies of a certain size it is often the case that **internal technical manuals** have been developed which specify how to design, distribute, install, operate, etc. the equipment used in its installations. The contents of the manuals can vary considerably, although they must always fulfil local and national legislation, and also the normal engineering standards. These are available compiled in the form of codes (ASME, ASTM, API, NFPA, TEMA, AD-Merkblatt, etc.), supplying complementary information to the documentation which a company may have obtained. In Chapter 10 some of the most important codes are quoted with the addresses of the organizations which publish them.

Therefore, in general, the first step is to use the available internal technical manuals, as well as the codes and engineering standards, in the evaluation of the acceptability of a design. If differences in a design are encountered with respect to what is considered standard practice it is necessary to examine them with every care, as sources of possible risk. In any event the reasons for not following the usual procedures should be investigated, and the designer should ask himself if a new design covers the risks at least to the same level. This is valid not only for the initial design of the plant but also for later modifications. A typical example is the Flixborough accident in 1974, where the carrying out of a 'temporary' modification (the bypass of a reactor) included the placing of a substandard design pipe, the failure of which caused the accident. The assessment of possible hazards in an installation should continue as the plant ages, carrying out frequent **safety audits** which allow the state of the material, instrumentation, operation procedures, emergency equipment, etc. to be judged and compared to the company's requirements for new plants. The subject of safety audits and reviews is discussed in detail in Chapter 8.

Another comparative method of hazard identification where use is also made of accumulated experience by an industrial organization is the **safety checklists**. A safety checklist is a useful reminder which, in general, has been developed over the years by different people and which, as in the previous case, allows comparison of the state of a system with an external reference, directly identifying in some cases lack of safety, or in others the areas which require a more in-depth study. The checklists can be applied to assessments of equipment, materials or procedures and the level of detail varies considerably from the general to those which are developed for very specific equipment, processes or procedures. An example is given of a general checklist in Table 2.1. As can be seen, the list provides a series of points for consideration and questions which call attention to aspects which may have been overlooked.

2.2.1 Historical record analysis

Historical record analysis is a tool for hazard identification which makes use of the data gathered on past industrial accidents. The advantage of this technique lies in that it refers to accidents which have already happened, so the dangers

Table 2.1   Summarized checklist (compiled from references [3–6].
Indicate what is right:

A = Already taken into account,
B = Not applicable,
C = Further study required

	A	B	C
LOCATION			
• Adequate plant layout: Has the separation of units been established according to a previous risk evaluation?			
• Accessibility: Are there any dangerous obstructions above or below ground?			
• Are there sufficient streets and corridors, with adequate signposts?			
• Are there entrances and emergency exits sufficient in width and number?			
• Is there enough space for elevated service lines (electricity, steam, water, compressed air, etc.)?			
• Has the proximity of sources of ignition been considered? The prevailing winds?			
• Flooring characteristics: Sufficient resistance to support the weights of the operation? Maximum loads indicated?			
• Adequate drainage? Control/protection against possible floods?			
• Adequate location of the loading and unloading facilities, off the main thoroughfares?			
BUILDINGS AND WAREHOUSES			
• Stairs, emergency exits, passages: Sufficient in number and width, free of obstructions and obstacles?			
• Cranes and elevators: well-designed, with adequate safeguards?			
• Adequate signs for head obstructions and other obstacles?			
• Adequate ventilation for the type of activity being carried out?			
• Adequate lighting for the type of activity being carried out?			
• Heating and cooling adequate for the type of activity being carried out?			
• Is access or stairway to the roof required?			
• Are fire doors necessary? Should fire resistant materials be used in certain areas of the building? Is fireproofed structural steel needed?			
• Is emergency equipment available and clearly indicated?			
• Is a design needed that takes into account the possibility of explosions?			
• Are heat and smoke detectors required?			
• Is there protection against electrical discharge (lightning conductors, earthing of equipment)?			
MATERIAL, EQUIPMENT AND PROCESSES			
• Has the possibility of interference bewteen adjacent operations been considered?			
• Has adequate storage of special or unstable materials been anticipated? Have the rest been segregated? Are there materials for which special handling equipment is needed?			
• Are there any products or materials which could be affected by extreme meteorological conditions?			

Table 2.1 *cont'd*

	A	B	C
• Are all raw materials and products properly classified and labelled?			
• Is the construction material used in process equipment adequate to meet operational conditions?			
• Is there a possibility of vapour confinement is certain areas?			
• Have *all* the dangerous characteristics of the substances in use been identified (autoignition temperatures, flash points, flammability limits, possibility of spontaneous decomposition, reactivity, impurity effects, possibility of runaway reactions, secondary reactions, corrosion and compatibility of materials, toxicity, etc.)?			
• Has possible exposure of plant personnel or the public to adverse agents been considered (chemical products via various entry routes: inhalation, oral or dermal, powder and smoke, noxious radiation, noise, biological agents, etc.).?			
• Are special hoods for fume, dust or vapour needed? Is personal protection equipment required?			
• Is there a possibility of static discharges? What is the conductivity of the materials used and what are their electrical charge accumulation characteristics? Is there an adequate earthing installed?			
• Is protection against explosions required? Have the pressure relief systems, explosion suppression systems, explosive atmosphere detectors, etc. been verified as adequate? Are the vents pointed in the right direction? Has the possibility of counterpressures been considered?			
• Are flame arrestors needed in the venting lines? Are other special precautions required for the type of material vented?			
• In the combined rupture disc/relief valve systems, have the valves been protected against the possibility of a blockage due to the fragments of rupture discs? Have pressure gauges been installed between them?			
• Is rapid emptying of tanks, reactors etc., required in emergencies?			
• Has an adequate design for the maximum operating pressure been completed?			
• Has corrosion allowance been considered?			
• Have safety factors for pressures, temperatures, flows, levels, and other process variables been considered?			
• Can dangerous conditions be caused by a mechanical failure?			
• What are the most likely opportunities for human errors, and what are their consequences?			
• Have the consequences of a failure in one or more utilities been considered? Electricity (agitation, circulation, instruments, controls, emergency systems, light, etc.), steam (heating, vacuum, pumps, etc.), air (instruments, pumps, etc.), water (cooling, fire extinction, reactions, etc.), gas (inerting, etc.)			
• Have guards been considered for drive belts, conveyors, pulleys, gears, mobile equipment in general, also for cutting edges and hot surfaces of all types?			
• Have pressure relief devices been installed in discharge lines and, when required, also in suction lines of the process pumps?			
• Has accessibility to all equipment been revised, especially to the critical elements?			
• Has protection and identification of the most fragile lines been considered? Is there sufficient support for the pipes? Has thermal contraction and expansion been considered?			

Table 2.1   *cont'd*

	A	B	C
• Does the possibility of gradual or sudden blockage of the process lines exist? Have the consequences been contemplated?			
• Have the quantities of toxic or flammable materials been revised ? Can they be reduced?			
• Can the equipment be placed in such a way that maintenance tasks can be performed with sufficient safeguards (complete electrical disconnection, line blanking, adequate room and facilities for rescue, etc.)?			
• Have procedures for taking samples been revised?			
• Are process diagrams kept up to date?			
• Are operating procedures kept up to date? Is there a standard procedure for this? Is there verification that procedures are followed?			
• Have spare parts for critical components or equipment been considered?			
• Is explosion proof electrical equipment required?			
• Does passive (e.g. tanks) or active (e.g. ball mills) equipment require inerting?			
• Are residues adequately eliminated ? Have drainage pipes been tested?			
• Is there a rigorous procedure for the control of quality and composition on reception of raw materials?			
• Is there an established communication procedure between supervisors when changing shifts?			
INSTRUMENTATION AND CONTROL			
• Have the valves, switches, instruments, etc. been appropriately identified?			
• Are the alarms, protection equipment, automatic start-up and shut-down equipment and instrumentation in general regurlarly checked? Is there regular verification of correct functioning of control panel lamps and indicators?			
• Do *all* control elements fail safe? Is there protection against automatic start after a stop?			
• Is there an adequate policy for the establishment and change of control parameters, also for the manual control of some operations?			
• Are the instruments and control equipment rated for the service in which they are used? Are they regularly checked, including electrical wiring?			
• Has the possibility of installing redundant instrumentation or protection systems been considered?			
• Have possible delays in the response of specific equipment been estimated? What are the consequences?			
• Are safeguards provided for process control in the case of removal of equipment for servicing?			
• Has the convenience of installing new interlocks been considered? Have all the consequences of the existing ones been taken into account?			
• What would be the shut-down procedure in case of loss of electricity or air to the instruments? How would it affect the plant?			

Table 2.1   *cont'd*

	A	B	C
EMERGENCY ACTIONS			
• Are emergency showers and eyewashes required?			
• Is personal emergency protection equipment required?			
• Switches and emergency valves: are they tested frequently? Are they clearly indicated? Are they accessible?			
• Is emergency power and lighting guaranteed?			
• Has the integrity of the control room during emergencies been considered?			
• Are extinguishers required? How many, what class and size?			
• Are automatic sprinklers required?			
• Does the fire detection and extinguishing equipment meet current legislation (water supply, including backup supply, pumps, hydrants, tanks, pipes, alarms, protection of fire-fighting equipment, etc.)? Are the materials used in fire-fighting equipment compatible with the process materials?			
• Is smoke, heat, or flammable vapour detection equipment required? For toxic vapours?			
• Has spill containment been anticipated?			
• Has the installation of new alarms been considered?			
• Is emergency equipment kept up to date? Also the emergency procedures? Are periodic practices held?			
• Has a system for controlling the exact number of people in the installation, including suppliers and visitors been considered?			
• Is the safety documentation on the different materials used kept up to date? Has the possibility of synergy effects been considered?			
• The emergency plan: Was it developed bearing in mind the worst reasonable scenarios with a sufficiently detailed risk analysis?			
• Have the emergency communications been verified, even during holiday periods?			

identified are undoubtedly real. On the other hand, its principal limitation also resides in this fact, as the analysis only refers to accidents which have taken place or on which there is information. The number of cases to be analysed is therefore finite, and does not cover all eventualities. It is necessary to bear in mind that the available information on accidents is limited and often biased, in the same way that many accidents and incidents are recorded in a restricted way or are not recorded at all. The latter is especially true in cases of incidents which could have had catastrophic consequences but which did not materialize or occurred in a limited way due to a combination of fortunate circumstances.

In spite of the above, historical accident analysis is a useful technique, which permits identification of specific hazards. It can at least indicate to the management of a company that in other similar plants or in plants that process similar substances a certain type of accident has occurred, which should be sufficient to initiate a risk analysis to assess whether or not the accident could take place at the plant in question. It is also a very valuable way to verify *a posteriori* the models which are available to predict the consequences of accidents.

Data on accidents which have happened in the past can come from very diverse sources, such as internal company information, newspapers, interviews with witnesses and reports from investigation committees. Obviously, not all of these sources are equally useful or equally reliable. In particular, media information is often unreliable, owing mainly to the fact that the person who writes it does not necessarily have an adequate technical background, and is not familiar with the characteristics of the accident. Also the habitual pressure of time in the editing of a newspaper article should be borne in mind, especially if it was written in the hours following an important accident, where precision regarding the consequences and characteristics could be flawed. However, there is no doubt that such articles frequently provide useful data, sometimes the only data available on an accident. With regard to the other sources mentioned, access to a company's own information is often difficult. It is also obviously difficult (and sometimes not practical) to try to interview the direct eyewitnesses to an accident. They could form part of the affected population, in which case their impressions would be incomplete and of limited use, or they may be specialists and plant personnel, more useful for their broader knowledge of the circumstances involved, but often worried about protecting those responsible. Lastly, the reports of official investigation committees are generally the best source of information, but are only available for relatively few events, so their usefulness is also limited.

Using sources such as those outlined above and others which are available (insurance company reports, scientific publications, judicial summaries, etc.), diverse public and private organizations have developed data banks for industrial accidents, in which the information available has been organized to facilitate consultation. The data gathered refer to the identification of the type of accident and the circumstances in which it took place, the nature and quantity of the substance or substances involved, location, causes and consequences, with an estimate of damage to people and property. Often the data available are sufficient to allow laying out patterns for certain types of accidents, such as the initiating events, the most frequently involved substances or the sequence of events. There are numerous data banks on accidents which contain relevant information for the chemical industry, such as CHAFINC (Chemical Accidents, Failure Incidents and Chemical Hazards Databank), CHI (Chemical Hazards in Industry), HARIS (Hazards and Reliability Information System), MHIDAS (Major Hazard Incident Data Service), NIOSH (National Institute for Occupational Safety and Health), SONATA (Summary of Notable Accidents in Technical Activities) and WOAD (Worldwide Offshore Accident Databank).

Example 2.1

A company which manufactures polyurethane foams is gathering the data necessary to carry out a risk analysis of their process. One of the hazardous substances involved is toluene diisocyanate (TDI). The results obtained in a limited search of the MHIDAS and NIOSH databanks on the accidents in which this substance was involved over a twenty-year span (1971–1991) are given below.

Both data banks give complementary information. The MHIDAS records contain, above all, data on major accidents, while the NIOSH data bank is oriented towards information on occupational hazards, although it also gives some data on accidents.

Seven accidents were found which related directly to TDI between 1971 and 1991. The consequences of these include one death, 23 hospitalized and six more affected to different extents. One of the cases included, besides the TDI, a variety of contributing substances.

The causes of the accidents were: transport (impact, derailment), followed by a leak and evaporation (four cases), one case of inhalation of vapour during repair work (and consequently the death of a worker), one case of mechanical failure in the storage plant and one case of a runaway reaction between the TDI and water in a tank. In each case more specific details are given on the circumstances of the accident. The losses of containment were as much instantaneous (container collapse), as semi-continuous. On at least two occasions it was necessary to evacuate a large number of people (1500 and 3000, respectively), and another case required decontamination work to prevent the TDI leakage from affecting underground water.

The information gathered on occupational hazards involves both studies carried out in the workplace and investigations on laboratory animals. Various references point out predictable damage by exposure to TDI: harm to the respiratory system, irritation by direct contact of liquid TDI with the skin, severe ocular irritation and diverse ocular affliction caused by exposure to TDI vapours. The use of protection is recommended especially on hands, arms, face and eyes, as well as good ventilation of the workplace, medical vigilance of workers exposed and rotation in the workplace as soon as any symptoms appear. Approximately one in 30 workers is hypersensitive to TDI. Various toxic indicators are discussed in the references given.

2.3 Risk indices

Risk indices, such as the Dow or the Mond indices, present a direct and relatively simple method of estimating the global risk associated with a process unit, as well as classifying the units as to their general level of risk. They are not, therefore, systems which are used to point out individual hazards, rather they provide a numerical value which permits the identification of areas in which the potential risk reaches a certain level. In these areas, if necessary, a more detailed risk analysis can be applied, so the obtained value of the risk index could be useful when deciding the extent of the study. In any event the risk indices are useful because they provide a quick and fairly reliable estimate of the order of magnitude of certain risks in a given unit.

The Dow fire and explosion index is widely used in the chemical industry. In its seventh edition [4] it covers aspects related to the intrinsic hazards of materials, the quantities handled, operating conditions, etc. These factors are weighted successively to obtain an estimate of the index value for the area which could be affected by an accident, the damage to property within this area and the working days lost as a result of the accident. Also, frequent use is made of the Mond index [7], similar in many repects to the Dow and which also specifically includes the toxic aspects of materials. However, the Dow index permits an estimate which is somewhat easier to visualize due to its preference for graphs instead of equations, and in its last edition takes into account, although in a marginal way, aspects of toxicity, including specific penalties. If a more detailed assessment of toxicity hazards is desired, there are specific indices, such as the Dow 'Chemical exposure index' [8].

Some aspects of the Dow index are described later, with the aim of illustrating the potential of the method and its ease of use. However, it should be borne in mind that what is shown here is a very summarized version, and that it is necessary to consult the Guide [4] for a correct application method.

2.3.1 Dow fire and explosion index (F&EI)

The method is applied in stages, and begins with the selection of the pertinent process units. It is usual to define as a process unit any primary equipment , such as a compressor, a pump, a storage tank, a heat exchanger, a reactor or a distillation column. In other cases, reduced groups of primary elements could be considered as process units whenever they clearly function as a unit and are situated in a restricted physical space. Thus in this way the Guide [4] also quotes as examples of process units the monomer feed preparation section or the styrene scrubber of a latex plant. Moreover, in any plant, there are numerous process units. The process units pertinent to application of the method are those which could have a relevant impact from the viewpoint of plant safety (for the materials processed, for the quantity of dangerous materials, for process conditions such as pressure and temperature or for any other reason, including the history of problems and incidents in the unit).

The Dow index should be calculated for all of the process units which have been identified as pertinent. The next step is determining the material factor (MF). The material factor is a number between 1 and 40 which is given to the substance being processed in the unit, according to its intrinsic potential to release energy in a fire or an explosion. The index Guide [4], in its appendix, lists more than 300 substances common in industry, for which the MF and other important parameters such as heat of reaction (normally of combustion), the NFPA indices of danger to health (N_h), flammability (N_f) and reactivity (N_r), the flash point (which is defined in Chapter 3) and boiling point are given. If the substances are not in the appendix, the MF value can be calculated using a simple equivalence table [4] from the values of N_f and N_r, given in the NFPA 325 M standard. If the compound cannot be found in the standard either, the N_f and N_r values can be obtained from various properties of the substance. For gases and liquids these are: flash point, boiling point, and the starting temperature of exothermic activity, obtained by DTA (differential thermal analysis) or DSC analysis. A selection of material factors and other substance properties, taken from the Dow F&EI Guide, are shown in Table 2.2.

The next stage consists of determining the concurrent hazard factors. They can be of two types: general process hazards, such as the presence of exothermic

Table 2.2. Material factors (*MF*) and other properties necessary for calculating the Dow F&EI, for some selected substances [4]

Substance	MF	$\Delta H_c \times 10^{-3}$ (Btu/lb)	NFPA N_h	NFPA N_f	NFPA N_r	Flash point (°F)	Boiling point (°F)
Acetaldehyde	24	10.5	2	4	2	−36	69
Acetic acid	14	5.6	2	2	1	103	244
Acetone	16	12.3	1	3	0	−4	133
Acetylene	29	20.7	0	4	3	Gas	−118
Acrolein	29	11.8	4	3	3	−15	127
Ammonia	4	8.0	3	1	0	Gas	−28
Benzene	16	17.3	2	3	0	12	176
Butane	21	19.7	1	4	0	−76	31
n-Butene	21	19.5	1	4	0	Gas	21
Carbon monoxide	21	4.3	3	4	0	Gas	−313
Chlorine	1	0	4	0	0	Gas	−29
Chloroform	1	1.5	2	0	0	−	143
Cumene	16	18.0	2	3	1	96	306
Cyclohexane	16	18.7	1	3	0	−4	179
Dimethylamine	21	15.2	3	4	0	Gas	44
Ethane	21	20.4	1	4	0	Gas	−128
Ethanol	16	11.5	0	3	0	55	173
Ethanolamine	10	9.5	2	2	0	185	339
Ethylene	24	20.8	1	4	2	Gas	−155
Ethylene glycol	4	7.3	1	1	0	232	387
Formaldehyde	21	8	3	4	0	Gas	−6
Gasoline	16	18.8	1	3	0	−45	100–400
Glycerine	4	6.9	1	1	0	390	554
Hydrogen	21	51.6	0	4	0	Gas	−423
Isopropanol	16	13.1	1	3	0	53	181
Methane	21	21.5	1	4	0	Gas	−258
Methyl alcohol	16	8.6	1	3	0	52	147
Phenol	10	13.4	4	2	0	175	358
Propane	21	19.9	1	4	0	Gas	−44
Pyridine	16	5.9	2	3	0	68	240
Styrene	24	17.4	2	3	2	88	293
Toluene	16	17.4	2	3	0	40	232
Triethylamine	16	17.8	3	3	0	16	193
p-Xylene	16	17.6	2	3	0	77	279

reactions or carrying out loading and unloading operations, and special process hazards, such as operations close to the flammable range or pressures different from atmospheric. The summing of the concurrent hazard factors in the process is done by assigning a penalty to each one of the sections. This permits defining the general process hazard factor F_1 and special process hazards factor F_2 of the process as follows:

$$F_1 = 1 + \Sigma \text{ (penalty for each one of the general process hazards)} \quad (2.1)$$

$$F_2 = 1 + \Sigma \text{ (penalty for each one of the special process hazards)} \quad (2.2)$$

To facilitate calculation of the penalties, the Dow F&EI form, which is shown in Figure 2.3, can be used. The applicable interval of each penalty is shown on the form, but the application of a specific value requires consulting the casuistic, explained in the Guide [4].

Once the F_1 and F_2 factors have been calculated, the unit hazard factor, F_3, can be obtained by as the multiplication of the other factors. F_1 and F_2 are multiplied instead of being added due to the interactive nature of the hazards involved. Thus, the effects of an explosion caused by the failure of equipment that works under pressure (section 2.E of the form) could be magnified if the unit is closed (section 1.D), or if the emergency teams have restricted access (section 1.E) . The hazard factor of the unit F_3, which is normally between 1 and 8, is used to find the fire and explosion index (F&EI), which is calculated as the product of the unit hazard factor and the material factor ($F_3 \times$ MF). Therefore, processes with equivalent penalties will give different F&EI values if they are applied to substances with different material factors.

Because the material factor varies between 1 and 40, the variation interval of the fire and explosion index is situated between the extreme cases of 1 and 320, although the majority of correlations (for example, the one used to calculate the radius of exposure) consider maximum values of 200. In earlier editions of the Guide a classification of risks was established from slight to critical, according to the F&EI value calculated, with a F&EI value higher than 159 considered as critical.

The area of exposure is an ideal circle which contains the equipment and installations which could be affected by a fire or explosion in the process unit being evaluated. Obviously this means a major simplification, as accidents rarely happen with totally symmetric effects, but gives an approximate measure of the radius affected by the accident. For a more precise estimate, the methods described in later chapters can be used. The radius of the ideal circle of exposure (R(m)) is calculated as follows:

$$R(m) = 0.256 \times \text{F\&EI} \quad (2.3)$$

In the F&EI Guide the value of the equipment contained within the exposure area is used to calculate the maximum probable property damage (MPPD), once

DOW FIRE AND EXPLOSION INDEX		Plant		Date	
Site	Process unit			Evaluated by	Revised by

Materials and process		
Materials in process unit		
State of operation		Basic material(s) for material factor
☐ Design ☐ Start-up ☐ Normal operation ☐ Shut-down		

MATERIAL FACTOR (see table I and appendices A and B)
Note requirements when unit temperature over 60° C

1. GENERAL PROCESS HAZARDS	Penalty factor range	Penalty factor used
Base factor	1.00	1.00
A. Exothermic chemical reactions	0.30–1.25	
B. Endothermic processes	0.20–0.40	
C. Material handling and transfer	0.25–1.05	
D. Enclosed or indoor process units	0.25–0.90	
E. Access	0.35	
F. Drainage and spill control. Volume…m³	0.25–0.50	
General process hazard factor (F₁)		1.00

2. SPECIAL PROCESS HAZARDS		
Base factor	1.00	1.00
A. Toxic materials	0.20–0.80	
B. Sub-atmospheric pressure (<500 mmHg)	0.50	
C. Operation in or near flammable range: ☐ Inerted ☐ Not inerted		
1. Tank farms storage flammable liquids	0.50	
2. Process upset or purge failure	0.30	
3. Always in flammable range	0.80	
D. Dust explosion (see table 3)	0.50–2.00	
E. Pressure (see figure 2). Operating pressure 5 bar a; Relief setting 7 bar a		
F. Low temperature	0.20–0.30	
G. Quantity of flammable/unstable material: Quantity 3000 kg. H꜀=20800 Btu/Lb		
1. Liquids or gases in process (see figure 3)		
2. Liquids or gases in storage (see figure 4)		
3. Combustible solids in storage. Dust in process (see figure 5)		
H. Corrosion and erosion	0.10–0.75	
I. Leakage: Joints and packing	0.10–1.50	
J. Use of fired equipment (see figure 6)		
K. Hot oil heat exchange system (see table 5)	0.15–1.15	
L. Rotating equipment	0.50	
Special process hazard factor (F₂)		1.00

Figure 2.3   Dow Fire & Explosion Index form.

the credit for installed safety measures has been noted. Likewise, in function of the MPPD value calculated one can estimate the days lost in repairing or replacing the damaged equipment. Therefore, a first estimation of material losses which could result from an accident can be made from the fire and explosion index value. However, the principal usefulness of the calculated F&EI value is in establishing a hierarchical structure of risks for the different units. It is therefore a relative measure that indicates in which installations the efforts to reduce risks should be concentrated.

Example 2.2

Calculate the Dow fire and explosion index for the storage of 25 000 MT of liquid ethylene at atmospheric pressure in a refrigerated tank (relief and design pressure equal at 0.06 bars relative, see Chapter 7).

To maintain the ethylene in the liquid phase, the vapours formed in the tank are returned to it after being compressed at 75 bars and condensed with sea water.

The tank can be filled by cisterns or through a line which carries ethylene vapour (after condensation in this case). It can also be sent at a pressure of 5 bars absolute, after being vaporized with sea water, to a plant where it is used at a rate of 18 000 kg/h. Alternatively ships can be loaded by two pumps and a fixed line which leads to the marine offshore floating loading dock (capacity 1500 m³/h).

For the application of the Dow index the installation can be considered as divided into three functional units, all surrounded by dikes whose implantation is shown in Figure 2.4(a):

1. *The storage tank in the strict sense.*
2. *The filling/loading area which comprises a pressurized reception tank where the ethylene gas coming from the pipes enters a compressor and a condenser (refrigerated by sea water). The condensed stream which leaves the condenser is returned to the reception tank, from where the liquid part is sent by a pump (with another in reserve) to the storage tank, and the*

ZONE 2

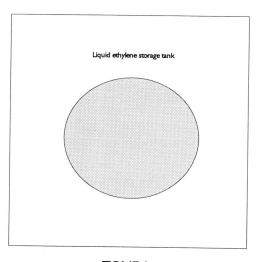

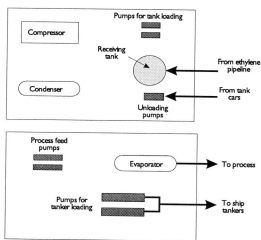

ZONE I   ZONE 3

Figure 2.4(a)   Equipment in the three zones considered in Example 2.2.

vapour part returns to be compressed and condensed again. The ethylene
unloaded from the cisterns also enters the reception tank directly.
3. *The pumping-out area, which comprises two large pumps for loading*
   tankers and another two pumps (one in reserve) and a re-vaporizer (heated
   with sea water) to send ethylene vapour to the plant.

To resolve the example the Dow F&EI Guide [4] is followed at all times. It is important to point out again that it is necessary to consult the Guide to argue the casuistry correctly, because what is shown here are only the study's conclusions.

We begin with an analysis in zone 1:

The normal boiling point of ethylene is $-103.8°C$ and its density is 577 kg/m^3 at $-110°C$, therefore, the volume of liquid in the tank will be approximately 44 000 m^3

The basic compound to be considered for the material factor is, obviously, the ethylene, with a value of 24 (Table 2.2). We shall analyse first the *general hazards* of the process.

A and B: Exothermic and endothermic chemical reactions. Not considered, because it is a storage tank in which all types of reaction are avoided.

C: We are studying storage in a tank, not in a warehouse and there are no disconnectable systems (they are in zones 2 and 3).

D and E: The system is in the open air, and we will initially assume that it has free access from at least two sides. There is no penalty.

F. For tanks inside dikes (we will assume the dike has sufficient capacity for the volumes mentioned in the Guide) the penalty is 0.5.

In this way the factor F_1, for general process hazards, is 1.5 (the sum of all the factors plus 1). Regarding the process *special hazards,* we have:

A: Because N_h is 1, the prescribed value in the Guide is 0.2.

B: There is no penalty as the tank does not operate under vacuum.

C: When the N_f value is 3 or 4, the Guide recommends a penalty of 0.5, unless there is an inert airtight vapour recovery system, which is not the case.

D: There is no penalty for dust explosions.

E: For the operating pressure of 1.01 bars absolute, and the design pressure of 1.07 bars absolute (in this case the same as the set pressure of the safety valves) a penalty of 0.2 is obtained from Figure 2 of the Guide (applying a factor of 1.3 as it is a liquefied flammable gas).

F: It is understood that a tank conceived for working at very low temperatures will be designed with adequate materials. There is no penalty in this section.

G: For ethylene, the heat of combustion is 20 800 Btu/lb (Table 2.2), and given that the storage capacity is 25 000 MT, from Figure 4 of the Guide (storage of liquids or gases, extrapolating curve A, for liquid gases) a penalty of 2 is obtained.

H: Given that the installation is close to the sea and there is probably some risk of corrosion (small, although we do not have specific data), we shall take a minimum penalty of 0.1.

I, J, K, L: As there are no joints or seals, furnaces with burners, oil heating systems, or rotary equipment, no penalties are assigned.

The factor F_2, for special hazards, adding those previous to 1, is 4.0. The hazard factor of the unit (F_3), the product of F_1 and F_2, will be 6.0, which, multiplied by the material factor, gives us a fire and explosion index of 144. Using equation (2.3) or Figure 8 of the Guide a radius of exposure of 37 m is calculated.

The calculations for areas 2 and 3 are carried out in a similar way. In Figures 2.4(b), (c) and (d) the respective forms, duly completed, from the Dow fire and explosion are shown for the storage tank and the other two sections.

The three units considered produce high values of the Dow index, which indicates that they should be separated from other process units, and, if possible,from each other, to avoid damage in the case of an accident. The separation between the units should be decided taking other aspects into consideration, such as the need for the emptying pumps to remain operative in the case of the tank catching fire.

2.4   Generalized methods of hazard identification

2.4.1   HAZard and OPerability (HAZOP) analysis

A HAZOP study serves to identify safety problems in a plant and is also useful to improve its operability. The implicit assumption of HAZOP studies is that the hazards or operability problems appear only as a consequence of deviations from the normal operating conditions in a given system, at any of the different process stages: start-up, operation at steady state, unsteady state operation (e.g. in batch plants), and plant shut-down. Consequently, whether the HAZOP analysis is applied in the design stage or whether it is carried out on an installation which is already built, the procedure consists of evaluating, line by line and vessel by vessel, the consequences of possible deviations in the operating conditions of a continuous process, or in the sequence of operations of a batch process. The HAZOP method has now been in operation for 25 years and can be considered to be firmly established. In fact, it has become a design requisite in large chemical companies such as ICI [9].

Before HAZOP and similar techniques, a common practice when introducing a modification in a chemical plant consisted of circulating the project or parts of it to the different departments or people who could give a critical opinion or make suggestions. Obviously, using this method it is quite probable that important aspects remain unanalysed, due to the lack of structure in the checking process. Neither is

DOW FIRE AND EXPLOSION INDEX		Plant *Etilento Español S.A.*	Date *May 1995*
Site	Process unit *Atmospheric ethylene storage tank*	Evaluated by	Revised by

Materials and process		
Materials in process unit	*Ethylene*	

State of operation □ Design □ Start-up □ Normal operation □ Shut-down	Basic material(s) for material factor *Ethylene*

MATERIAL FACTOR (see table I and appendices A and B) Note requirements when unit temperature over 60° C		24

1. GENERAL PROCESS HAZARDS	Penalty factor range	Penalty factor used
Base factor	1.00	1.00
A. Exothermic chemical reactions	0.30–1.25	
B. Endothermic processes	0.20–0.40	
C. Material handling and transfer	0.25–1.05	0.00
D. Enclosed or indoor process units	0.25–0.90	0.00
E. Access	0.35	0.00
F. Drainage and spill control. Volume...m^3	0.25–0.50	0.50
General process hazard factor (F_1)		1.50

2. SPECIAL PROCESS HAZARDS		
Base factor	1.00	1.00
A. Toxic materials	0.20–0.80	0.20
B. Sub-atmospheric pressure (<500 mmHg)	0.50	
C. Operation in or near flammable range: □ Inerted □ Not inerted		
1. Tank farms storage flammable liquids	0.50	0.50
2. Process upset or purge failure	0.30	0.00
3. Always in flammable range	0.80	0.50
D. Dust explosion (see table 3)	0.50–2.00	
E. Pressure (see figure 2). Operating pressure 1.013 bar a; Relief setting 1.073 bar a		0.21
F. Low temperature	0.20–0.30	
G. Quantity of flammable/unstable material: Quantity 25000 kg. H_c=20800 Btu/Lb		
1. Liquids or gases in process (see figure 3)		
2. Liquids or gases in storage (see figure 4)		1.85
3. Combustible solids in storage. Dust in process (see figure 5)		
H. Corrosion and erosion	0.10–0.75	0.10
I. Leakage: Joints and packing	0.10–1.50	
J. Use of fired equipment (see figure 6)		
K. Hot oil heat exchange system (see table 5)	0.15–1.15	
L. Rotating equipment	0.50	
Special process hazard factor (F_2)		3.86

Process unit hazard factor (F_1 x F_2 = F_3)	5.79
DOW fire and explosion index (F_3 x MF = F&EI)	**139**
Radius of exposure (m)	**36**

Figure 2.4(b) Calculation of Dow Fire & Explosion Index for zone 1 in Example 2.2.

it possible with this procedure to consider all the relevant aspects of the process, because an important requisite for this is the simultaneous interaction of participants with different professional experience and points of view. The HAZOP method attempts to improve both aspects, based on the following points:

1. The systematic character of the analysis: An examination based on the successive application of a series of guide words is carried out [1], with the objective of

◆DOW◆ **DOW FIRE AND EXPLOSION INDEX**	Plant *Etilento Español S.A.*		Date *May 1995*
Site	Process unit *Ethylene compression and liquefaction*	Evaluated by	Revised by

Materials and process		
Materials in process unit *Ethylene* *Sea water*		

State of operation ☐ Design ☐ Start-up ☐ Normal operation ☐ Shut-down	Basic material(s) for material factor *Ethylene*

MATERIAL FACTOR (see table I and appendices A and B) Note requirements when unit temperature over 60° C			24

1. GENERAL PROCESS HAZARDS		Penalty factor range	Penalty factor used
Base factor		1.00	1.00
A. Exothermic chemical reactions		0.30–1.25	
B. Endothermic processes		0.20–0.40	
C. Material handling and transfer		0.25–1.05	0.50
D. Enclosed or indoor process units		0.25–0.90	
E. Access		0.35	
F. Drainage and spill control. Volume…m³		0.25–0.50	0.50
General process hazard factor (F_1)			2.00

2. SPECIAL PROCESS HAZARDS			
Base factor		1.00	1.00
A. Toxic materials		0.20–0.80	0.20
B. Sub-atmospheric pressure (<500 mmHg)		0.50	
C. Operation in or near flammable range: ☐ Inerted ☐ Not inerted			
1. Tank farms storage flammable liquids		0.50	
2. Process upset or purge failure		0.30	0.30
3. Always in flammable range		0.80	
D. Dust explosion (see table 3)		0.50–2.00	
E. Pressure (see figure 2). Operating pressure 75 bar a; Relief setting 90 bar a			1.10
F. Low temperature		0.20–0.30	
G. Quantity of flammable/unstable material: Quantity 3000 kg. H_c=20800 Btu/Lb			
1. Liquids or gases in process (see figure 3)			
2. Liquids or gases in storage (see figure 4)			0.25
3. Combustible solids in storage. Dust in process (see figure 5)			
H. Corrosion and erosion		0.10–0.75	0.10
I. Leakage: Joints and packing		0.10–1.50	0.30
J. Use of fired equipment (see figure 6)			
K. Hot oil heat exchange system (see table 5)		0.15–1.15	
L. Rotating equipment		0.50	
Special process hazard factor (F_2)			3.25

Process unit hazard factor (F_1 x F_2 = F_3)			6.50
DOW fire and explosion index (F_3 x MF = F&EI)			**156**
Radius of exposure (m)			**40**

Figure 2.4(c)   Calculation of Dow Fire & Explosion Index for zone 2 in Example 2.2.

providing a reasoning procedure, capable of facilitating the identification of deviations. Each time a reasonable deviation is identified, the causes and consequences are analysed as well as the possible corrective actions, making an orderly record of the results of the analysis. Although obviously there is no guarantee that all the possible deviations are identified, it is a very considerable improvement over the previous procedure.

2. Its multidisciplinary nature: The HAZOP analysis is applied by a team, which should be formed by people from different disciplines, from inside and often from outside the company or plant. The method is based on the principle that

DOW DOW FIRE AND EXPLOSION INDEX	Plant *Etilento Español S.A.*		Date *May 1995*
Site	Process unit *Ethylene pumping and vaporisation*	Evaluated by	Revised by

Materials and process		
Materials in process unit	*Ethylene* *Sea water*	

State of operation □ Design □ Start-up □ Normal operation □ Shut-down	Basic material(s) for material factor *Ethylene*

MATERIAL FACTOR (see table I and appendices A and B) Note requirements when unit temperature over 60° C		24

1. GENERAL PROCESS HAZARDS	Penalty factor range	Penalty factor used
Base factor	1.00	1.00
A. Exothermic chemical reactions	0.30–1.25	
B. Endothermic processes	0.20–0.40	
C. Material handling and transfer	0.25–1.05	0.50
D. Enclosed or indoor process units	0.25–0.90	
E. Access	0.35	
F. Drainage and spill control. Volume…m³	0.25–0.50	0.50
General process hazard factor (F₁)		1.50

2. SPECIAL PROCESS HAZARDS		
Base factor	1.00	1.00
A. Toxic materials	0.20–0.80	0.20
B. Sub-atmospheric pressure (<500 mmHg)	0.50	
C. Operation in or near flammable range: □ Inerted □ Not inerted		
1. Tank farms storage flammable liquids	0.50	
2. Process upset or purge failure	0.30	0.30
3. Always in flammable range	0.80	
D. Dust explosion (see table 3)	0.50–2.00	
E. Pressure (see figure 2). Operating pressure 5 bar a; Relief setting 7 bar a		0.28
F. Low temperature	0.20–0.30	
G. Quantity of flammable/unstable material: Quantity 3000 kg. H$_c$=20800 Btu/Lb		
1. Liquids or gases in process (see figure 3)		
2. Liquids or gases in storage (see figure 4)		0.25
3. Combustible solids in storage. Dust in process (see figure 5)		
H. Corrosion and erosion	0.10–0.75	0.10
I. Leakage: Joints and packing	0.10–1.50	0.10
J. Use of fired equipment (see figure 6)		
K. Hot oil heat exchange system (see table 5)	0.15–1.15	
L. Rotating equipment	0.50	0.50
Special process hazard factor (F₂)		2.73
Process unit hazard factor (F₁ x F₂ = F₃)		4.10
DOW fire and explosion index (F₃ x MF = F&EI)		**98**
Radius of exposure (m)		**25**

Figure 2.4(d)   Calculation of Dow Fire & Explosion Index for zone 3 in Example 2.2

people with distinct experiences and training can interact better and identify more problems when they work together than when they work separately and combine the results later. The varied approaches to a problem are what make the HAZOP analysis a tool which stimulates the generation of ideas. In particular, the method presupposes that the members of the HAZOP team do not hesitate to discuss the deviations, causes, consequences and solutions which occur to them, although at first sight these may appear to be rather unreasonable or even impossible, as it could stimulate other members of the team to think of deviations, etc., which are similar but possible. To achieve this objective it is necessary that all the members expound their ideas freely, and at the same time avoid excessive criticism of those shown by others, so as not to inhibit participation.

2.4.2 Methodology of HAZOP analysis

As previously indicated, a HAZOP analysis can be carried out on operating plants or on one which still has to be built. For the latter, the HAZOP analysis as discussed in this section requires that the design is well-defined, so that the necessary information is available. In particular, the application of the method demands that the piping and instrumentation diagrams (P&IDs) are complete. From this moment on, the HAZOP analysis should be carried out as soon as possible, with the aim of being able to implement the necessary changes at a minimum cost. However, it is important to also point out that in later stages this or other types of analysis may be carried out. Therefore, as soon as the raw materials and intermediate products have been identified, a critical review should be carried out to evaluate the possibilities of changes in the process towards routes of greater intrinsic safety, and also to identify the areas where more information is needed on the dangerous properties of the substances involved [10].

In accordance with the general guidelines explained, the development of a HAZOP analysis requires, as a preliminary step, the formation of a competent team, in which people with different functions participate. For a new design, typically there should be at least one project engineer (normally the person responsible for completing the project within budget) [1], a process engineer and an instrumentation/control engineer, as well as one or more persons from the production area (production engineer, foremen, operatives), and sometimes a chemist (who may come from the research and development department, especially if the materials handled have new characteristics or if it is thought that unusual reactions could take place). Apart from the above, it is fundamental that someone with considerable previous experience in HAZOP analysis directs the study, who does not need to know the plant in question. His mission consists of acting as coordinator, ensuring that the correct procedure is followed and that no significant aspects are left out, stimulating discussion, etc. He would normally put forward the views of the safety department (otherwise, a member of this department should also form part of the team).

The organization of the HAZOP analysis is indicated in Figure 2.5. From the **guide words** given in Figure 2.6, or other similar words, the procedure in Figure 2.5 is initiated, applying the guide words to each one of the process lines which enter or exit from a particular unit in the plant. The guide words are applied to actions (reaction, transfer, etc.) as well as to specific parameters (pressure, temperature, etc.).

As a first step, for each one of the process lines the **intention** is usually specified, i.e. the purpose that it fulfils in the plant, under normal operational conditions (for example, provide heating steam to reactor R12 at a certain pressure, temperature and flow rate). From here on, the application of the guide words help

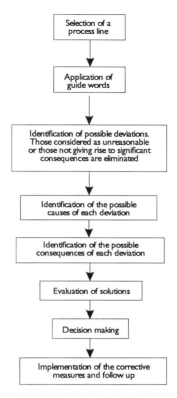

Figure 2.5 Schematic of HAZOP analysis.

identify *deviations*, that is, circumstances where the defined intention is not met (for example, the pressure in the line is too high, there are variations in the flow, interruptions in the vapour supply, etc.). The deviations produce **consequences** (for example, excessive heating of the R12 reactor), and at the same time have **causes** which give rise to them (for example, human error, control valve failure, etc.). For a deviation to be considered in the analysis it must have significant consequences and reasonable causes. Once a deviation with these characteristics is identified, the next step consists of proposing corrective solutions, and assessing their cost. In some cases a more in-depth analysis is required including for example, a computer simulation of the accident to estimate its consequences, and/or a formal study of the probability of the event. In this case, the HAZOP team may recommend an in-depth study before suggesting the implementation of specific measures. In many other cases the HAZOP analysis is sufficient to decide the application of corrective measures or even changes in the design.

Lastly, it is essential to guarantee a systematic record of the HAZOP analysis results, which is usually carried out using the traditional format in columns shown in Example 2.3. There is commercial software which helps to produce acceptable

HAZOP analysis reports, with built-in functions for rapid examination of earlier records. A computerized file of previous HAZOP analysis results helps to carry out new analyses with greater effectiveness, as often the ideas, deviations identified and solutions proposed by other teams for other installations are applicable to that which is under analysis. A well-organized record also simplifies the preparation of summaries and notes which could be circulated to the team members before future meetings, as well as lists of pending questions, etc.

2.4.3 Application of HAZOP analysis to sequences of operations and discontinuous processes

The guide words shown in Figure 2.6 are some of those regularly used in continuous processes. If the plant operates in a discontinuous way, the method of analysis is subject to some modifications. In a continuous process it is assumed that the condition of a certain equipment or installation does not vary during 'normal' operation. Conversely, in a discontinuous process there is an inherent temporary variation, and, therefore, an installation passes through different states. Another characteristic which complicates the analysis of these processes is that often the same discontinuous installation is used for distinct purposes, thus a discontinuous reactor could operate in different ways to obtain distinct products, or a distillation column could be used to purify a considerable variety of products. Obviously this increases the possibilities of cross-contamination and interference between successive operations.

In a discontinuous process the P&ID of the installation does not indicate its current state and so complementary information is required. On the other hand,

No	The anticipated intentions of the design are not achieved, e.g. no flow in a line
More/less	Qualitative increases or decreases from the values intended, e.g. higher temperature, lower reaction rate, higher viscosity, etc.
Besides	The design intentions are achieved, but something else also happens, e.g., the steam heats the reactor but also increases the temperature of a nearby flange
Part of	Only part of the events occur as anticipated, e.g. the desired feed rate is achieved but the feed composition has changed.
Inversion	An effect opposite to that desired occurs, e.g., the flow takes place in the reverse direction, the system cools down instead of heating up, the reverse reaction takes place, etc.
Other than	The desired result is not achieved. Instead, something completely different occurs, e.g., catalyst change, spurious valve operation, unanticipated stop, etc.

Figure 2.6 Frequently used HAZOP guide words.

multiple steps are often used in discontinuous processes. Thus, in a discontinuous chemical reactor, deviations are possible during loading, heating to operation temperature, reaction with controlled heat elimination, cooling to unloading temperature, unloading and cleaning of the reactor.

In addition to the process lines, Kletz [1] recommends applying guide words to the operations to be carried out in a batch process. For instance, in the operation of loading a tonne of a reactive A to a reactor, the team should consider deviations such as:

- A is not loaded
- More A is loaded
- Less A is loaded
- In addition to A, something else is loaded
- Instead of A, something else is loaded
- Only part of A is loaded (if A is a mixture)
- Reverse loading (i.e. flow from the reactor to the container of A)
- A is added too soon.
- A is added too late.
- A is added too quickly.
- A is added too slowly.

In the same way, discontinuous operations that are carried out in relation to a continuous process (for example, maintenance work on process equipment, changes of catalyst, catalyst regeneration, start-up, shut-down, etc.) should be studied, examining the proposed order of operations, and applying adequate guide words to each stage.

The above is also applied to the instructions contained in computer programs which control a specific automatic action (for example, the corrective actions in the presence of an increase in pressure or concentration above a certain value, the actions to follow for an automatic stop, etc.). In this case an examination of the computer's response for each of the possible deviations is required. In addition, specific circumstances such as power cuts or loss of the signal from an instrument in the plant should also be analysed.

Example 2.3

The reactor of Figure 2.7(a) is used in the production of butadiene from the dehydrogenation of a feed consisting fundamentally of butane and butene. The reaction is carried out at about 590 °C, with a Cr_2O_3/Al_2O_3 catalyst diluted with inert solids in a 4:1 ratio. As a whole, the process is endothermic despite the exothermicity of the coke deposition which takes place simultaneously with the dehydrogenation of the reactants. When the level of coke deposited in the reactor reaches a certain value, the regeneration of the reactor by combustion of the coke deposits becomes necessary. This is a strongly exothermic process, and is used to compensate for the decrease in *temperature during the dehydrogenation, so that the complete dehydrogenation/regeneration cycle is practically autothermic. Both stages are carried out in the same fixed bed reactor.*

There are two main problems associated with the regeneration process: The first consists of ensuring that the evacuation of flammable hydrogen vapours and hydrocarbon vapours (reactants, products and volatile fractions of the coke deposits) is completed before proceeding to introduce oxygen. The second problem is in carrying out the combustion of the coke deposited on the catalyst while at the same time avoiding the formation of hot spots in the reactor, which could cause irreversible deactivation of the catalyst by sintering, as well as damage to the reactor. The temperature is maintained within acceptable limits by exercising a careful control of the quantity of oxygen fed to the reactor. The normal procedure during regeneration is the following:

1. *Begin by purging the reactor with steam at 450°C, to drag out all the flammable vapours which are present.*
2. *Once the concentration of hydrocarbons in the exit stream is not detectable, oxygen is gradually introduced. For this, the air flow through line L2 is regulated to give an oxygen concentration of 2–3% at the reactor inlet (after mixing with steam). This raises the reactor temperature to around 600°C. Continue feeding this proportion until the temperature starts to decrease, which indicates the end of the combustion of the easily accessible coke and the start of the diffusion-controlled process.*
3. *Continue increasing the concentration of air in stream L3, while maintaining the maximum temperature of the reactor at 600°C, until the air feed reaches 100%. When the temperature falls below 500°C, the regeneration is concluded.*
4. *Purge again with steam to eliminate the oxygen from the reactor. When the concentration of oxygen in the exit stream is very low or zero, the dehydrogenation process starts again.*

Carry out a HAZOP analysis of the regeneration process as described above.

The regeneration operation described corresponds to an unsteady state process, i.e. although there is a continuous flow of gases, the process is discontinuous for the solid. It could thus be considered as an intermediate case between the two possible situations in the application of HAZOP analyses. However, the method for a continuous process is used here, as it is more appropriate for the case studied. Next, a HAZOP analysis is carried out for the L1 to L4 streams. The application of the method to line L5 requires a description of the equipment downstream from the reactor.

It should be borne in mind that the study presented here is necessarily simplified due to reasons of space. Only the most obvious deviations are pointed out with the aim of illustrating the application of the method. On the other hand, Figure 2.7(a) shows an incomplete P&ID, in which the process lines that are used in the dehydrogenation stage do not appear, although they should obviously be considered in a HAZOP study.

The completed HAZOP form is shown in Figure 2.8. It can be seen that a very significant increase in the initial instrumentation is recommended. This reduces considerably the probability of accidents, especially of an explosion or of a runaway reaction. In addition, the application of the HAZOP analysis introduces important improvements in the operability of the reactor. Thus the new P&ID shown in Figure 2.7(b) corresponds to a reactor which is not only safer, but is also more efficient.

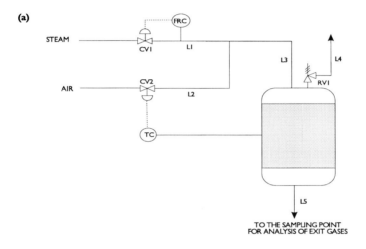

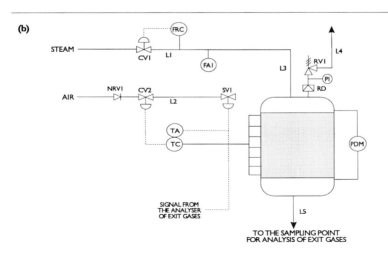

Figure 2.7 Piping and instrumentation diagrams (P & IDs) of the installation in Example 2.3 (a) before and (b) after the HAZOP analysis.

As already indicated, although there is a continuous flow of gas, the operation of the reactor of Figure 2.7 is carried out under non-stationary conditions. In these cases it is useful to also check the guide words for the operation of discontinuous systems and see whether they are applicable. For example, the guide sentences 'instead of A load B', 'A is loaded too soon', 'A is loaded too late', etc., would also have helped to identify some of the deviations shown in Figure 2.8.

2.4.4 'What if' analysis

'What if' analysis is much less structured than HAZOP analysis discussed above, although its application presents some obvious similarities. Owing to this lack of structure, members of the team carrying out the analysis need to have considerable experience, otherwise important omissions are likely.

The objective of a 'what if' analysis is to consider the negative consequences of possible unexpected events. The 'what if' approach uses the question 'What would happen if...?' applied to deviations in the design, construction, modification and operation of industrial installations. The questions refer to specific areas (for example, electrical safety, fire protection, instrumentation of given equipment, storage, handling of materials, etc.) and are addressed by a team of several experts who have detailed documentation of the installation and operating procedures as well as access to plant personnel to obtain complementary information. In general, from the application of the question 'What would happen if...?' suggestions of initiating events and possible failures are obtained, from which a dangerous deviation could happen.

For example, in a batch reactor we could ask: 'What would happen if the proportions of the loaded reactants were incorrect?', and after the corresponding debate arrive at the conclusion that, in the case that the concentration of one of the reactants was too high, a runaway reaction could happen, with a dangerous increase in pressure and temperature. The analysis does not end here, but goes on to examine the possible corrective actions, such as modification of the emergency system (for example, taking measures to introduce a quenching agent in the reactor to achieve a rapid extinction of the reaction), or modification of the operational procedures to decrease the probability of a failure of this type (for example, analysis of the reactor load before preheating, etc.).

2.4.5 Fault tree analysis (FTA)

The chemical industry started to use fault tree analysis (FTA) in the 1960s, following the development of this technique by Bell Laboratories. Fault tree analysis assumes that an unwanted event (an accident or a dangerous deviation of any kind), called the **top event**, has already happened, and looks for its causes and the sequence of events which could bring it about. As in previous cases, not all of the possible causes and sequences of events are necessarily identified, so it is advisable to combine FTA analysis with other techniques which increase the reliability of identification. The general principle that unidentified hazards are uncontrolled hazards should always be borne in mind.

Line L1
Intention: Supply steam at 450°C to the reactor, with a flow rate of up to 10 000 kg/h

Guide word	Deviation	Causes	Consequences	Action required
NO	No flow	Valve CV1 fails closed Line fracture Line blockage Failure of upstream equipment	a) If there is no steam flow during the purge step flammable vapours could remain in the reactor: Possible explosion at the start of the regeneration stage b) During regeneration, oxygen concentration would be too high: Formation of hot spots and possible runaway reaction	a) Install a low flow alarm, FA1 b) Covered under 'More flow' (air)
MORE	More pressure	Failure of CV1	If the pressure drop through L3 is sufficiently high, reverse flow could take place in L2 during the purge stage	Install a non-return valve in L2, NRV1
LESS	Less flow	CV1 failure Pressure drop in the reactor increases due to interparticle deposit of catalyst fines or to coke growth	a) Same as 'No flow' b) Pressure drop too high	a) Covered under 'No flow' a,b) Install pressure differential gauge PDM to measure the pressure drop in the bed. Record the readings to assess gradual plugging over long periods of time
	Less pressure	Unreliable steam supply CV1 failure	As in 'No flow'	Covered under 'No flow'

Figure 2.8 Results of HAZOP analysis for example 2.3

Figure 2.8   *cont'd.*

Line L2
Intention: Supply air at ambient temperature to the reactor, with a flow rate of up to 10 000 kg/h

Guide word	Deviation	Causes	Consequences	Action required
NO	No flow	Blower failure CV2 fails closed NRV1 blocked Line fracture	No air enters the reactor, complete regeneration is not achieved (the regeneration is limited to the fractions of coke which can be gasified with steam)	No specific action required. The failure has consequences regarding operability, but it is evident (the expected increases in temperature do not occur), and can be corrected
MORE	More flow	Failure in CV2 Failure in temperature control system TC	Too much air enters the reactor. Rapid temperature increase, runaway reaction	A single temperature measuring point is insufficient, as steep temperature gradients are likely. Install a set of thermocouples at different bed positions. Implement *auctioning control* (the thermocouple registering the highest temperature provides the control signal for CV2) Install a high temperature alarm, TA, interlocked to a plug valve for rapid flow interruption, SV1
LESS	Less flow	Similar to those discussed under 'No flow' (in this case, partial rather than complete failures)	Similar to those discussed under 'No flow'	No specific action required
	Less pressure	Blower failure	a) As in 'No flow' b) Possibility of reverse flow in L2	b) Covered by the installation of NRV1
PART OF	Dust particles in air flow	Damaged filter at the compressor air intake	Interparticle fouling in catalytic reactor, partial blockage of air flow	Frequent filter maintenance

Figure 2.8 *cont'd.*

Line L3
Intention: Supply steam, air or steam/air mixtures to the reactor, at the desired temperature, pressure and flow

Guide word	Deviation	Causes	Consequences	Action required
NO	No flow	See causes for 'No flow' in lines L1 and L2 L3 fracture	Regeneration is not accomplished	Covered by previous sections
PART OF	Too much air in the feed	Fault in the air flow and/or steam regulator	As in 'More flow' (air) and 'Less flow' (steam)	Covered by previous sections
OTHER THAN	In the purge stage, air is introduced instead of steam	Failure in the valve opening sequence, due to failure of the control system or to a human error during manual operation	Air is introduced into a reactor full of flammable vapour. Probable explosion.	Install appropriate interlocks so that valve SV1 remains closed as long as the continuous gas analyser in L5 detects the presence of flammable vapours in the reactor exit stream

Line L4
Intention: Provide the required relief flow in case of reactor overpressure, directing it to the flare by way of a sufficiently wide duct

Guide word	Deviation	Causes	Consequences	Action required
NO/LESS	No flow or less than anticipated flow through RV1	RV1 failure. Partial obstruction of the valve clearance due to coke deposition	The valve does not provide the necessary relief flow. Reactor over pressuring is likely	Frequent valve maintenance. Consider the possibility of installing a rupture disc RD in series with RV1. Disc material should not favour coke formation
NO	Rupture disc RD does not open (this assumes that RD has been installed)	The rupture disc leaks, causing the pressure to balance on both sides	The relief system does not work as planned. Reactor overpressuring is likely	Install a pressure indicator, PI, between the rupture disc and the safety valve

Fault tree analysis is, therefore, a deductive process which helps to determine how a particular event could take place. As a risk analysis method it is one of the most structured, and can be applied to a single system or to interconnected systems. It is, moreover, one of the few techniques capable of dealing adequately with common cause failures, which are discussed in Chapter 6. However, the application of the FTA analysis to complex systems may disguise considerable mathematical difficulties for the inexperienced. This has given rise to extensions of the method such as HARA analysis (Hazard Assessment by Risk Analysis), which is more easily applied [11].

In relation to the techniques discussed up to now, FTA analysis has the additional advantage of serving not only to identify process hazards, but to quantify the risks involved. Fault tree analysis breaks down an accident into its contributing factors, whether human errors, plant equipment failures, external events, etc. The result is a logical representation of the different sequences of events capable of generating the top event, at the apex of the tree. The following simplified example illustrates the application of the analysis.

Example 2.4

A chemical engineering student plans a well-deserved graduation journey to various countries. As he thinks he will be away for 6 weeks, he decides to install an automatic watering system, with the aim of keeping his plants alive. The system is shown in Figure 2.9. As an additional precaution he leaves his keys with a neighbour, asking that he checks the health of the plants and the functioning of the watering system. The system in Figure 2.9 consists of a tank with sufficient capacity to cover the demand for water (the student has measured the volume of water he uses to water plants, averaging it for the last 4 weeks, and has added a safety factor of 30%), which he has placed on the table, at a sufficient height according to his calculations, a restriction in the feed line to help regulate the flow of water, an electrovalve and a timer which opens for15 minutes each week. If the plants remain without water for more than 1 week they will dry out. Before leaving, the student has carried out tests on the system, and is satisfied with its operation. Carry out a fault tree analysis for the top event 'The plants dry through lack of water'.

The fault tree is shown in Figure 2.9. It contains some basic symbols which are commonly used, such as OR gates and AND gates. The logic of an OR gate is that the output event is always verified when at least one of the input events occurs. On the contrary, the AND gate demands that all the input events are verified for the output event to take place. Other symbols used in fault tree diagrams appear in Figure 2.10. In our case, the AND gate indicates that for the top event to occur not only must the irrigation system fail, but also the periodic revisions should fail. In accordance with the logic

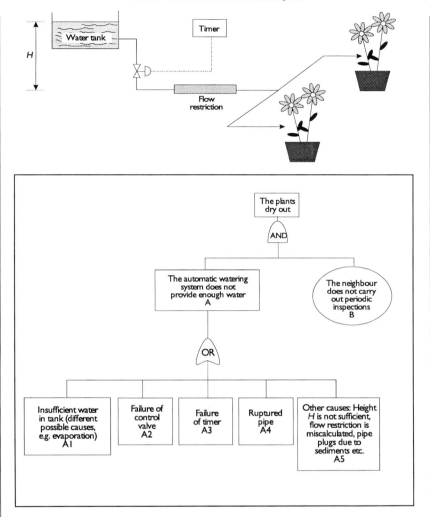

Figure 2.9   Watering system and fault tree for the top event 'the plants dry out' in Example 2.4.

applied to fault trees, which is discussed more generally in Chapter 6, the top event which corresponds to the output of the AND gate exit has a probability equal to the product of the probabilities of events A and B. In addition, A is the output of an OR gate with several inputs. The probability of event A with two inputs M and N is given by

$$P(A) = P(M \text{ or } N) = 1 - \{1 - P(M)\} \{1 - P(N)\} \tag{2.4}$$

$$P(A) = P(M) + P(N) - P(M)P(N) \tag{2.5}$$

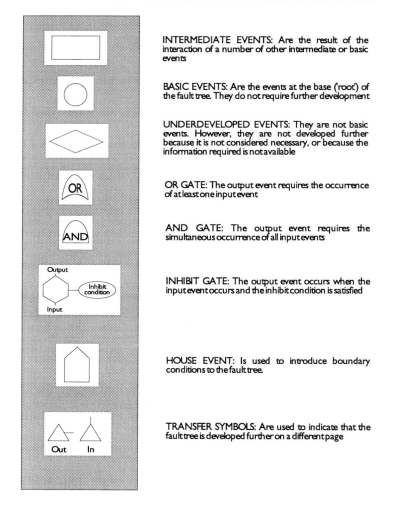

INTERMEDIATE EVENTS: Are the result of the interaction of a number of other intermediate or basic events

BASIC EVENTS: Are the events at the base ('root') of the fault tree. They do not require further development

UNDERDEVELOPED EVENTS: They are not basic events. However, they are not developed further because it is not considered necessary, or because the information required is not available

OR GATE: The output event requires the occurrence of at least one input event

AND GATE: The output event requires the simultaneous occurrence of all input events

INHIBIT GATE: The output event occurs when the input event occurs and the inhibit condition is satisfied

HOUSE EVENT: Is used to introduce boundary conditions to the fault tree.

TRANSFER SYMBOLS: Are used to indicate that the fault tree is developed further on a different page

Figure 2.10   Commonly used symbols in fault tree analysis.

For a low value of the probability of individual events the term $P(M)P(N)$, containing the product of the probabilities of two independent events, can be neglected, and equation (2.5) can be approximated by

$$P(M \text{ or } N) = P(M) + P(N) \tag{2.6}$$

In a similar way, in the case of various inputs of low probability, the probability of the output event of an OR gate can be written approximately as

$$P(M \text{ or } N \text{ or } P \text{ or}... S) = P(M) + P(N) + P(O) + ... P(S) \qquad (2.7)$$

In the case considered, the OR gate has five entrances, and in accordance with equation (2.7), the probability (or frequency) of event A will be the sum of the probabilities (or frequencies) of events A1 to A5. Let us assume that our information establishes that events A1, A3, A4 and A5 have a frequency of 2, 1, 0.003 and 0.2 times per year respectively, and that the control valve has a probability of failure on demand (event A2), of 0.01 (i.e. it will fail to operate once every 100 times that action for this particular valve is requested). This poses a problem of units. Probabilities or frequencies can be added separately but not probabilities and frequencies. Likewise, frequencies can be multiplied by probabilities, or probabilities between themselves, but multiplication of frequencies between themselves is not advisable. To solve the problem of adding probabilities and frequencies it is sufficient to change A2 to frequency units. If we assume that the demand rate of once a week on a control valve is maintained, there will be 52 demands throughout the year, and the failure rate will be $0.01 \times 52 = 0.52$ times per year. In this way, the frequency of event A will be:

$$F(A) = 2 + 0.52 + 1 + 0.003 + 0.2 = 3.723 \text{ times/year}$$

If we rate the probability that his neighbour forgets to carry out the weekly inspections for this period at 3% (he is a highly responsible neighbour), the frequency of the top event (AND gate) is given by the product of F(A) and P(B), i.e. 0.112 times/year. As the student is going to be absent for 6 weeks (0.115 years), the probability of him finding his plants dry on his return is 0.013, or 1.3%. Another way of viewing this is as follows: If the student goes on a 6-week trip once a year and the fault rates previously described are constant, for approximately 1 out of every 77 journeys he would find the plants dry on his return. Note that in the previous example, human errors (B), equipment failures (A2, A3, A4) and maintenance failures (A1, A5) are included. In the example, arbitrary values of frequency and failure probabilities for each of the different events have been assigned. In a real case, each one of the fault tree branches should be developed for the basic events or until reaching a point at which a frequency or probability can be assigned with a reasonable certainty. Thus, for example, event A2 could be developed further through an OR gate which would branch to several events: An electrical power failure (which would produce a common cause failure in the timer), a mechanical failure, etc.

The previous example constitutes a simplified application of the fault tree logic. Many other examples of FTA applications can be found in the references given at the end of this chapter [1, 3, 5, 12–16]. Following is another example of FTA taken from a paper by Aelion and Powers [14].

Example 2.5

To avoid reverse flow when starting operation of the centrifugal pump corresponding to the system in Figure 2.11, the following operation procedure is used:

1. *With the discharge valve initially closed, start pump.*
2. *Observe pressure build-up.*
3. *When this is sufficient, open the discharge valve.*
4. *Carry out flow control starting with the control valve in closed position. The control can be done manually or automatically. During start-up the flow control loop is on manual.*
5. *Check that the pump is not left running for too long with the exit valve closed, to avoid overheating.*

Carry out an FTA analysis taking 'reverse flow' as the top event in the system described.

The fault tree corresponding to this example is shown in Figure 2.12. For reverse flow to occur when starting the feed to the mixer, all the events at the input of the AND gate should happen simultaneously: Both valves should be open and in addition there should be a driving force for the reverse flow. When the tree branch corresponding to the flow control valve is developed, we find an OR gate with two inputs: The control valve is stuck open (equipment failure) or it is prematurely opened by the operator (human error). On the other hand, the existence of a reverse pressure gradient could result from the loss of pressure in the feed or from an excessively high pressure in the mixer. Note that none of the events at the base of the tree is further developed. Instead, a probability value is assigned and the downwards tree construction is interrupted. These underdeveloped events are identified with a diamond. Logically, in most cases, it is necessary to develop the tree down to the basic events to assign this value. Thus, the high pressure in the mixer could be due to a number of causes (excessive temperature, chemical reaction, failure in the control of the other stream entering the mixer, etc.), which would have to be examined independently.

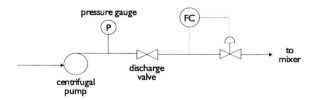

Figure 2.11   Detail of feed system to the mixer (Example 2.5).

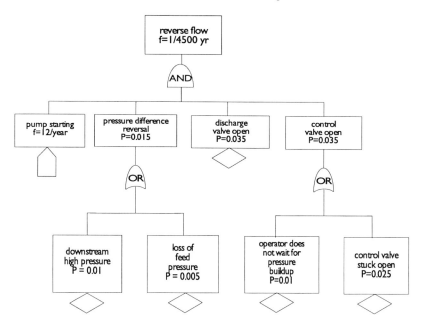

Figure 2.12 Fault tree analysis for Example 2.5.

The two examples given are very simplified. Rigorous fault tree analysis in complex systems requires considerable time and expertise in its application. Despite this, mathematical techniques have been developed which permit simplification of fault trees, and there are computer programs on the market which notably facilitate their construction. It is also important to identify the **minimal cut sets**, which is the name given to those sets of basic events which are sufficient to cause the top event (in Example 2.4 it would be the groups formed by event B and any of the events Al to A5). Once the minimal cut sets are identified they must be classified, assessing their importance by taking into account both the number of events involved and their probability. In Chapter 6 the terms used are defined with more precision, and sources of data on failure rates of equipment and individual components are discussed. Also, some of the simpler rules used to carry out formal FTA and studies of the reliability of equipment and components are explained.

2.4.6 Event tree analysis (ETA)

The fault tree analysis just discussed starts at a certain event and investigates reasonable mechanisms through which this could take place. Contrary to the procedure followed in FTA studies, event tree analysis (ETA) evaluates the consequences which could take place after a certain event. Studying how the initiating event could occur does not concern ETA, but rather what are its possible results. Therefore, in ETA emphasis is placed on the initial event which is assumed to have happened, and a logic tree is constructed which connects the said initial event to the final effects, where each tree branch represents a possible line of evolution leading to a final effect (or to the absence of one if a sequence of favourable circumstances avoids harmful consequences).

Event tree analysis is especially suitable for studying possible sequences of events after an accident. This permits analysis of the possible scenarios and the establishment of a hierarchical structure for them, taking into account their severity and probability, selecting emergency scenarios for their quantitative evaluation and preparation of adequate responses for them. ETA is developed in accordance with the following scheme [3]:

1. Identification of relevant initiating events.
2. Identification of the safety functions designed to respond to a given initiating event.
3. Construction of event tree.
4. Description of the sequence of resulting events.

In the context of the analysis proposed here, the initiating event could be any important deviation provoked by equipment failure or human error. This initial event could have very different consequences, depending on the safeguards implemented in the system, on the workers' reaction, and the concomitant circumstances. For the application of the event tree analysis to make sense, the initiating event should not be too close to the final effects. Thus it would not be sufficient to define as initiating event the explosion of a reactor and investigate from then on its possible effects such as magnitude of the shock wave, domino effect, etc. In this case, the most suitable approach would consist of carrying out an FTA analysis, which takes the explosion of a reactor as the top event and investigates its causes. Each sequence of events (cause – intermediate events – explosion) would therefore give rise to an accident (explosion) of different characteristics, and, therefore, to different effects. On the contrary, to apply event tree analysis one must select a deviation which does not directly lead to a specific final accident. For example, 'the refrigeration system is insufficient' or 'the dosage of the feed to the reactor is incorrect', and from that, analyse the system response.

As already explained, the initiating event may give rise to different sequences of events. Thus, in the case mentioned before about the feed dosage being incorrect, only in some cases (increase in the rate of an exothermic reaction causing runaway, reaction mixture within explosion limits, etc.) will dangerous circumstances appear. In others the system will be capable of self-regulation or the accident will simply not give rise to a hazardous situation, even though it may imply a loss of efficiency in the operation. In this way, for example, an incorrect dosage of the feed would lead in many cases to extinction of the reaction through lack of one of the reactants. On the other hand, the design of process equipment is expected to include a sound control system and built-in safety responses capable of handling a majority of common deviations. Thus, even in the case of an unforeseen deviation causing a hazardous situation, the system should have control elements and safety functions (feed stop, explosion suppression system, pressure relief systems, etc.) capable of correcting the deviation before catastrophic consequences happen.

Once the first two stages have been carried out (identification of initiating events and safety functions), the event tree can be developed to the final effects. The estimation of the magnitude of these generally requires the use of quantitative consequence analysis, using models capable of estimating the final effects for a given scenario.

Example 2.6

A mixture of ethylene, acetylene and hydrogen is fed into a fixed bed catalytic reactor with the aim of selectively hydrogenating the acetylene fraction (Figure 2.13). The reaction is exothermic, and to obtain a good selectivity a supported Pd catalyst is used, on which the reaction takes place at about 90°C. Construct an event tree for the case of a leak occurring in the control valve at the exit of the reactor (frequency = 0.001/year).

In accordance with the characteristics of the above scenario, the initiating event consists of a leak of flammable gases below their autoignition temperature. Therefore, as will be shown in Chapter 3, although a flammable mixture with air is produced, an external source of ignition is required for an explosion to occur. The external ignition could take place through sources in the process (open flames, hot surfaces, electric motors, etc.) or through external sources (motor vehicles, cigarettes, etc.). Likewise, it could happen almost immediately after the leak, in which case the total quantity of flammable mixture formed with the surrounding air will be small and there will not be significant pressure effects, or it could occur after a certain time, which would give rise to a substantial accumulation of flammable material and to an unconfined vapour cloud explosion (UVCE), with a high destruction potential. Lastly, the leak could be detected at an early stage, and controlled by closing the corresponding isolation valves (for example, V-2), or in very favourable circumstances, even if the leak is not detected rapidly, ignition might not occur, so the vapours emitted would disperse in the atmosphere without igniting.

The previous possibilities are reflected in the event tree diagram shown in Figure 2.14, in which arbitrary probability values have been assigned to each of the events which form the branches. In Chapter 3 there is a more extended discussion of the possible consequences of a leak of this type. Note that other possibilities could have been included, or the ones shown could have been developed further (for example, considering whether or not the flammable cloud moves towards populated areas, considering domino effects, etc.).

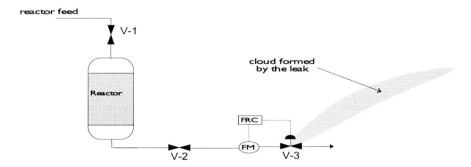

Figure 2.13   Schematic of the installation in Example 2.6.

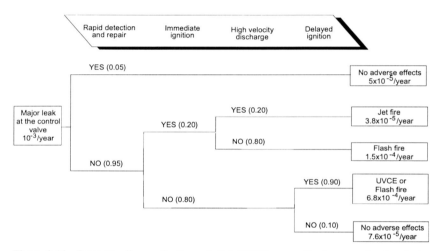

Figure 2.14   Event tree diagram for Example 2.6. UVCE = unconfined vapour cloud explosion.

2.4.7   Failure modes and effects analysis (FMEA)

Failure modes and effects analysis (FMEA) consists of an examination of individual components with the object of assessing the effect that their failure could have on the behaviour of the system. It is a systematic analysis, often of considerable length, which is mainly concerned with component failures rather than human errors. It has been used extensively in the nuclear industry for steady-state studies, in preference to other techniques such as HAZOP analysis, which is preferentially used to study the hazards associated with start-up and shut-down operations [1].

In the context of this analysis, a failure mode is 'a symptom, condition or fashion in which hardware fails. A mode can be identified as a loss of function, premature function (function without demand), an out of tolerance condition, or a simple physical characteristic such as a leak (incipient failure mode), observed during inspection' [17]. In FMEA all of the known component failure modes are considered in turn and the consequences of each failure are analysed and recorded.

FMEA is carried out by a team and requires considerable documentation including process and instrumentation diagrams, electrical diagrams, operation procedures, instrument logic diagrams, information on controls and inter-dependences, etc. [18]. The team which carries out the analysis should have sufficient information to understand the design and the operation of a component, and its interaction with the system of which it forms a part. The team leader should have previous experience in FMEA studies, knowledge of engineering systems, including process control and mechanical and electrical design, as well as experience regarding failures of equipment and transient operation.

The rest of the team should include complementary experiences, at least one control specialist and a system engineer familiar with the design and operation of the installation under study. It is essential that the team is capable of analysing not only the direct effects of failure mode, but also its influence on the operational parameters of the system and the response of the control system during the transient state.

The different steps in the development of FMEA analysis begin with the definition of the system and the level of detail in the study. The definition of a group of components at the P&ID description level can hardly ever be considered complete for an FMEA study, as the operation of the system requires external inputs (energy sources, water, information from the plant control system, etc.). Therefore, the first mission of the analysis team is to establish the functional limits of the study. As to the degree of detail, different levels can be considered [3]. If a study is carried out at the plant level, the FMEA should be focused on the individual systems (such as the feed system, mixing system, reaction system (reactor plus auxiliary systems), separation systems, utilities, etc.), and the effects of their possible failure modes on the operation at plant level. If an analysis is done on a system or subsystem level, the FMEA is carried out on individual equipment (feed pump, oxidation reactor, refrigeration circuit pump, refrigeration circuit control valve, temperature sensor and alarm, etc.). On occasion, if the team has sufficient experience and it is required, an FMEA study at subcomponent level could be tackled, although normally in a typical chemical industry FMEA the failure of subcomponents is considered as the component failure modes are being analysed.

The next stage consists of defining an adequate format for the study. The goal is to obtain a coherent analysis, and the way to achieve this is to have a standard format. A typical FMEA table includes formats of the type presented in Figure 2.15. When the last column of criticality ranking is included the analysis is usually called FMECA (Failure Modes, Effects and Criticality Analysis). A scale of 1 (without adverse effects) to 4 (immediate danger to personnel and installation, requiring emergency shut-down) has been suggested to rate the severity of the failure mode [3], with levels 2 and 3 corresponding respectively to low-risk situations, which do not require shut-down, and those of higher risk levels, which require normal shut-down.

Date:			Reference:	Page:	
Plant:			Evaluated by:		
System:			Supervised by:		
Item	Description/ comments	Failure mode	Detection of failure	Effects	Criticality ranking

Figure 2.15   Example of FMECA (Failure Modes, Effects and Criticality Analysis) form (adapted from [3] and [4]).

The most important step when filling in the FMECA form (Figure 2.14) is obviously the identification of all the relevant failure modes and the effects they produce. Thus, a pump which is operating normally could fail to stop on demand, could stop spuriously, the hydraulic pump seals could leak, etc. Each one of these failure modes can give rise to different effects, with different criticality ranking.

The method of failure detection is also very important. Consider a batch reactor used to carry out an exothermic reaction with a potential for runaway in the event of coolant failure. The reactor has a refrigeration system to control the reaction temperature, alarms for high reactor temperature and low coolant flow rate, with automatic activation of a valve for fast reactor unloading in both cases, leading to a safe reactor shut-down. The worst-case scenario corresponds to runaway reaction (failure of the temperature control) and failure of the protection systems (the temperature alarm fails off, or the dumping valve fails closed). The FMEA shows [10] that if the dumping valve fails closed, the failure remains undiscovered until the end of the cycle when the reactor is emptied, thus giving rise to a highly hazardous situation.

FMEA is therefore another tool for hazard identification and risk analysis which complements those already described. As in previous cases, the analysis does not end when the form is completed. Remaining for discussion are all those cases which require further study (which in many cases will demand a quantitative analysis). On the other hand, the relevant failure modes identified must be followed by corrective actions, which the team proposes or recommends for study by other experts. Normally, after a certain period of time, new team meetings are carried out or a follow-up committee convened, with the object of assessing the state of implementation of the recommendations.

2.5   Questions and problems

2.1   A plant for the production of acetone uses two storage tanks of 229 m^3 to store acetone. The tanks are vertical cylinders, made of carbon steel, and are usually 75% full. There is a containment dike which surrounds the tank on three sides and directs any spillage towards a pool located a sufficient distance away. The acetone is stored in the tanks at atmospheric pressure and ambient temperature. Inerting of the vapour space has not been anticipated. The tank is fed through a 2" pipe, using two alternating pumps. The tank unloads to cistern lorries in the installation next to the tanks, using another double set of pumps, with the necessary valves. Apply the Dow fire and explosion index to the acetone storage and loading facilities. Calculate the FEI value, the radius of exposure and the damage factor to the unit.

2.2   A benzene/toluene mixture (5:3 ratio), is fed into a 1.5 m diameter continuous distillation column. The unit has 18 valve plates, a total condenser which uses water as a coolant, and a 'kettle' type boiler heated by steam. The distillate contains 98% of benzene.

(a) Estimate the flammable liquid contents in the column.
(b) Apply the Dow fire and explosion index to the column. Calculate the FEI value, the radius of exposure and the damage factor to the unit.

2.3   Assume basic instrumentation and carry out a HAZOP analysis on the distillation column of the previous problem. Recommend adequate supplementary instrumentation/control in view of the results.

2.4   The feed of a fixed bed catalytic reactor is a mixture of ethylene, acetylene and hydrogen at 2 atmospheres. Selective hydrogenation of the acetylene fraction is desired (see Example 2.6). The reaction is exothermic, and the reactor works at a low temperature (around 100°C), with the aim of obtaining good selectivity. Because of the exothermic nature of the hydrogenation process the reactor requires cooling.

(a) After setting up a P&ID of the installation, carry out a HAZOP analysis, discussing the principal hazards of the operation.
(b) Recommend supplementary instrumentation/control (if any) to be installed.
(c) Select the most suitable refrigeration system and coolant. Discuss safeguards against the failure of the refrigeration system.

2.5   Carry out a qualitative FTA, taking as the top event the explosion of the reactor from the previous example.
2.6   Carry out an ETA for a leak of hydrogen cyanide (normal boiling point 26.1°C, flash point −18°C), through a fissure in a welded part of the storage tanks for this compound (276 m^3, 75% full) built in 304 stainless steel, insulated and maintained at 5°C and atmospheric pressure.

2.7   Carry out an FTA for the top event 'obtain the wrong product in a tobacco vending machine'. Analyse the different failure modes at the system level and carry out an FMEA.

2.8   What are the main hazards in a laboratory where certain liquid and gas chemical products are handled? Answer the question from the viewpoint of the possibility of loss of life because of an accident. When answering think of a specific laboratory of which you know the materials stored, preferably an accessible one, which can be inspected. After discussion, elaborate a safety checklist for the laboratory, taking as a basis the general one shown in this chapter. Carry out a safety analysis on the laboratory using the said checklist. Make recommendations to improve safety.

2.6   References

1. Kletz, T. (1992) *Hazop and Hazan. Identifying and Assessing Process Industry Hazards*, 3rd edn, The Institution of Chemical Engineers, Rugby.
2. Lee, S., Batres, R., Lu, M. L. and Naka, Y. (1996) *Study on Safety Evaluation for Startup*, Proceedings of the 5th World Congress of Chemical Engineering, San Diego, Vol. II, pp. 969–94.
3. Battelle Columbus Division-AIChE/CCPS (1985) *Guidelines for Hazard Evaluation Procedures*, American Institute of Chemical Engineers, New York.
4. American Institute of Chemical Engineers-Dow Chemical Company (1994) *Dow´s Fire and Explosion Index Hazard Classification Guide*, 7th edn, American Institute of Chemical Engineers, New York.
5. Crowl, D. A. and Louvar, J. F. (1990) *Chemical Process Safety, Fundamentals with Applications*, Prentice Hall, Englewood Cliffs.
6. Lees, F. P. (1980) *Loss Prevention in the Process Industries*, Butterworth-Heinemann, London.
7. ICI (1985) *The Mond Index*, 2nd edn, Imperial Chemical Industries Explosions Hazards Section, Technical Department, Winnington.
8. American Institute of Chemical Engineers-Dow Chemical Company (1994) *Dow´s Chemical Exposure Index Guide*, 1st edn, American Institute of Chemical Engineers, New York.
9. Swann, C. D. and Preston, M. L. (1995) Twenty-Five years of HAZOPs. *J. Loss Prev. Process Industries*, **8**, 349–54.
10. European Federation of Chemical Engineering (1985) *Report of the International Study Group on Risk Analysis. Risk Analysis in the Process Industries*, The Institution of Chemical Engineers, Rugby.
11. Goodner, H. W. (1993) A new way of quantifying risks. *Chem. Eng.*, **100**(10), 114–20.
12. Hauptmanns, U. (1986) *Análisis de Arboles de Fallos*, Ediciones Bellaterra, Barcelona.
13. Prugh, R. W. (1980) Application of fault tree analysis. *Chem. Eng. Prog.*, **76**, 59.
14. Aelion, V. and Powers, G. J. (1992) Risk reduction of operating procedures and process flowsheets. *Ind. Eng. Chem. Res.*, **32**, 82.
15. Greenberg, H. R. and Salter, B. (1991) Fault tree and event tree analysis, in *Risk Assessment and Risk Management for the Chemical Process Industry* (eds H. R. Greenberg and J. J. Cramer), Van Nostrand Reinhold, New York.
16. CCPS (Center for Chemical Process Safety) (1989) *Guidelines for Chemical Process Quantitative Risk Analysis*, American Institute of Chemical Engineers, New York.
17. CCPS (Center for Chemical Process Safety) (1989) *Guidelines for Process Equipment Reliability Data*, American Institute of Chemical Engineers, New York.
18. O'Mara, R. L. (1991) Failure modes and effects analysis, in *Risk Assessment and Risk Management for the Chemical Process Industry* (eds H. R. Greenberg and J. J. Cramer), Van Nostrand Reinhold, New York.

3 Consequence analysis: fires and explosions

Courage friends! These are the lights of victory!

Gonzalo Fernández de Córdoba, called 'the Grand Captain', to his troops at the battle of Ceriñola (1503) immediately after an enemy shell provoked an explosion in his gunpowder store.

3.1 Introduction

A look around us is sufficient to realize the large quantity of flammable products such as plastics, textiles, paper, construction materials, etc., that surround us. The majority of these have been manufactured by process industries, but when one thinks about their flammability characteristics only their use is considered and not their prior manufacture. In this chapter we shall consider the flammability of various materials as far as they intervene in chemical industry manufacturing processes. We are interested, therefore, to ascertain the risks of fire, explosion of gases, liquids and their vapours, powders, aerosols, etc., that are used in processing. Other elements, such as construction materials used by the industry, which also require precautions against fire, fall outside the scope of this book.

Fires and explosions, in this order, are the most frequent kind of accident in the chemical industry, followed by the release of toxic substances. This is not surprising given the quantity and characteristics of the substances normally processed. Table 3.1 shows a percentage distribution of industrial fires, according to the material involved [1]. From the point of view of risk analysis, the evaluation of the consequences of fires and explosions, which this chapter is about, requires a definition of the scenario in which the fire or explosion would take place. We need to know, for example, how much material within the flammability limits exists in a cloud at the moment of explosion, or how much flammable liquid exists in the leak which has caught fire. For this, the consequence assessment models used in this chapter usually require as an input the results of the emission models described in Chapter 4.

Any practising chemical engineer should have information that allows him to:

- Be familiar with the fire and explosion characteristics of different materials.
- Estimate the consequences of a fire or explosion, in a given scenario.
- Propose procedures to reduce the risk of fires and explosions or, should they occur, to reduce their effects.

Table 3.1   Distribution of industrial fires according to the material involved (% of number of cases).

(a) Industry as a whole

Material	(%)
Wood or paper	27.9
Flammable liquids	22.1
Chemical products, plastics and metals	15.7
Textiles	10.3
Natural products	9.6
Flammable gas	6.4
Volatile solids	5.4
Equipment containing oil	2.2
Other/unknown	0.4

(b) Chemical industry

According to the physical state of material		
	Gas	13
	Vapour	20
	Liquid	25
	Solid	29
	Unknown	13
According to the type of material		
Hydrocarbons: 29.5	Gas	4
	Liquid/vapour	23
	Solid	2.5
Other products: 70.5	Organic liquid/vapour	20
	Organic solids	9
	Cellulosic materials	8
	Hydrogen	9
	Steel	2.5
	Sulphur	1
	Unknown	21

The general structure of this chapter follows the order of the first two of these points, while some aspects concerning the third are discussed in Chapters 7 and 9.

3.2   Flammability characteristics

The term 'flammability' refers to the greater or lesser ease with which a substance can burn in air or in other gases that can serve as an oxidizer. Combustion is a chemical reaction in which energy is released due to the oxidation of a particular material, and fire is, under certain circumstances, a visible consequence of this combustion. In most cases, conventional combustion occurs in the vapour phase, which means that liquids must evaporate and solids pyrolyse before combustion takes place.

The elements required to produce a fire are shown in the so-called 'fire triangle' reproduced in Figure 3.1. If one element is missing, then a fire cannot occur. Therefore, a fire cannot occur if there is no fuel, or if it is not present in the proportion required, if there is no oxygen or other oxidizer in adequate quantities, or if there is no source of ignition of sufficient power.

Fires and explosions have many similar characteristics, the main difference being the velocity at which energy is released during combustion of the material, which is much lower in fires than in explosions, whether these are deflagrations or detonations. There exists, moreover, a practical relationship of cause and effect, and it frequently occurs that a fire causes an explosion and vice versa.

The **flash point** of a substance t_f, is the minimum temperature at which sufficient vapour is produced to form, close to the surface of a combustible liquid, a mixture with air which is within the flammability limits. The term flash point only has meaning when applied to liquids which are stable in air, or to solids that melt before burning. There are standardized methods for determining flash points, that differ according to the characteristics of the liquid under study (see the US standard ASTM E502). Generally, methods using closed vessels (usually employed with low flash point substances) give values of t_f less than those obtained with open vessels, although the differences are small. In all cases the procedure involves a slow heating of the liquid in contact with air, applying a source of ignition at predetermined intervals and recording the moment at which burning occurs. The flash point generally increases with an increase in total pressure. A rough estimate of the flash point of hydrocarbons can be achieved using the equation [2]:

$$t_f = 0.683 \, t_b - 71.7 \qquad (3.1)$$

where both t_f and, the boiling temperature, t_b are in degrees centigrade.

The **autoignition temperature**, t_a, is the temperature at which a flammable substance is capable of burning in air, without an 'external' source of ignition. Autoignition occurs, therefore, without the necessity of a flame or spark, due to the very thermal level of the gaseous mixture, or due to contact with a hot surface. In the latter case the autoignition temperature is more difficult to predict due to the nature of the solid surface or of materials deposited on it (dust, etc.). A catalytic effect may be produced which lowers the energy required to initiate the process. This introduces a supplementary risk when working with materials with a low

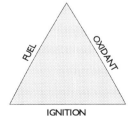

IGNITION

Figure 3.1 The fire triangle.

autoignition temperature, in which case not only surfaces with temperatures higher than t_a must be avoided, but also catalytic effects (which can lower the autoignition temperature by more than 100°C), must be limited by ensuring the cleanliness of surfaces.

The autoignition temperature depends on several variables. Thus, apart from the possible catalytic effect already mentioned, it has been established that an increase in the system volume, in the total operating pressure, or in the concentration of oxygen reduces the t_a value, whilst a variation in the fuel concentration has a more complex effect. For this reason it is essential to determine t_a in conditions as similar as possible to those of the process.

The **flammability limits** provide us with the range of fuel concentration (normally in percentage volume), within which a gaseous mixture can ignite and burn. Below the Lower Flammability Limit (LFL), there is not enough fuel to cause ignition. Thus, for example, if a mixture of 1% of methane in air at atmospheric pressure and ambient temperature is placed in contact with a naked flame, the methane will be gradually consumed as the molecules cross the high-temperature regions close to the flame, but the combustion will not propagate to the gaseous mixture, as would occur if the mixture were within the flammability limit. Similarly, with fuel concentrations greater than the Upper Flammability Limit (UFL), there is insufficient oxygen for the reaction to propagate away from the source of ignition. The flammability limits of various substances are listed in Table 3.2, together with their autoignition temperatures and flash points.

It is often the case that the stoichiometric mixture containing the exact amount of oxygen (in air) necessary for complete combustion is approximately at the geometric mean of the flammability limits. Also, it is often found that, for a number

Table 3.2 Flammability characteristics of some selected substances

Substance	LFL in air (%)*	UFL in air (%)*	Autoignition temperature t_a (°C)	Flash point t_f (°C)
Acetaldehyde	4	60.0	175	−38
Acetone	2.5	13.0	538	−18
Acetylene	2.5	80–100	305	−
Benzene	1.3	7.9	562	−11.1
Butane	1.6	8.4	405	−60
Ethane	3	12.5	515	−135
Ethanol	3.3	19.0	423	12.8
Ethylene	2.7	36.0	490	−121
Formaldehyde	7.0	73.0	430	−67.2
Hexane	1.1	7.5	225	−26
Hydrogen	4.0	75.0	400	−
Methane	5.0	15.0	538	−188
Propane	2.1	9.5	450	< -104
Propylene	2.4	11.0	460	−108
Styrene	1.1	6.1	490	31.1
Toluene	1.2	7.0	536	4.4

* The flammability intervals correspond to standard conditions (1 atm, 25°C), in air.

of hydrocarbons, the ratio of the UFL to the LFL is between 3 and 5. However, there are substances with exceptionally wide flammability limits, such as hydrogen or acetylene. In the latter case, the upper limit is 100%, because acetylene can decompose explosively in the absence of air (although in this case the explosion is not caused by combustion of the acetylene). A precise determination of the flammability limits requires the use of a standardized apparatus and conditions (see, for example, ASTM E681), with the aim of eliminating variability due to the geometry of the apparatus, strength of the ignition source, temperature, degree of mixing, etc.

When the flammability limits cannot be determined experimentally, empirical equations for their determination are available. It should be noted, however, that the experimental measurement should always be attempted using the conditions closest to those in which the substances are to be used. The empirical equations and the rules of thumb such as those described in the previous paragraph are subject to error, often considerable. As an example of equations for the estimation of flammability limits in air, those of Jones are frequently cited. [2, 3]:

$$LFL = 0.55 \, C_{est} \qquad (3.2)$$

$$UFL = 3.50 \, C_{est} \qquad (3.3)$$

where C_{est} is the stoichiometric concentration of the flammable product for complete combustion in air. The estimate for a general compound $C_nH_xO_y$ is obtained by considering complete combustion to carbon dioxide and air:

$$C_nH_xO_y + (n + x/4 - y/2)O_2 \rightarrow nCO_2 + (x/2)H_2O \qquad (3.4)$$

The concentration in air is calculated by correcting the oxygen calculated from the above equation for the accompanying nitrogen.

Another empirical relation frequently used for the prediction of the lower flammability limit in air is that of Spakowski:

$$LFL \times (-\Delta H_{comb}) = 4.354 \times 10^3 \qquad (3.5)$$

where $(-\Delta H_{comb})$ is the standard upper heat of combustion, and is expressed as kJ/mol, whilst the LFL is obtained from the previous equation as a volume percentage.

Example 3.1

Estimate the flammability limits for methane, propane, ethanol and benzene using the equations of Jones and Spakowski, and compare them with the experimental values given in Table 3.2.

The stoichiometric equation for complete combustion of methane is

$$CH_4 + 2O_2 \rightarrow CO_2 + 2H_2O$$

Therefore, for complete combustion a mole of methane requires 2 moles of oxygen accompanied by $2(79/21) = 7.52$ moles of nitrogen. The stoichiometric concentration is, therefore, $1/10.52 = 0.095$, or 9.5%. According to equations (3.2) and (3.3) the flammability limits would be calculated as LFL $= 5.2\%$, and UFL $= 33.2\%$.

$(-\Delta H_{comb}) = 889.5$ kJ/mol [4]. Therefore, from equation (3.5) one obtains LFL $= 4.9\%$.

Calculations for the remaining compounds are carried out in the same way. The results are shown in the table below.

Hydrocarbon	LFL (%) Equation (3.2)	LFL (%) Equation (3.5)	UFL (%) Equation (3.3)	LFL (%) Experimental	UFL (%) Experimental
Methane	5.2	4.9	33.2	5.0	15.0
Propane	2.2	2.0	14.1	2.1	9.5
Ethanol	3.6	3.5	22.9	3.3	19.0
Benzene	1.5	1.3	9.5	1.3	7.9

The previous examples show a greater reliability in the lower limit estimation, while significant errors are observed in the values of UFL, which again demonstrates the need to use experimental values whenever possible.

3.2.1 Modifications to the flammability limits

The flammability limits can be modified by ambient conditions. Of special importance is the dependence on temperature, where an increase widens the flammability interval. Thus it is estimated that an increase in temperature of some 100°C raises the upper limit by approximately 8%, and lowers the lower limit by the same amount. The Burgess–Wheeler equation is often quoted [3,5] for the estimation of the flammability limits of hydrocarbon vapours at temperatures different from the ambient temperature:

$$LFL_{(t)} = LFL_{(25°C)}[1 - 0.75(t - 25)/(-\Delta H_{comb})] \qquad (3.6)$$

$$UFL_{(t)} = UFL_{(25°C)}[1 + 0.75(t - 25)/(-\Delta H_{comb})] \qquad (3.7)$$

where t is the temperature in degrees centigrade and $(-\Delta H_{comb})$ is the lower standard heat of combustion, given in kcal/mol.

Variations in pressure have little effect on LFL until they reach values less than 5 kPa, where in any case, the flame front only propagates with difficulty. On the contrary, the upper limit can increase considerably with an increase in pressure, with the resulting widening of the flammability zone. Zabetakis [6] gives the following empirical expression for this dependence:

$$UFL_{(P)} = UFL_{(1atm)} + 20.6(\log P + 1) \qquad (3.8)$$

where P is given in MPa.

With respect to the concentration of oxygen, the lower flammability limits in oxygen and in air are very similar, since oxygen is in excess at the concentrations corresponding to the LFL. On the other hand, the upper flammability limit usually increases considerably with the concentration of oxygen. Thus for propane, the upper flammability limit increases from 9.5 to 55% when determined in oxygen instead of air, while the lower limit hardly varies (2.1 against 2.3%) [2]. If in place of oxygen one uses other oxidants, the limits change considerably. Data on the flammability limits of different substances in chlorine, in nitrogen oxides, etc. is available in literature dealing with explosions and flammability.

3.2.2 Minimum oxygen for combustion

The estimation of flammability limits based on the concentration of combustibles in air has been discussed. However, whatever the concentration of flammable gas or vapour, the ignition of the mixture can be prevented if the level of oxygen is sufficiently reduced. Below the concentration of oxygen (%) termed the **minimum oxygen for combustion** (O_{min}), the reaction does not generate enough energy to raise the temperature of the mixture (including the inerts present), to allow the propagation of the flame. The use of this concept is obvious, as it permits the elimination of the possibility of explosions independently of the concentration of combustible. Thus O_{min} is a fundamental parameter when designing inerting systems for process vessels and storage tanks that may contain flammable mixtures.

If it is not possible to obtain experimental values for a particular system, Bodhurtha [2] recommends making an approximate estimate of O_{min} (for combustion with nitrogen as the inert gas) as the stoichiometric oxygen necessary for complete combustion at the lower flammability limit.

If instead of nitrogen, a gas or vapour of greater heat capacity is used as an inert, for example, carbon dioxide or steam (avoiding conditions where condensation can arise), the values of O_{min} increase correspondingly. Thus the values of O_{min} for inerting of mixtures of ethyl alcohol–air with nitrogen, steam and carbon dioxide are, respectively, 10.5, 12.3 and 13% [2].

Example 3.2

Estimate the values of O_{min} for inerting with nitrogen of mixtures of acetone in air and butane in air, respectively. Compare them with the experimental values (13.5% for the acetone and 12% for butane).
The corresponding stoichiometric equations are

$$CH_3COCH_3 + 4O_2 \rightarrow 3CO_2 + 3H_2O$$

$$C_4H_{10} + (13/2)O_2 \rightarrow 4CO_2 + 5H_2O$$

As shown in Table 3.2, the lower flammability limit for acetone in air is 2.5%. Because 4 moles of oxygen are needed for each mole of acetone, O_{min} can be estimated at 10%, lower than the experimental value. Similarly, the lower limit for butane is 1.6%, so that O_{min} is 10.4%, which again gives a conservative estimate. This is a frequent result, but cannot be generalized for all cases. For example, the method predicts a value of O_{min} of 11% for butadiene, higher than the experimental value of 10%.

3.2.3   Flammability diagrams

Figure 3.2 is a concentration–temperature diagram in which the flammability limits for a hypothetical mixture of gases are shown. Point A corresponds to the flash point, i.e. a saturated vapour mixture at a temperature at which the resulting concentration corresponds to the lower flammability limit (LFL). Analogously, at point B there is a saturated mixture at the concentration corresponding to the upper flammability limit (UFL).

As stated previously, and as can be seen in Figure 3.2, an increase in temperature widens the flammability interval. As a consequence, depending on the initial position in the diagram, an increase in temperature can displace a mixture such as C, initially non-flammable, into the flammability zone (C').

The difference between the temperatures corresponding to points A and B in Figure 3.2 gives us the temperature interval at which a saturated system would be within the flammability limits. Thus, as King [7] demonstrates, in atmospheric tanks with fixed tops vented to the atmosphere flammable mixtures can exist if the temperature is within the limits shown, and so, whenever possible, tanks with

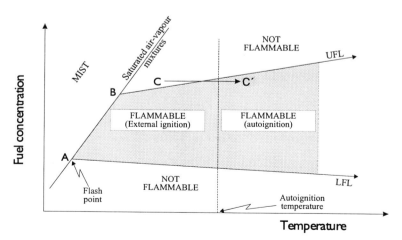

Figure 3.2   Concentration–temperature diagram showing different flammability characteristics. UFL = upper flammability limit; LFL = lower flammability limit.

floating tops should be used for this type of application. The diagram is divided
into two zones, either side of the autoignition temperature. Below this temperature
an external source is required for ignition of the flammable mixture. Above it, the
molecules that constitute the mixture already have sufficient energy for ignition.

Another way of placing a mixture outside the flammability limits is to make it
inert. Figure 3.3 shows the effects of the addition of air, combustible or inert
(nitrogen), to a given gaseous mixture. On line AD of this figure are all the possible
binary combustible–air mixtures. The addition of fuel at B (LFL), or the addition
of air at C (UFL), places the system in the flammability zone, whilst the addition
of nitrogen either at B or C places it outside. If a point representing a given mixture
is M, the system can be placed outside the flammability limits by inerting it with

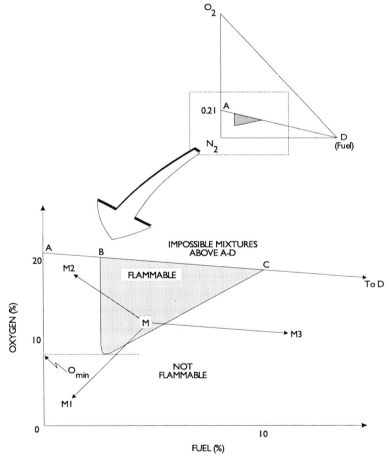

Figure 3.3   Effects of the addition of air, fuel or inert gas (nitrogen) on the flammability of a mixture
(adapted from Bodhurtha [2]).

nitrogen (trajectory M – M1), by the addition of air (trajectory M → M2), or by making the mixture richer through the addition of fuel (trajectory M → M3).

3.2.4   Mixtures of flammable vapours

There are relatively few reports in the literature giving flammability intervals for mixtures with various combustible products. The LFL values for mixtures of flammable products can be estimated using the Le Chatelier equation:

$$\text{LFL}_{\text{mixture}} = \frac{1}{\Sigma\left(Y_{i,\text{comb}} / \text{LFL}_i\right)} \tag{3.9}$$

where $Y_{i,\text{comb}}$ is the mole fraction of each of the flammable components with respect to the total quantity of fuel, and LFL_i its lower flammability limit, also expressed as a mole fraction. From this it is deduced that a mixture can be within the flammability limits, even though each of the individual components is below the lower limit. The Le Chatelier rule is an empirical relationship which has exceptions and should be applied with caution. Some of the limits can be found in [8].

Example 3.3

Determine whether a mixture of 4% methane and 1% propane in air (standard conditions) is within the flammability limits.

The methane represents 80% of the mixture's combustible content, and the propane 20%. Their lower flammability limits are, respectively, 5 and 2.1%. Applying the Le Chatelier rule, equation (3.9),

$$\text{LFL} = \frac{1}{\dfrac{0.8}{0.05} + \dfrac{0.2}{0.021}} = 0.039$$

i.e. a total concentration of combustible of 3.9%. Therefore the mixture is within the flammability interval (5% > 3.9%), although neither of the components individually exceeds the lower flammability limit.

3.2.5   Flammability degree of different materials

A unique parameter capable of characterizing the flammability of different materials does not exist, but rather various properties with different degrees of importance interact. Some of them have already been discussed: The flash point (of great importance to the flammability of combustible liquids), flammability limits and autoignition temperature; but also important are the energy of ignition, combustion velocity, heat of combustion, melting point of solids, viscosity, carbon/

hydrogen relationship, etc. The NFPA [9] classifies different materials according to their flammability characteristics, dividing them into five categories:

- *Degree of flammability 0*: Corresponds to materials that do not burn when exposed to temperatures of 815°C in air for 5 minutes. Examples of this type of material are aluminium chloride or ammonium nitrate.
- *Degree of flammability 1*: Assigned to materials that need considerable preheating to burn, whatever the ambient conditions. In this group are materials that burn in under 5 minutes when exposed to air at 815°C and combustible liquids, solids and semisolids with a flash point higher than 93.4°C. Examples are diethylene glycol and benzyl alcohol.
- *Degree of flammability 2*: The materials in this section do not form dangerous atmospheres on contact with air under normal conditions, but are able to do so after moderate heating, or when exposed to high temperatures. Within this group are those liquids with a flash point above 37.8°C and below 93.4°C, as well as solids and semisolids that easily produce flammable vapour. Examples are acetic acid and aniline.
- *Degree of flammability 3*: Corresponds to liquids and solids that, upon ignition, can burn in ambient conditions or close to them. Grade 3 liquids give rise to flammable atmospheres in air, in practically all the usual conditions. Solids belonging to this group are those fibres, e.g. cotton, or relatively coarse dusts which, although not normally forming explosive atmospheres in air can burn easily, as well as other solids that can burn easily because they contain oxygen molecules, e.g. dry nitrocellulose. Other examples of products belonging to this group are acetone and diethylamine.
- *Degree of flammability 4*: Belonging to this group are those materials that vaporize quickly at ambient conditions and burn readily. Included are gases, cryogenic materials, flammable liquids with a flash point below 22.8°C and materials that, due to their physical form or properties can easily disperse in air, forming explosive mixtures, e.g. fine dusts of combustible solids and mists of flammable liquids. Coal, in spite of its low volatility, can have an NFPA index of 3 or even 4, when ground to a very fine particle size. Other examples of substances in this group are acetaldehyde and butane.

3.3 Sources of ignition

It was previously indicated that the presence of a source of ignition is the third necessary element in the fire triangle. The inverse is also true, in the sense that, without sources of ignition, fire is prevented.

For ignition to occur it is necessary to supply the minimum energy for ignition, sufficient to start the burning of a mixture. All materials possess a characteristic minimum energy for ignition, which varies with ambient conditions such as pressure and composition of the combustible mixture. The typical minimum ignition energies for hydrocarbons are about 0.25 mJ, there being some substances with considerably less energy requirement (e.g. hydrogen at about 0.03 mJ). To give an idea of the order of magnitude represented by these energies, and the ease with which they can be achieved, the electrostatic charge accumulated by a person walking on a carpet on a dry day can cause a discharge of 22 mJ, and the electric spark from an ordinary switch is usually between 20 and 30 mJ.

The possible sources of ignition are numerous (see, e.g., [10]). Table 3.3 lists a percentage distribution of ignition sources. Autoignition has already been mentioned, and does not require an external source. A related case, although of a different nature, is self-heating, the process of slow oxidation that releases heat until the autoignition temperature is reached if the heat cannot be adequately dispersed. It is usually associated with liquids of low volatility (because high volatility gives off a significant amount of energy as latent heat), and at moderate ventilation. It occurs on heat insulating materials on which oil or liquid polymers have been spilt.

As can be seen from Table 3.3, the most frequent sources of ignition in the chemical industry are hot surfaces and open flames. This is a specific characteristic related to the type of activity and contrasts with the statistics on fires for general industry, where the first places in the list of heat sources causing ignition are occupied by manufacturing machinery and electrical distribution equipment [1] or by smokers [3].

Ignition from electrical causes occurs frequently in the chemical industry as is shown in Table 3.3. Here one should consider not only ignition caused by equipment or electrical installations (13.6% of identified causes), but also those derived from electrostatic discharges (2.3%). In the first case, electrical equipment is widely used in process industries and can be a cause of ignition if strict control over its characteristics is not exercised. There are substantial regulations on electrical equipment to be used for different degrees of hazard from the atmosphere (see Chapter 7).

Nothing like so well-defined is the manner of confronting ignition caused by electrostatic discharges. Static electricity is an important cause of ignition in process plants, aggravated by the fact that it is not so well understood, nor are all of its potential hazards usually considered. Its generation is associated with contact and

Table 3.3   Distribution of ignition sources causing fires in the chemical and petrochemical industries [11] (% number of cases). Data from a study of the Fire Protection Association during the period 1971–73

Sources of ignition	%	Modified %*
Hot surfaces	10.1	18.2
Burner flames	10.1	18.2
Electrical equipment	7.6	13.6
Spontaneous ignition	7.6	13.6
Sparks and heat from friction	7.6	13.6
Flame cutting	5.1	9.1
Children with matches	3.8	6.8
Malicious ignition	2.5	4.5
Static electricity	1.3	2.3
Unknown causes	44.3	–

* The modified % have been calculated excluding fires with unknown cause of ignition.

separation of different types of materials, which imparts an excess of electrons to one of the materials after separation. The accumulated charge can easily be eliminated if the contacting bodies are conductors, but if one or both have low electrical conductivity, this can give rise to differences in potential of thousands of volts, which eventually produces an electrical discharge.

The effects of static charge accumulation are important in many systems in which the previously mentioned sequence (contact, separation and accumulation of charge) takes place. Thus, as examples in solid–fluid systems, we can cite flow through pipes or filters, filling of tanks, dispersion of liquids through sprays, etc. as well as fluidization operations and pneumatic transport. In fluid–fluid systems examples are the coexistence of non-miscible liquids, sedimentation of a phase dispersed in drops within a continuous one, cleaning with steam, leaks of vapour, etc. Finally, important causes of static charge generation in solid–solid systems includes among other examples conveyor belts, rotor drive belts, the handling of bobbins with plastic or paper, the accumulation of charge in the human body, etc.

Ignition can also be produced in a flammable mixture when it is compressed using high compression ratios without sufficient refrigeration. It is a well-known fact, and used in practice, that ignition of combustibles can be caused by compressing the mixture of fuel plus air until the autoignition temperature is reached. The increase in temperature resulting from the adiabatic compression of an ideal gas can be calculated from

$$T_{final} = T_{initial} \left(\frac{P_{final}}{P_{initial}} \right)^{\frac{k-1}{k}}$$ (3.10)

where k is the ratio of heat capacities, the temperatures are expressed in degrees Kelvin and the pressures are absolute. A significant number of explosions are known to have been caused by the entrance of flammable vapours into compressors not specifically designed for this. Vapours emanating from the lubricating oils of compressors can also cause explosions.

Other sources that frequently cause the ignition of flammable mixtures are: (a) open flames, whether intentional (torches, burners, etc.) or accidental (as a consequence of explosions or inadequate venting, fires, flames produced by negligence of personnel, etc.); (b) ignition from mechanical causes: collision of metal surfaces, cutting operations and moulding of metals, etc.; (c) welding operations, not only from the arc or the welder flame, but also the from metal heated in the process; (d) motor vehicles; (e) others: hot surfaces, microwaves, chemical energy (e.g. in pyrophoric materials), etc.

3.4 Explosions

An explosion liberates energy in a sudden and violent manner. The causes can be various, but generally explosions are classified according to the type of energy of origin. In this chapter we are concerned with explosions produced by energy liberated by physical pressure or by chemical energy. In the former, the energy of a compressed gas is suddenly released, generally due to a mechanical failure followed by the collapse of the container. It occurs, for example, with the catastrophic rupture of a pressurized gas cylinder. This category also includes the sudden depressurization of a liquefied gas stored under pressure at a temperature above its normal boiling point (which causes its rapid evaporation along with an increase in pressure due to an increase in the number of moles in the gas phase), and explosions due to an increase in the internal pressure of a vessel due to external heating (fire).

As to explosions due to the liberation of chemical energy, they are caused by a chemical reaction which raises the temperature and/or the number of moles in the gas phase. The energy liberated in a chemical explosion depends on the nature and physical state of the reactants and products. In this respect it is important to distinguish between explosions of gaseous mixtures containing combustible vapours and an oxidizer (normally air), which are of immediate concern for the safety in the chemical industry, and those produced by solid explosives such as TNT, etc. These have different characteristics, because the solid contains sufficient oxygen and, therefore, the explosion can take place in the absence of air.

As indicated previously, in the case of a combustible mixture, the velocity at which the release of potential chemical energy takes place is the main difference between fires and explosions. As with fires, explosions can be avoided by working outside the flammability limits, lessening the concentration of oxidizer to the corresponding value or eliminating sources of ignition. As will be seen later, the application of intrinsic safety criteria from the initial stages of design sometimes all but eliminates the hazards derived from explosions or, alternatively, allows the design of equipment that is able to resist explosions, without suffering damage or propagating its effects. On the other hand, extrinsic safety can be added using equipment capable of suppressing explosions at an early stage of their development, with pressure relief systems, etc.

If a homogeneous vapour cloud within the flammability limits ignites in the absence of external restrictions, the result is a spherical flame which rapidly propagates to the rest of the cloud. The combustion causes an increase in the temperature, and normally also in the number of moles. The velocity of the process causes an increase in local pressure which does not equilibrate with its surroundings, which is different to what happens in a fire, where the process is sufficiently slow to dissipate the increase in pressure.

The flame front in expansion can be visualized as a permeable piston that causes pressure waves in the unreacted mixture. These waves generally propagate at the local speed of sound. If the velocity of the reactive front is sufficiently high, it produces the well known superposition of frontal waves which gives rise to the formation of a shock wave. Explosions can be either a deflagration or a detonation, depending on the relative velocities of sound and of the combustion front in the unreacted mixture. If the velocity of the flame front is less than the propagation of sound in the unburned gas mixture, a deflagration is produced, whilst on the contrary the detonation of the flammable mixture takes place. Combustion in a petrol engine is a deflagration, although produced in about 1/300th of a second. A detonation requires that the process is completed in approximately 1/10 000th second [3]. The typical velocities of a deflagration are of some hundreds of metres per second, whilst in a detonation velocities greater by an order of magnitude may be reached. Therefore, it is accepted [12] that in a deflagration conventional mechanisms of heat transfer operate, whilst in a detonation the increase in temperature is due primarily to the shock wave formed. The conditions necessary for a detonation to occur are stricter than those for flammability, detonation intervals being narrower. The detonation of mixtures of combustible gas with air usually requires a certain degree of confinement. Detonation can occur directly or can also take place by transition from a deflagration. As stated previously, in this case a significant acceleration of the flame front is required, which can occur in pipes, but is unlikely in process vessels. The pressures reached in detonations are higher than those of deflagrations, and their effects much more destructive. The great majority of explosions of flammable mixtures in the chemical industry are deflagrations. However, detonations have been reported under different circumstances (e.g. layered dust explosions in mine tunnels [13]).

3.5   Effects of fires and explosions

The final result of a fire or an explosion in the chemical industry depends upon the intrinsic nature of the accident and the conditions under which it occurs, which can increase or decrease its effects. The nature of an accident, i.e. the type of accident that eventually takes place, is a function of the sequence of events that configure the accident, in this case a fire or an explosion.

Figure 3.4 shows different chains of events with the results which can be expected: formation of pressure waves, projectiles, or of thermal radiation. Thus, in physical explosions (i.e. when there are no chemical reactions contributing to the effects of the explosion, or if there are, they are of minor importance), when there is only the gas phase present, the possible effects are reduced to the formation of shock waves and projectiles, provided that no ignition of the mixture is produced. An explosion which is initially physical, can become a chemical explosion (i.e. in a process whose effects are determined by a chemical reaction of combustion of the gas mixture). For this it is necessary that the gas involved is of a combustible nature, that it forms with air a mixture within the flammability interval and that ignition takes place. As indicated in Figure 3.5, from this point either an unconfined vapour cloud explosion (UVCE) or a flash fire can occur. Figure 3.4 indicates the formation of a pressure wave and of projectiles as the final effects of the UVCE. Thermal effects have been left out because, although they often exist, they are usually of less importance.

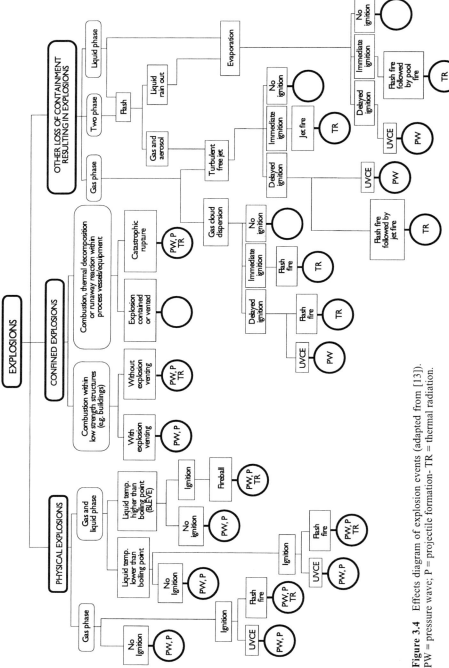

Figure 3.4 Effects diagram of explosion events (adapted from [13]).
PW = pressure wave; P = projectile formation- TR = thermal radiation.

When liquid and vapour are present in a physical explosion the consequences are different. If the liquid is below its boiling temperature, only the material which is already in the vapour phase (plus any supplementary fraction evaporated in the brief period of the process) takes part in the explosion. Under these circumstances the chain of evolution is parallel to the previous case. On the contrary, if the liquid is stored under pressure at a temperature above its boiling point, the initial physical explosion that breaks the receptacle produces a sudden decompression, giving rise to a massive evaporation of the superheated liquid. This is known as a BLEVE (Boiling Liquid Expanding Vapour Explosion). BLEVE explosions are of great destructive power due to the high increase in pressure caused by the sudden incorporation of liquid into the gas phase. The ignition of a BLEVE produces a mass of gases at high temperature known as a 'fireball', with significant thermal effects.

Another kind of explosion considered in Figure 3.4 is confined explosions. They include deflagrations initially constricted by receptacles or buildings. The term 'structures of low resistance' used in Figure 3.4 refers principally to buildings

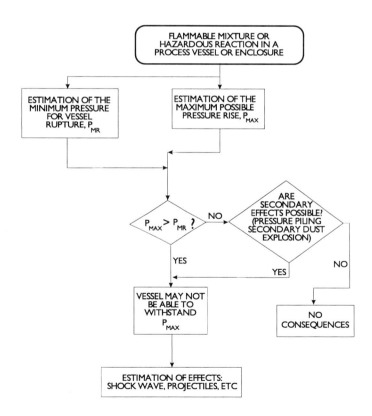

Figure 3.5   Flow chart for the case of a confined explosion in a receptacle (adapted from [14]).

or silos, while it is assumed that process receptacles possess a greater resistance. In this case, if the structure allows the containment or adequate venting of the explosion there are no further consequences, if not, one or more of the following adverse effects may be produced: pressure waves, formation of projectiles and thermal radiation.

Lastly, as shown in Figure 3.4, explosions and fires can be caused as a consequence of a loss of containment. Whether the loss of containment is of a gas or a liquid capable of significant evaporation, or if a two-phase emission is produced, the result is the formation of a cloud which, upon ignition, can cause a UVCE or a flash fire.

In other cases, as will be discussed later, the result is the formation of a jet fire, a pool fire or dispersion without consequences if no ignition takes place. As can be seen in the figure, the moment of ignition has great importance in determining the type of accident finally produced.

Following the above effects diagram, methods for estimating the effects of confined explosions, unconfined explosions, vessel rupture (including physical explosions), pool fires, jet fires and BLEVEs will be dealt with successively.

3.5.1   Confined explosions

According to the definition given previously, in this section we will discuss the development characteristics of explosions in receptacles, and more specifically the two fundamental parameters: the maximum pressure and the rate of pressure increase. The flow chart for the case of a confined explosion in a receptacle is shown in Figure 3.5. When the maximum pressure exceeds the resistance of the container one must calculate the overpressure produced as a function of the distance, and also estimate the effects of the projectiles formed. This calculation can be performed following the guidelines given in the section 'vessel rupture', although the specific characteristics of the confined explosion must be taken into account. Thus, when the rupture is provoked by an increase in pressure caused by an internal explosion, the inertia of the process is such that the velocity of increase in pressure is, generally, greater than the velocity of rupture of the receptacle, so that the effective rupture pressure is between the pressure at which the mechanical resistance of the receptacle is exceeded and the maximum pressure obtained if the explosion had been totally confined.

Maximum pressure in confined explosions

When a deflagration takes place inside a vessel, the maximum pressure (absolute) that can be reached depends on the initial pressure (absolute), the initial and final number of moles and the initial and final temperatures (K), according to

$$P_{MAX} = P_{initial} \frac{n_{final} T_{final}}{n_{initial} T_{initial}} \tag{3.11}$$

The above equation assumes that the ideal gas equations are followed, although if the final pressure is high, the behaviour of the gas can deviate considerably from the ideal gas law. As is shown, the maximum pressure varies linearly with the initial pressure, which corresponds to the larger number of moles of reactants present at the start of the reaction as the initial pressure increases. In receptacles under pressure (not in atmospheric tanks), designed according to the usual chemical industry codes, the minimum pressure for vessel rupture P_{MR} is at least 3 to 4 times the Maximum Allowable Working Pressure (MAWP). A value of 2.5 times this pressure can be used as a guide for comparison between P_{MAX} and P_{MR}, i.e. if P_{MAX} is 2.5 times the MAWP or greater, rupture of the receptacle can be considered probable. The maximum pressure reached in a vessel containing a combustible gas mixture depends on the composition of the mixture, reaching a maximum at concentrations approaching the stoichiometric limit. For deflagrations in air, the maximum pressure that can be reached is about 8 times the initial pressure, whereas in oxygen the maximum is some 16 times the initial. In detonations, the maximum pressure can reach some 20 times the initial, or more if there exist significant reflection effects in the receptacle walls.

The above discussion implicitly assumes that the mixture in the receptacle is homogeneous. However, an important case which frequently occurred is that in which a small quantity of combustible gas accidentally enters a receptacle full of air. If ignition takes place before the mixture has had time to equilibrate, an explosion may occur, even if the number of moles introduced, considering the whole receptacle, would give a concentration lower than the flammability limit. Bodhurtha [2] calculates that 180 g of propane at ambient temperature introduced into a receptacle of $10 \, m^3$ (which corresponds to 1% average concentration, i.e. below the flammability range), can cause an explosion which generates an overpressure of more than 140 kPa.

As stated before, P_{MAX} increases if the initial pressure increases. A similar reasoning leads us to conclude that if the initial temperature increases the maximum pressure decreases for a given vessel, because the number of moles initially present decreases.

Other factors such as turbulence or strength of the ignition source have little influence on the value of P_{MAX}, and the same occurs with the volume and shape of the receptacle, unless the process is so slow that heat losses become significant, in which case the surface/volume ratio appreciably influences the maximum pressure reached.

Rate of pressure increase

The rate of pressure increase (dP/dt) has great importance in the design of pressure relief systems in process vessels. If the pressure inside a receptacle is to be maintained stable, it is necessary to design the relief system in such a way that a sufficient number of moles is evacuated to prevent the pressure increase reaching a certain value. The typical variation of the pressure curve with time is given in Figure 3.6. It can be seen that there is an initial period during which the pressure

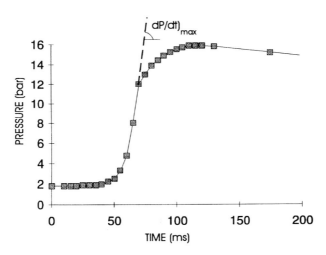

Figure 3.6   Example of pressure evolution in a confined explosion.

increases slowly, followed by a period in which the pressure increases faster, and a maximum value of the rate of pressure increase $(dP/dt)_{max}$ is reached. The rate of pressure increase then decreases and eventually becomes zero on reaching the maximum pressure.

In contrast to that observed with maximum pressure in confined explosions, the value $(dP/dt)_{max}$ grows when the temperature is increased, because the rate of reaction increases with temperature. As for other factors, it has been found that $(dP/dt)_{max}$ increases linearly with the initial pressure. It also increases considerably with increased turbulence, due to the higher degree of mixing in the vessel.

The type of ignition source has a more complex influence on $(dP/dt)_{max}$ [2]. Stronger igniters can increase the value of $(dP/dt)_{max}$ although in some cases there is no change. Ignition close to the centre of the receptacle increases the value of $(dP/dt)_{max}$ with respect to that obtained close to the walls. Greater values of the maximum rate of pressure increase are obtained when there are multiple ignition sources, although this case is unlikely in practice, unless it is specifically desired, as happens in some types of automobile engine. Where the receptacle geometry is concerned, it has been observed that, for similarly shaped receptacles and for the same degree of turbulence and ignition potency, the variation of $(dP/dt)_{max}$ with the receptacle volume follows the equation known as the cubic law, which establishes lower rates of pressure increase for larger receptacles:

$$\left(\frac{dP}{dt}\right)_{max} V^{1/3} = K_{st} \tag{3.12}$$

where V is the receptacle volume and K_{st} is a constant termed the deflagration index. The equation can be applied to combustible gases/vapours or solids, using different values of K_{st}. Equation (3.12) can be used to predict rates of pressure increase for industrial size vessels using the results of laboratory experiments with similar vessels. As an example of values obtained for K_{st}, Bartknecht [11] gives values of 75 and 550 bar.m/s for propane and hydrogen, respectively (standard conditions, with a 10 J ignition source), and different intervals for combustible powders: 59–165 bar.m/s for sugar, 83–211 bar.m/s for wood dust, 56–229 bar.m/s for pulverized cellulose, etc.

Example 3.4

In the laboratory a relatively simple experimental system can be used to measure the characteristics of explosions. Basic requirements are a vessel capable of resisting without damage the maximum pressure of an explosion, an adequate ignition system, a stirrer to produce different degrees of turbulence, temperature measurement and control systems to fix the temperature of the receptacle before the explosion, flow systems to load the receptacle with the reaction atmosphere and to discharge it after the explosion, and high-speed pressure transducers and adequate data logging systems to follow the rapid changes in pressure produced during the process. A system of this type with a 15-litre vessel was used to determine the explosion characteristics of a new industrial product, A. The process uses mixtures of A with air at a pressure of 1.9 bars. Figure 3.6 shows the data for pressure evolution obtained in one of the experiments. Determine the main explosion characteristics under the conditions used. How would these be modified in a 500-litre industrial vessel?

Figure 3.6 shows the characteristic curve of confined explosions. The induction period in this case is extended to some 40 ms, and the maximum pressure is reached at about 115 ms. The maximum pressure value is 15.9 bars, a little more than eight times the initial pressure.

The figure also shows the tangent to the curve corresponding to the maximum rate of pressure increase $(dP/dt)_{max}$. The slope of this line is 820 bar/s. The value of K_{st} can be obtained using this value and the receptacle volume, using equation (3.12):

$$K_{st} = (820)(0.015)^{1/3} = 202 \text{ bar.m/s}$$

For a 500-litre vessel no major variations are expected in the maximum pressure if the rest of the conditions are comparable. The maximum rate of pressure increase is given by the cubic law, of which we now know K_{st}. Using a volume of 500 litres in equation (3.12) a maximum rate of pressure increase of 254.5 bar/s is obtained.

Pressure accumulation

In systems with interconnected receptacles it is possible that higher pressures than those mentioned previously are obtained, because of accumulation of pressure.

Let us consider two process vessels connected by a pipe. After ignition in the first, part of the flammable mixture, still not reached by the combustion front, is pushed by the pressure front towards the second receptacle via the connecting pipe. This causes a pressure increase in this receptacle before ignition occurs. When ignition takes place, on arrival of the flame front at the second container, the explosion starts at a higher 'initial' pressure, so that the final pressure is also higher, as seen previously. The pressure increase of the explosion in the second receptacle can be very considerable (reaching $(P_{MAX})^2$ in the worst case), so that the possibility of pressure accumulation should be borne in mind in the design of interconnected equipment.

The final pressure in the second receptacle depends on the system's geometry. Especially important are the length and diameter of the connecting pipe and the presence of restrictions in it. Bodhurtha [2] cites experiments in which a receptacle is divided into two compartments by means of a panel with a hole in it. Moderate decreases in the size of the hole produce significant increases in the maximum explosion pressure obtained in the system after ignition in one of the compartments.

3.5.2   Unconfined explosions

These are explosions that occur outside buildings or process receptacles. Within this group are included unconfined vapour cloud explosions (UVCE), that have caused some of the worst accidents in the chemical industry. A UVCE requires firstly the formation of the cloud, due for instance to the collapse of a vessel containing a volatile flammable liquid or to a leak of flammable gas in a pipe. The elapsed time to ignition is a critical factor in determining the destructive power of the explosion. Thus with early ignition the size of the flammable cloud is still sufficiently small that the effects are of small magnitude. As the time before ignition increases, the effects increase, as a consequence of accumulation of material in the cloud. Lastly, if ignition is sufficiently delayed, most or all of the materials emitted will be diluted to concentrations below the flammability limit, so that the effects would be small or non-existent. In the following chapter, methods for estimating the size of the cloud formed under different circumstances are discussed.

The deflagration of a cloud of gas or combustible powder produces a reaction front that moves away from the point of ignition, preceded by a shock wave or pressure front. This shock wave continues after the material in the cloud has been consumed, its magnitude decreasing as it exchanges momentum with the surroundings. Figure 3.7 represents the evolution of overpressure profiles as the wave distances itself from the ignition point. The exact shape of the pressure profile in the initial moments depends on the type of explosion. In any case, at a certain distance from the point of origin, the region of positive pressure (overpressure) is usually followed by a rarefied zone, in which there is a weak negative pressure with respect to the atmosphere, that generally does not exceed 0.25 bars absolute. In spite of this, its destructive effects can be very significant,

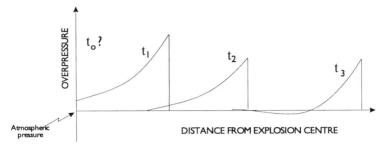

Figure 3.7 Evolution of overpressure profiles in a UVCE (unconfined vapour cloud explosion) (adapted from Lees [5]).

due to the fact that buildings generally are not designed to resist greater pressures within than without.

If one considers the pressure changes that take place with time at a fixed distance from the explosion's origin, the overpressure variation has the form shown in Figure 3.8. When the shock wave impacts an object, the pressure on the surfaces parallel to the direction of wave propagation increases almost instantaneously to the maximum pressure value P^o. Other important parameters which characterize the pressure variation at a particular point are the time of arrival, t_a, the duration of the positive phase, t_+ i.e. the time during which the pressure on a surface parallel to the propagation is greater than atmospheric pressure, and the pressure decay parameter α, that describes the shape of the curve from the moment when the maximum overpressure is reached. In the literature [2, 5, 14] charts are given to obtain the above parameters as a function of the distance from the explosion point.

The following table shows the variation of the three parameters for the expansion wave resulting from the explosion of a ton of TNT [5]:

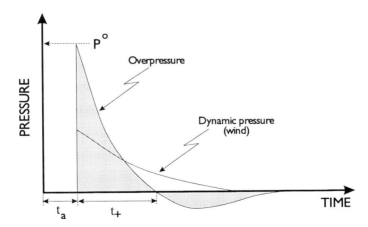

Figure 3.8 Variation of incident overpressure and dynamic pressure with time at a fixed location (from Bodhurtha [2]).

Distance (ft)	30	60	80	120	200
t_a (ms)	4	18	37	59	123
t_+ (ms)	5.5	12	16	18	22.5
α	3.5	1.08	0.87	0.93	1.15

To describe mathematically the positive phase of the overpressure curve shown in Figure 3.8 a modified Friedlander equation [5] can be used:

$$P = P^o \left(1 - \frac{t}{t_+} \right) \exp(-\alpha\, t/t_+) \qquad (3.13)$$

where t starts from the moment the pressure wave arrives. Other important properties are the shock wave velocity, the maximum dynamic pressure and the maximum reflected overpressure, which can be expressed [5] as shown below. The shock wave velocity in air, U, is:

$$U = C_o \left(1 + \frac{6P^o}{7P} \right)^{1/2} \qquad (3.14)$$

where C_0 is the speed of sound in air, and P and P^o respectively are atmospheric pressure and maximum overpressure. The term **dynamic pressure** refers to the transformation of the kinetic energy of the wind generated in the explosion to pressure energy when encountering a solid surface in its path, and its evolution is also shown in Figure 3.8. For explosions in air it can be expressed as

$$q^o = \frac{5}{2} \frac{\left(P^o \right)^2}{7P + P^o} \qquad (3.15)$$

where q^o is the maximum dynamic pressure. Lastly, it is important to consider the maximum overpressure due to wave reflection. When the pressure wave hits a solid surface not parallel to the propagation direction a reflection is produced, and the reflected pressure varies not only with the value of P^o, but also with the angle of incidence. The maximum overpressure takes place when the pressure wave hits a surface perpendicular to its direction of propagation. In this case the relation between the overpressure experienced and that which takes place on a surface parallel to the direction of propagation is given by:

$$\left(P^o \right)_r = 2P^o \left(\frac{7P + 4P^o}{7P + P^o} \right) \qquad (3.16)$$

where $(P^o)_r$ is the overpressure produced on a surface perpendicular to the direction of propagation as a consequence of the reflection (where r denotes the 'reflected' overpressure). Equation (3.16) shows that the maximum reflected overpressure is at least double P^o (for weak explosions where P^o is small compared to atmospheric pressure), and could become 8 times greater (for explosions with a high value of P^o where the opposite occurs).

3.5.3   Estimation of the effects of an explosion as a function of the distance

Although other factors (such as the rate of pressure increase, the duration of the positive phase and the magnitude of the negative pressure zone) also have an influence, the damage produced by an explosion depends mainly on the maximum overpressure. Thus, an overpressure of 0.15 psi is cited [15] as typical for the breaking of glass, 1 psi for the partial demolition of low buildings and 10 psi for the total destruction of buildings. A more complete listing is given in Table 5.3.

In spite of the limitations due to its simplified nature, the TNT model is still widely used for the prediction of overpressures at a given distance from the centre of an explosion. This model is based on an empirical law, established from trials done using explosives. This law establishes equivalent effects for explosions occurring at the same normalized distance, expressed as :

$$z = R /(W_{TNT})^{1/3} \tag{3.17}$$

where z is the normalized distance (m.kg$^{-1/3}$), R is the real distance (m) and W_{TNT} is the amount of TNT used in the explosion, or the theoretical amount of TNT that would release the same energy. Strictly speaking, the scale-up factor is the cube root of energy rather than the mass of explosive, but for a given material it can be assumed that the energy released is proportional to the mass involved. From equation (3.17) it can be deduced that the distances at which two explosions produce the same effects are given by:

$$\frac{R_1}{R_2} = \left(\frac{W_{TNT,1}}{W_{TNT,2}} \right)^{1/3} \tag{3.18}$$

Figure 3.9 represents overpressure as a function of the normalized distance. It should be noted that an inversion layer in the atmosphere can disturb the propagation of a shock wave. As a consequence, for P^o values below approximately 0.25 atmospheres, a given overpressure can occur at distances considerably greater than those obtained in Figure 3.9 without taking into account atmospheric inversion. In the literature, the explosion parameters are usually plotted on different graphs, depending on whether the explosions are elevated or at ground level. However, for the case of overpressure the difference is almost always less than 20%, for which reason both types of explosions are presented together in Figure 3.9.

The procedure for establishing the effects of an explosion at a specified distance begins with the calculation of the energy involved, expressed as an equivalent mass of TNT. The normalized distance can now be calculated by dividing by $(W_{TNT})^{1/3}$ and the overpressure is estimated using Figure 3.9. Once the overpressure at a specified distance is obtained, the vulnerability of persons and installations can be estimated using the methods described in Chapter 5.

When estimating some effects, e.g., deaths or injuries from impacts received from the explosion, the relevant parameter for correlation is the impulse, I_p, which is defined as

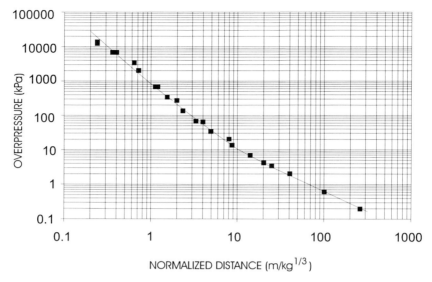

Figure 3.9   Application of TNT model. Overpressure (P°) versus normalized distance. The figure contains data from several references ([2, 5, 14]), corresponding to both spherical (elevated), and hemispherical (ground level) explosions.

$$I_\mathrm{p} = \int_0^{t_+} P(t)\mathrm{d}t \tag{3.19}$$

After substitution of equation (3.13) and integration the impulse is obtained as a function of the parameters of the modified Friedlander equation

$$I_\mathrm{p} = P^\circ t_+\left[\frac{1}{\alpha} - \frac{1}{\alpha^2}\left(1 - e^{-\alpha}\right)\right] \tag{3.20}$$

Example 3.5

Estimate, for distances between 100 m and 1 km, the overpressure which will be reached in explosions equivalent to 500 and 5000 kg of TNT respectively. Calculate in both cases the radius within which it is probable that buildings become uninhabitable (partial demolition) as a consequence of the explosion.

The normalized distances are calculated by dividing the actual distance by $(W_{\mathrm{TNT}})^{1/3}$. Once the normalized distance is calculated, Figure 3.9 gives the overpressure. Below are given the calculations for 500 kg of TNT:

Distance (m)	100	200	400	700	1000
Normalized distance (m/kg$^{1/3}$)	12.6	25.2	50.4	88.2	126.0
Overpressure (kPa)	7.5	3.0	1.3	0.72	0.45

Figure 3.10 shows the variation of overpressure with distance. As shown in Chapter 5, in order for an explosion to cause partial demolition of houses, some 7 kPa is sufficient. This occurs at distances less than 110 m for an explosion equivalent to 500 kg of TNT, and at some 240 m for an explosion of 5000 kg.

3.5.4   *Unconfined Vapour Cloud Explosions (UVCE)*

Let us assume that a leak of propane from a storage deposit gives rise to a cloud with concentrations within the flammability limits. As it advances the cloud may encounter sources of ignition such as open flames, fired ovens, electrical apparatus or lighted cigarettes. If ignition takes place, in most cases it will occur close to the edge of the cloud, as it advances. As already indicated, if this happens immediately the leak occurs, the results of the accident would be a flash fire or an explosion of small magnitude. As time passes, the material accumulated in the cloud becomes sufficient for the explosion to cause important damage. Lees [5] revised the discussion on the threshold value of the cloud which would give rise to an explosion instead of a flash fire. Although discrepancies exist on this point, the majority of authors consider that below 1 to 15 tonnes of flammable vapour in the cloud, explosions are unlikely, although lesser values have been quoted, in the order of some tens of kilos, for very reactive species, such as hydrogen. In this respect, one should keep in mind that the quantity of material that can accumulate in a short period can be very large if the flow of material is also high. For example, it is calculated that in the explosion of the cloud of cyclohexane at Flixborough (the largest explosion registered in Great Britain in time of peace), the ignition delay was about 50 seconds [7]. Finally, if there is no encounter with a source of ignition or if it occurs very late, the dispersion of the propane into the atmosphere will reduce its concentration to below the lower flammability limit (2.2%), and no explosion will occur.

A recent analysis of past vapour cloud explosions [16] shows that vapour cloud ignition predominantly resulted from (a) sparking electric devices, (b) hot surfaces

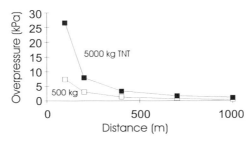

Figure 3.10   Overpressure as a function of distance for explosions equivalent to 500 and 5000 kg of TNT.

such as extruders and hot steam lines, (c) friction between moving parts of machines and d) open flames. Also, in most cases the ignition sources were fairly close to the source of the leak.

In order to apply the TNT model to the estimation of the effects of a UVCE, the fraction of total energy of the explosion used in the shock wave must be calculated first. Once the corresponding value is estimated, it is converted into the equivalent mass of TNT:

$$W_{TNT} = \frac{\eta M E_c}{E_{TNT}}$$
(3.21)

where W_{TNT} is the equivalent mass of TNT (kg) that would produce the same effects as the explosion, η represents the explosion yield, i.e. the quotient between the energy in the shock wave and the theoretical energy available in the explosion, M is the total mass of flammable material in the cloud (kg), E_c the lower heat of combustion of the material (kJ/kg), and E_{TNT} the heat of combustion of TNT (approximately 4680 kJ/kg). It should be noted that obtaining the value of M is no trivial matter. One possibility is to assume that all the material emitted forms part of the flammable region of the cloud, which can lead to considerable overestimations of the explosion's effects. A more realistic procedure consists of applying a dispersion model (Chapter 4) to the emission, and estimating the volume of the cloud within the flammability region at a given instant.

From the value of W_{TNT} obtained the overpressure is calculated using Figure 3.9. Considerable difficulties exist in estimating the value of η, and it is also difficult to correlate it with data from other accidents due to the uncertainty in the precise quantities of vapour involved in previous UVCEs. It is generally accepted [5, 14] that, taking as a basis for calculation the total quantity of vapour in the cloud, the value of η is between 1% and 10% for most explosions. Exceptions have been quoted in which values of 25% or 50% have been reached, especially for strongly asymmetric clouds, although the majority of available data place the yield of explosions within the above limits.

One must also bear in mind that, in spite of extensive use of the TNT model, explosions of flammable vapour clouds present some distinctive characteristics with respect to high explosives. The main differences can be reduced to four points: (a) A flammable vapour cloud extends over a large volume compared to that occupied by the equivalent explosive, (b) the actual UVCE gives values of overpressure close to its epicentre which are lower than those predicted by the TNT model, (c) initially the shape of the shock wave is different in a UVCE, (d) in an UVCE the time duration of the shock wave is greater. The most important difference is (b), which means that the TNT model can only be applied to calculate the effects of an explosion outside the area of the cloud.

TNO correlation model

As indicated in the previous section, although the TNT model is well-established and predicts with reasonable accuracy the results of TNT explosions and other high explosives, its application to unconfined vapour cloud explosions is restricted by the difficulty of estimating an adequate yield value. The TNO correlation model [12] is an alternative attempt to relate the total energy released from the combustion of flammable vapour in the cloud with the effects of the explosion, also using a very simplified model, which in this case is based on data from previous UVCEs. This model was developed for total energies in a cloud between 5×10^9 J (which corresponds to a cloud of some 100 kg of hydrocarbons) and 5×10^{12} J. Below this range one expects that the effects of the explosion are of low magnitude, and above (clouds of some 100 tonnes of hydrocarbons), the extrapolation is too great and the predictions unreliable. The TNO model is not applied to materials classified as 'highly explosive' (such as acetylene, hydrogen and ethylene oxide), nor to those of 'low explosive' (methane, ammonia). It is applied to materials in the 'middle group', which includes hydrocarbons such as ethane, ethylene, propane, butane, isobutane, pentane and cyclohexane [12]. An explosion yield must also be estimated for this model, although values are suggested for it. The output of an explosion is considered as the product of two factors

$$\eta = \eta_c \eta_m \qquad (3.22)$$

where η_c indicates the efficiency factor due to the non-stoichiometric nature of the cloud, arbitrarily fixing its value at 0.3, and η_m represents the mechanical efficiency factor. The values suggested for η_m are 0.33 for an isocoric combustion and 0.18 for an isobaric combustion. Since it seems likely that some form of confinement will exist in an explosion, the isocoric approach is often used which results in a value of η of approximately 10%, unless the particular circumstances of the explosion advise using other values for η_c and η_m.

The TNO model establishes damage circles according to the expression

$$R_i = C_i (\eta ME_c)^{1/3} \qquad (3.23)$$

where R_i is the *maximum* distance, in metres, within which one can expect a certain type of damage i, and the product ME_c corresponds to the energy, in joules, obtained from the combustion of all the material in the cloud. This energy is calculated as the product of the total mass of flammable vapour by the lower heat of combustion. The values of the constant are shown in Table 3.4.

Table 3.4 Value of constant C_i for application of the TNO correlation model [12]

C_i (m.J $^{-1/3}$)	Type of damage caused
0.03	Significant damage to buildings and process equipment
0.06	Repairable damage to buildings. Damage to house facades
0.15	Breakage of glass, injuries likely
0.4	Threshold of glass breakage (10% of total)

Example 3.6

A pipe ruptures, giving rise to a cloud of some 3000 kg of propane, which ignites. Estimate the distance at which the breakage of 10% of windows (which corresponds to an overpressure of approximately 0.02 bars, according to Table 5.3), is likely to take place. Use the TNT model and the TNO correlation model.

Application of the TNT model

From Figure 3.9, an overpressure of 0.02 bars (2 kPa) corresponds to a normalized distance of 35 m.kg $^{-1/3}$. The total energy in the cloud is calculated taking into account the heat of combustion of propane (19 900 Btu/lb, see Table 2.2). $ME_c = 1.39 \times 10^{11}$ J. If a yield of 10% for the explosion is accepted, the equivalent weight of TNT is 2970 kg (equation (3.21)). Thus the distance for an overpressure of 0.02 bars would be

$$R = (35)(2970)^{1/3} = 503 \text{ m}$$

If the yield was only 1%, $R = (35)(297)^{1/3} = 234$ m. As indicated previously, the majority of explosions would be within these limits, although there are important exceptions.

Application of the TNO correlation model

The value 1.39×10^{11} J is within the applicability limits and propane is classified as a compound of intermediate reactivity, thus the model can be applied. From Table 3.4 the value of C_i would be 0.4. According to equation (3.23)

$$R = 0.4(0.1 \times 1.39 \times 10^{11})^{1/3} = 962 \text{ m}$$

The differences in the estimates are important (a factor between 2 and 4). This is principally due to the uncertainty in the yield factors applied. Also, one must take into account the fact that the TNO model gives maximum distances for a given type of damage.

3.5.5 Flash fires

In the context of this chapter we will follow the normal convention in the literature on safety in the chemical industry, describing a flash fire as the very rapid combustion of a mixture of a flammable vapour with air with characteristics (mass, combustion heat, flammability, etc.), such that the pressure effects are slight, and the only damage to be considered is that corresponding to thermal radiation.

As previously observed, the transition point between a flash fire and a UVCE is not known with precision. On the other hand, the model of flash fires is not well-developed. Attempts have been made to simulate the effects based on the thermal radiation, calculated from the flame temperature. Lees [5] briefly explains the proposed treatment. However, the inaccuracies in the estimates of this temperature can cause significant errors, given the dependence of the radiated energy on temperature (T^4). Other models for flash fires are simpler and use a gas dispersion model to determine the cloud area (generally the region corresponding to the lower flammability limit), limiting consideration of the thermal effects to within this area.

3.5.6 Vessel rupture

The rupture of a vessel containing a pressurized fluid occurs when its mechanical resistance is exceeded. There can be various immediate causes, including the failure of the equipment for control and relief of pressure, design or construction defects, reduction in the wall thickness due to corrosion, erosion or chemical attack, and reduction in resistance due to overheating or overcooling. It is important to consider the cause of the rupture because, depending on this, the failure will occur at approximately the operating pressure (e.g. when the cause is a reduction in the wall thickness due to corrosion), or at high pressures (e.g. when the pressure control and pressure relief systems fail). Whatever the cause, the energy stored is released, which causes the formation of the corresponding shock wave and accelerates the fragments resulting from the rupture. These are converted into dangerous projectiles and can cause additional damage. Also, as previously shown, although initially only the physical explosion exists, if the gas involved is combustible it can be ignited, wherefore other factors enter into play beyond the characteristics of a physical explosion.

Energy of the explosion

The maximum energy that can be liberated by the rupture of a receptacle whose mechanical resistance has been exceeded corresponds to the expansion of the gas contained within it. In most cases, the process can be approximated by an isentropic expansion. For an ideal gas we can write

$$ E = \frac{P_1 V_1}{k-1} \left[1 - \left(\frac{P_2}{P_1} \right)^{(k-1)/k} \right] \tag{3.24} $$

where P_1 and V_1 respectively represent the initial pressure and volume, P_2 is the final pressure at which the gas expands (normally atmospheric), and k is the ratio of specific heats.

Just as occurs with unconfined explosions, not all the maximum energy is available for the formation of the shock wave. As well as other minor losses, in a real explosion part of this energy is invested in supplying kinetic and potential energy to the fragments resulting from the explosion. Thus, in the fragile rupture of a receptacle into small fragments up to 80% of the energy can be expended in the shock wave, whilst in the case where large fragments are formed, this value can drop to 40%, the remaining 60% being used mainly to impart energy to the fragments [14].

Formation of projectiles

Provided that the energy stored is not exceptionally great, the major danger with the explosion of a container is due to the projectiles formed. Apart from the intrinsic danger that the impact of a fragment at high speed represents for people and installations, the projectiles can cause a chain of accidents, the so-called domino effect. Thus, for example, a relatively small explosion which causes the formation of projectiles can result in one of these projectiles perforating a receptacle containing combustible gas, which can then form a flammable cloud and cause a major explosion.

With explosions that cause formation of projectiles it is important to estimate the range of these fragments. The initial velocity of the projectiles depends on the energy of the explosion and on the mass of the fragment and its dimensions, which influence the friction resistance encountered. Moore [17] gives a simple equation for the estimation of initial velocity:

$$u = 2.05\left(\frac{PD^3}{W}\right)^{0.5}$$

(3.25)

where u is the initial velocity of the fragment (ft/s), P is the rupture pressure (psig), D the fragment diameter (inches) and W its mass (lbs). Due to the resistance of the air to the displacement of the projectiles, the estimation of the initial velocity obtained in this equation should be corrected downwards when speeds are calculated at significant distances from the explosion.

The range of the projectiles is often estimated from the empirical correlation of data obtained from explosions of TNT. Figure 3.11 shows the correlation obtained from the data of Clancey [15], with only one parameter, the mass of explosive. As to penetration, there are numerous equations given in the references to predict the penetration of fragments into various surfaces, almost always in perpendicular impacts. Lees [5] gives various expressions for calculating penetration, depending on the velocity of the fragment, its mass and the impact area.

Effects of pressure

Overpressure at a given distance can be calculated using the TNT model but only when the influence of the receptacle over the propagation of the pressure wave is known. The method proposed in a recent CCPS publication [14] assumes that the expansion of the gas initially contained in the receptacle occurs isothermally. Thus, the energy of the explosion can be expressed as

$$W = 0.0219\, P_1 V_1 \ln\left(\frac{P_1}{P_2}\right)$$

(3.26)

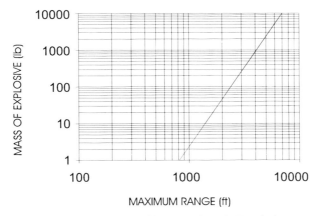

Figure 3.11   Maximum horizontal range of fragments from TNT explosions at ground level.

where W is expressed in mass of TNT in grams, the pressure in atmospheres and the volume in litres. The gas does not expand freely, but is contained in a receptacle that collapses. To take into account the influence of the receptacle walls a virtual distance to the centre of the explosion is calculated, which requires a prior estimation of the pressure on the surface of the receptacle, P_s. This is obtained from the equation of Prugh [18]:

$$P_B = P_s\left[1 - \frac{3.5(\gamma-1)(P_s-1)}{\left[(\gamma T/M)(1+5.9P_s)\right]^{0.5}}\right]^{-2\gamma/(\gamma-1)} \qquad (3.27)$$

where P_B and P_s are, respectively, the pressure at which the explosion of the receptacle is produced and the estimated pressure on its surface (both in bars absolute), γ is the ratio of specific heats, T is the absolute temperature (K), and M the molecular weight of the gas. The above equation must be solved iteratively because it is implicit for P_s. Once the pressure on the receptacle surface P_s is known, Figure 3.9 can be used together with equation (3.17) to calculate the distance R at which the pressure P_s is produced, using the value of W calculated in equation (3.26). The virtual distance mentioned before is calculated by subtracting the real radius of the receptacle, which is known, from the calculated distance R, and the value obtained is used to correct the distances in the estimates of the effects of pressure.

Example 3.7

Due to undetected structural stress, a 0.5 m³ spherical vessel which contains air at 412 atmospheres absolute and 20°C undergoes catastrophic rupture. Estimate the overpressure produced at a distance of 20 metres.

The value of W is obtained substituting $P_1 = 412\,\text{atm}$, $P_2 = 1\,\text{atm}$ and $V_1 = 500\,\text{litres}$ in equation (3.26). $W = 27.2\,\text{kg}$ of TNT is obtained. The pressure on the vessel surface is estimated using equation (3.27), with $P_B = 412\,\text{atm}$ or $417.3\,\text{bars}$ absolute, $M = 28.9$, $\gamma = 1.4$ and $T = 293.15\,\text{K}$. After iteration, one obtains for P_s a value of $9.48\,\text{bars}$ absolute, or $8.47\,\text{bars}$ of overpressure.

The next step is to calculate the distance at which an explosion of 27.2 kg of TNT would produce an overpressure of $8.47\,\text{bars}$ or $847\,\text{kPa}$. In Figure 3.9 this overpressure corresponds to a normalized distance of approximately $z = 1$, i.e. a real distance of 3 m.

The effect of the vessel wall is to reduce the overpressure at a given point with respect to that obtained from the unconfined explosion of an equivalent mass of explosive. Therefore, the existence of the vessel wall implies that to obtain the same effects the distances should be less than in an unconfined explosion. A spherical deposit of $0.5\,\text{m}^3$ capacity has a radius of 0.59 m. Because P_s is the pressure on the surface of the vessel, i.e., at a distance of 0.49 m, the difference $3.0 - 0.49 = 2.51$ is the virtual distance that must be *added* to provide the estimates of overpressure.

The effects obtained at 20 m are, therefore, those that would be obtained in an unconfined explosion at 22.51 m. For a normalized distance equal to 7.5 the overpressure is approximately 17 kPa.

3.5.7   Pool fires

The vapours emitted by flammable liquids can give rise to a flammable mixture with air if the temperature of the liquid is above the flash point. This can take place in a deposit of combustible (tank fire), or after a spillage or leak to the exterior (pool fire), and in the latter case the spillage can be confined (by a dike, or by the topography of the nearby terrain), or non-confined. Fires of this type tend to be local in their effects, very often the effects of propagation being more severe than those deriving directly from the event. Thus for instance, burning of the spilled liquid can affect the same or other deposits, heating its contents (with the resulting increase in pressure) and its metal wall, which will cause a decrease in its mechanical resistance. This constitutes one of the most frequent causes of BLEVEs, which will be discussed later.

The consequence analysis in this section is mainly aimed at estimating the radiation that would be emitted in different scenarios of tank fires and pool fires. The results of the analysis have obvious implications in the design of installations intended to contain flammable liquids, as well as the distances to be maintained between different units of the plant.

The flow chart for calculation is given in Figure 3.12. In spite of its name of pool fire, what really burns is the material that evaporates from the liquid which is at the base of the flames. In the case of confined fires, the size of the fire is determined by the size of the deposit or the containment dike, taking also into

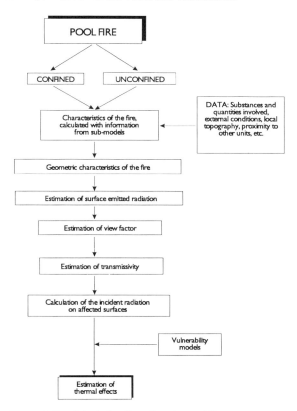

Figure 3.12   Calculation flow chart for pool fires.

account the tilt and shape of the flame. In the case of unconfined fires, a significant variation in size can be produced as the spilled liquid spreads. Whatever the case, it is necessary to establish the size of the fire over time if it is not constant. For this, prior calculations are necessary, such as those described in the following chapter. Thus if the spillage results from a perforation in the wall of a deposit, a sub-model for liquid spillage is used that establishes the flow rate over time. The consideration of this flow rate and the losses by evaporation and combustion of the liquid during the fire, together with the topographical information of the area close to the spillage, allow estimation of the pool fire dimensions at a given instant. The steps in Figure 3.12 lead to an estimate of the radiation received by neighbouring bodies. This information may be used together with the vulnerability models described in Chapter 5 for the estimation of damage to persons and property.

Flame dimensions

In many cases, the size of the fire is determined by physical barriers, like local accidents, or the presence of drains to storage areas away from the deposits. Often, a circular shape for the base of the fire can be assumed, and in cases where the containment dikes give rise to square or rectangular pools an equivalent diameter is used.

The most complex case is that of a semi-continuous spillage when the local topography is not able to contain the spilt liquid. In this case, the maximum diameter is reached when the mass flow rate of liquid leaked equals that consumed by the fire. An example of this type of analysis applied to the spillage of evaporating liquids can be found in the work of Opschoor [19], which calculates the variation of the pool radius as

$$\frac{dr}{dt} = \left[Cg(h - h_{min}) \right]^{0.5}$$

(3.28)

where r is the pool radius, C a constant, g the acceleration due to gravity, h the height of the layer of liquid at a given instant and h_{min} the minimum thickness of the layer of liquid. This approximation is only valid for thin layers of liquid, in which C has a value of between 1 and 4. The driving force for the spread of the spilt liquid is the difference between the height of the liquid at a given moment and the minimum thickness of the liquid layer, which depends on the type of surface on which the spillage occurs. A value of 5 mm for h_{min} is recommended for smooth surfaces, whilst for rougher surfaces the value of h_{min} can be several centimetres.

Examples of empirical equations for estimating the size of a liquid spill are given by Elia [20]. For the instantaneous spillage of a mass (m kg) of liquid, the radius during the expansion phase of the pool is given by

$$r = \left(\frac{t}{B} \right)^{0.5}$$

(3.29)

with

$$B = \left[\frac{\pi \rho_L}{8gm} \right]^{0.5}$$

(3.30)

where t is the time (seconds), ρ_L the density of the liquid (kg/m³), and g the acceleration due to gravity (m/s²). In the case of a continuous spillage the corresponding equations would be

$$r = \left(\frac{t}{B} \right)^{0.75}$$

(3.31)

with

$$B = \left[\frac{9\pi \rho_L}{32gm''} \right]^{0.333}$$

(3.32)

where m'' is the spillage flow rate (kg/s).

For a fire of circular base and a given radius, there are equations to estimate the height of the flame. The best known is that of Thomas [21], which gives the relationship height/diameter (H/D), without the presence of wind:

$$\frac{H}{D} = 42\left[\frac{M_b}{\rho_a\sqrt{gD}}\right]^{0.61} \tag{3.33}$$

where M_b is the velocity of combustion per unit of surface area (kg/m²s), whose calculation will be described later, ρ_a is the density of the ambient air (kg/m³), and g the acceleration due to gravity (m/s²). The height of a fire of cylindrical shape is usually two to three times the diameter of the base.

The presence of wind can alter the shape of the flame by tilting it, and with stronger winds, the base of the flame can be deformed in the wind direction. Generally it is sufficient to calculate the inclination of the flame, and even to ignore it in most cases of consequence analysis unless the inclination directs the flame towards another unit. From the point of view of thermal effects, in the absence of direct flame impingement the presence of wind can have opposite effects: On the one hand, the surfaces nearby and downwind from the flame can receive more radiation due to its inclination, but, on the other hand, a stronger wind contributes to a certain extent to cooling the surface receiving the radiation. The inclination of the flame (Figure 3.13) can be calculated according to [22]:

$$\cos\beta = 1 \qquad \text{for } u^*_w < 1 \tag{3.34a}$$

$$\cos\beta = \frac{1}{\left(u^*_w\right)^{0.5}} \text{ for } u^*_w > 1 \tag{3.34b}$$

where

$$u^*_w = u_w\left(\frac{\rho_v}{gDM_b}\right)^{1/3} \tag{3.35}$$

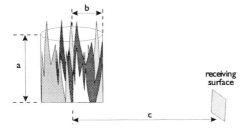

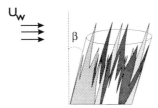

Figure 3.13 Geometric parameters of a cylindrical pool fire.

where u_w is the wind speed (m/s), $u*_w$ the dimensionless wind speed and ρ_v the vapour density (kg/ m^3). The rest of the parameters have been defined previously.

Combustion rate

The rate at which a pool fire burns fuel is usually expressed in m/s (m^3 of liquid consumed per m^2 of pool area per second). Large fires burn at an approximately constant rate, depending on the type of material. Because combustion takes place in the vapour phase, the rate of combustion is determined by the evaporation rate of the liquid. Knowledge of the combustion rate gives the rate of heat generation per unit surface of liquid, and also serves to estimate the duration of the fire. It is necessary to distinguish two situations:

1. Liquids with a normal boiling temperature less than ambient temperature

When a spillage of this kind occurs, if the liquid was stored under pressure at ambient temperature, after the spillage there is very short transient period, in which part of the spillage evaporates, cooling the remaining liquid until its boiling temperature is reached (Chapter 4), followed by a period in which the bulk of the liquid is at its boiling point. The calculations for the fire are done in this second period. In it, the heat necessary for evaporation comes fundamentally from the radiation of the flames, although there can also be a significant contribution from the cooling of the ground in the initial moments of spillage. Hoftijzer [22] calculates that for a spillage of liquid propane, after approximately one minute, more than 80% of the heat for evaporation is supplied by the radiation of the flame.

Estimation of the rate of combustion of liquids whose boiling point is less than the ambient temperature can be made using the following equation [22]:

$$M_b = 0.001 \frac{(-\Delta H_{comb})}{\Delta H_v} \qquad (3.36)$$

where the rate of combustion M_b is given in kg/m^{2}s, and $(-\Delta H_{comb})$ and ΔH_v are respectively the heats of combustion and evaporation of the substance involved.

2. Liquids with a boiling point higher than ambient temperature

In this case evaporation at the liquid surface occurs at the start, without liquid boiling. The heat coming from the radiation is spent not only in evaporating the liquid, but also taking it to its boiling point. This can be taken into account by substituting ΔH_v in the denominator of equation (3.36) for $C_p \Delta T + \Delta H_v$, where ΔT represents the difference between the initial temperature of the liquid and its boiling point.

Intensity of the radiation emitted

Once the rate of combustion is known, the next step is to estimate the radiation emitted. The radiation flux from the flames of a fire can be calculated using the Stefan–Boltzmann equation, knowing the temperature of the flame. The problem with this procedure is the uncertainty of the flame temperature, which, as has been pointed out previously, can give rise to important errors, because of the sensitivity of the Stefan–Boltzmann equation to the temperature. Therefore, other approximations are normally used.

One possibility is to use the values of radiation measured in fires of different materials. The intensity of radiation from a pool fire depends on the type of fuel, as well as the size of the fire, as the diameter of the fire affects the turbulence of the flame and the possibility of smoke forming, which in turn influences the radiation emitted. Thus Hoftijzer [22] cites experiments with LNG in which the formation of smoke and soot reduces the intensity of radiation emitted from some 100 kW/m² down to 36–52 kW/m² depending on the diameter of the fire. The maximum values of radiation intensity of flames usually given in the literature are between 170 and 200 kW/m². However, in real fires much lower values have been measured [14], about 20–60 kW/m².

The calculation of the radiation from the fire should be based, whenever possible, on experimental measurements in equivalent circumstances. However, due to the scarcity of experimental measurements of surface radiation intensity for large fires and the difficulty of extrapolating them to other circumstances, an empirical factor is introduced, the radiation fraction F_R, to calculate the emitted radiation. F_R is defined as the ratio between the energy emitted by radiation and the total energy released by the combustion. The values of F_R vary between 0.15 and 0.35–0.40 for most hydrocarbons. Values above 0.35–0.40 are considered conservative, because they are obtained [22] without taking into account the obscuring of the flame by the smoke formed. These values are close to 0.3, cited as conservative by Sigales [23].

According to the above, to estimate the radiation emitted per external unit surface of a pool fire, I, the total radiated energy must be calculated as the product $F_R M_b (\pi/4) D^2 (-\Delta H_{comb})$, and the value found is divided by the external surface area of the fire. To this end, the lateral surface of the ideal emitting cylinder is considered, of which the height H and the diameter D are known. Thus,

$$I = \frac{F_R M_b D(-\Delta H_{comb})}{4H} \tag{3.37}$$

View factor

In the absence of absorption by the atmosphere, the radiation received by a surface outside the perimeter of a fire of known characteristics can be calculated once the geometric view factor F_{vg} is known. This can be done knowing the geometry of the fire and the receiving surface and their relative positions. Of special interest in this section are the view factors between a cylinder and the ground, according to the diagram presented in Figure 3.13. The view factors for vertical and horizontal receiving surfaces, and for the maximum radiation intensity, with an inclined surface are given in Table 3.5. View factors for other configurations (between two flat surfaces, between spherical and flat surfaces, etc.) are given in Hoftijzer [22].

Table 3.5   View factor F_{vg} between a vertical cylinder and a surface on the ground [22]. For a, b and c the geometric relationships of Figure 3.13 apply

(A) VIEW FACTOR FOR A HORIZONTAL RECEIVING SURFACE (F_H)

		a/b							
		0.1	0.2	0.5	1.0	2.0	5.0	10.0	20.0
	1.1	0.132	0.242	0.332	0.354	0.360	0.362	0.363	0.363
	1.2	0.044	0.120	0.243	0.291	0.307	0.312	0.313	0.313
	1.5	0.005	0.024	0.097	0.170	0.212	0.228	0.231	0.232
c/b	2.0	0.001	0.005	0.027	0.073	0.126	0.158	0.164	0.166
	4.0	0.000	0.000	0.001	0.007	0.022	0.057	0.073	0.078
	10.0	0.000	0.000	0.000	0.000	0.001	0.007	0.017	0.026
	20.0	0.000	0.000	0.000	0.000	0.000	0.001	0.003	0.008

(B) VIEW FACTOR FOR A VERTICAL RECEIVING SURFACE (F_v)

		a/b							
		0.1	0.2	0.5	1.0	2.0	5.0	10.0	20.0
	1.1	0.330	0.415	0.449	0.453	0.454	0.454	0.454	0.454
	1.2	0.196	0.308	0.397	0.413	0.416	0.416	0.416	0.416
	1.5	0.071	0.135	0.253	0.312	0.329	0.333	0.333	0.333
c/b	2.0	0.028	0.056	0.126	0.194	0.236	0.248	0.249	0.249
	4.0	0.005	0.010	0.024	0.047	0.080	0.115	0.123	0.124
	10.0	0.000	0.001	0.003	0.006	0.013	0.029	0.042	0.048
	20.0	0.000	0.000	0.000	0.001	0.003	0.007	0.014	0.020

(C) MAXIMUM VIEW FACTOR (INCLINED SURFACES) (F_M)

$$F_M = \sqrt{F_v^2 + F_H^2}$$

For a deeper discussion of the significance and the view factor calculation any conventional text on heat transfer may be consulted.

In the case where the radiation received at a sufficiently large distance from the emitting surface is considered, the real dimensions of the source lose importance, and a point source can be assumed instead. In accordance with this scenario, the radiation emitted is distributed in a homogeneous manner over the surface of an ideal sphere, in such a way that a solid surface of one square metre at a determined distance would intercept a fraction $1/(4\pi X^2)$ of the emitted radiation, where X (m) is the distance to the receiving surface from the point of emission. It should be taken into account that if the hypothesis of a point source is used the factor $1/(4\pi X^2)$ should be applied over the total radiation emitted, and not on the radiation emitted per unit surface.

Atmospheric transmissivity

At a distance of 100 m, the atmosphere can absorb or disperse about 20–40% of the radiated energy emitted by the fire. This is due to absorption by CO_2 and above all by the water vapour present in the atmosphere. The atmospheric transmissivity τ is defined as the fraction of energy transmitted, and can be calculated approximately [24] considering only the water vapour:

$$\tau = 2.02 \, (P_w \, X)^{-0.09} \tag{3.38}$$

where P_w is the partial pressure of the water vapour (Pa) and X the distance in metres between the emitting and receiving surfaces.

Thermal radiation received

The radiation flux over the receiving surface can be expressed as

$$I_R = I \, \tau F_{vg} \tag{3.39}$$

where each of the factors is calculated as previously explained. Equation (3.39) assumes that there is no reflection on the receiving surface, a conservative hypothesis which is adequate in most cases. Once the radiation received as a function of the distance is known, an irradiation map can be established that shows the radiation received at a determined point. The application of vulnerability models which are discussed in Chapter 5 permits the estimation of the damage for a level of radiation and a given exposure time. When the thermal radiation received from the fire is moderate, it may be necessary to add to the value of I_R calculated, that corresponding to solar radiation, or to radiation from other sources, to obtain a better estimate of the total radiation received. Estimates for the solar radiation as a function of the time of the day, the season and the latitude can be found in the literature (e.g. [25]).

Example 3.8

A spillage of n-pentane in a circular pool of 20 m diameter burns on a summer day without wind, with an ambient temperature of 32°C. Estimate the radiation received by a vertical surface at 40 m from the centre of the fire. The partial pressure of water vapour in the atmosphere at the time of the fire is 2500 Pa.

The distance of the fire from the receiving surface is of the order of the dimensions of the fire, so one cannot use the point source hypothesis, and it is necessary to calculate the view factor taking into account the dimensions of the fire. The first step is to obtain an estimate of the combustion velocity using equation (3.36). The heats of combustion and vaporization applicable are, respectively, 10751.5 and 87.5 cal/g. Given the proximity of the boiling point of n-pentane (36.3°C) to the ambient temperature, it is not necessary to correct for the initially cold liquid. Using equation (3.36) $M_b = 0.123$ kg/m²s is obtained.

The height of the flames can now be calculated using equation (3.33), assuming a cylindrical fire, resulting in $H = 42.8$ m. Using this value together with $D = 20$ m, $(-\Delta H_{comb}) = 44\,941$ kJ/kg, $M_b = 0.123$ and $F_R = 0.35$ (a conservative value), in equation (3.37) one obtains the value of the emitted radiation per unit surface, $I = 226$ kW/m^2 .

The view factor is calculated using Table 3.5, for the dimensions (refer to Figure 3.13), $a/b = 4.3$, $c/b = 4.0$. Taking $c/b = 4.0$ the view factors for a/b equal to 2 and 5 are, respectively, 0.08 and 0.115. Interpolating for $a/b = 4.2$ a view factor of $F_{vg} = 0.106$ is obtained. The transmissivity is calculated using equation (3.38) with $x = 30$, giving a value of 0.736.

With the estimated values for intensity of emitted radiation, transmissivity and view factor, equation (3.39) yields an intensity of radiation on the receiving surface of 17.6 kW/m^2.

3.5.8   Jet fires

When a pressurized gas escapes through an orifice or a narrow conduit, a jet type discharge is produced, with a maximum exit speed that can be equal to that of sound if the ratio of the atmospheric pressure to the pressure in the receptacle is lower than the critical value (Chapter 4). After the orifice a decrease in speed takes place due to the widening of the passage. If a discharge of combustible gas ignites, the characteristic jet fire is formed, which has the shape indicated in Figure 3.14.

The calculation of the radiation from a jet fire can be made as in the previous case, providing the characteristic dimensions of the flame are known. If it is assumed that the shape of the jet fire is roughly equivalent to a cylinder, as shown in Figure 3.14, the relationship between its length and the orifice diameter is given by [26]:

$$\frac{L}{d_o} = \frac{5.3}{C_t} \sqrt{\frac{T_f\left[C_t + (1 - C_t)M_a/M_f\right]}{N_T T_{amb}}} \tag{3.40}$$

where L and d_o are, respectively, the length of the jet and the diameter of the orifice, N_T the quotient between the number of moles of reactants and products

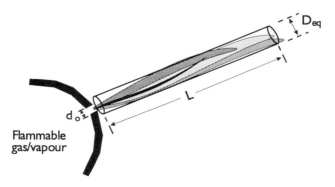

Figure 3.14   Characteristic dimensions of a jet fire.

according to the reaction stoichiometry, T_f and T_{amb} the flame and ambient temperatures (K), and M_a and M_f the molecular weights of the air and the fuel. The value of C_t is given by

$$C_t = (1 + rM_f/M_a)^{-1} \qquad (3.41)$$

r being the stoichiometric air/fuel ratio. For the calculation of the equivalent diameter Romano et al [26] recommend the equation of Raj and Kalelkar [27], based on experimental measurements:

$$\frac{D_{eq}}{d_o} = \sec\left(\frac{\psi}{2}\right) + \frac{L}{d_o}\sin\left(\frac{\psi}{2}\right)\sec^2\left(\frac{\psi}{2}\right) \qquad (3.42)$$

where ψ is the angle of the flame aperture, normally between 10 and 20°. A more detailed description of the emission of gases at high speed that can result in a jet fire can be found in the work of Hoftijzer [21]. Just as with some methods of estimation of the effects of pool fires, it is also necessary to know the temperature of the flame. However, in this case the influence of errors committed in the estimation is much less. The calculation of the total radiation emitted by the jet is done in the same way as for pool fires, using radiation factors F_R of 0.15 for hydrogen, 0.2 for methane and 0.3 for other hydrocarbons [14]. To this value it is necessary to apply a second coefficient of 0.67 to take into account the effects of incomplete combustion.

In estimating the effects of jet fires the main concern is direct impingement of the jet upon other surfaces, in which case the heat transfer effects can considerably exceed those due to radiation, whose calculation has been discussed. In this respect equation (3.40) can be employed, using reasonable scenarios to approximately estimate the maximum distances at which direct impingement of the flame on other surfaces is likely.

3.5.9 BLEVEs and fireballs

The term BLEVE, Boiling Liquid Expanding Vapour Explosion, which has been referred to previously, is regularly used in relation to the sudden rupture of a receptacle containing a liquefied gas under pressure. Because the liquid is stored at a temperature above its normal boiling point, the rupture of the receptacle causes the sudden evaporation of the liquid. This gives rise to a shock wave of great destructive power, accompanied by projectiles, which are often very large, formed from the same receptacle. In the frequent case that the stored material is flammable, the process is almost always accompanied by ignition of the cloud formed. Historically, BLEVEs have been produced with some frequency, and have almost always caused human casualties. Prugh [28] lists 50 BLEVE cases between 1926 and 1986, involving a total of 1300 deaths.

The most frequent cause of a BLEVE is the external fire, which too often originates from small leaks of the material stored. As it receives heat from the external fire, ever-increasing quantities of liquid vaporize with a consequent increase in pressure. At the same time, the radiation and also direct flame impingement from the external fire heats the wall of the receptacle. In the zone of the wall above the level of the liquid, the heat transfer to the interior is slower, which causes the wall temperature to increase rapidly, with the consequent decrease in its strength. The process eventually causes the collapse of the receptacle, the depressurization of the gas and liquid remaining, and the BLEVE of the whole. In this respect, it should be stated that the existence of a pressure relief valve is not sufficient to ensure that a BLEVE will not take place. Moreover, the flames of the fire guarantee the immediate ignition of the mixture in expansion if this is flammable.

On other occasions the rupture of the receptacle occurs simply because its strength has been exceeded, due to accidental over pressurization, construction defects, runaway reactions, etc., without the intervention of external fire. In this case ignition could anyway be produced during the rupture of the receptacle, or it could be delayed until the expanding cloud finds an appropriate ignition source, but the effects of pressure are similar to the previous case.

Often, after the rupture of the receptacle, one or both halves of it can be blown large distances as a consequence of the initial explosion and the impulse produced by the evaporation of the liquid remaining in them. King [7] quoted the accident at Mexico City (Table 1.1), in which parts of a cylindrical tank of 2×13 metres reached distances of 1200 m before destroying two houses, and the flinging of the major part of a 2 tonne absorption tower containing propane at more than 1 km in an accident at Romeoville, Illinois, in 1984. Casal [29] states that in the accident at Los Alfaques, the tractor of the tanker (6.5 tonnes) was found 170 m from the centre of the explosion, and a large fragment of the cistern (of about 5 tonnes), at more than 200 m. In Figures 3.15(a) and (b) are shown the state of the Los Alfaques camp site after the explosion, and the above-mentioned part of the cistern after impacting into a nearby house.

Calculation of BLEVE effects

The effects of a BLEVE are, as in previous cases, of three types: shock wave, projectiles and thermal radiation (caused by the associated fireball). However, given the special characteristics of this type of explosion, empirical expressions have been developed that are specific to BLEVEs.

Given the way BLEVEs are generated, the overpressures are difficult to predict, because the specific evolution of the vaporization and pressurization prior to the collapse of the receptacle, and the duration of the rupture–depressurization process have an important influence on the explosion effects. In addition, the consequences are different depending on what quantity of liquid is vaporized sufficiently quickly to take part in the initial explosion. Lastly, in the case of BLEVEs produced by external fires the estimate of the pressure at which the receptacle rupture is produced is uncertain, mainly because of the decreased strength produced by the high temperatures. In any case, the pressure effects are often limited, and so the main danger where the shock wave is concerned comes from the propagation to close by process units (domino effect).

Figure 3.15(a) Part of the cistern impacts a house, more than 200 m away from the explosion. (Courtesy of El Pais newspaper, Spain. Photo: Chema Conesa.)

Figure 3.15(b) General view of Los Alfaques campsite, after the BLEVE (Boiling Liquid Expanding Vapour Explosion) that occurred on July 11th 1978. (Courtesy of El Pais newspaper, Spain. Photo Chema Conesa.)

An approximate way of calculating the TNT equivalent of a BLEVE has been described by Prugh [28], and consists of: (a) estimate the quantity of liquid that would undergo sudden evaporation by reducing the pressure to that of the atmosphere, (b) calculate the supplementary volume of vapour (at the receptacle pressure before explosion) that this quantity of evaporated liquid would involve, and add this fictitious volume to the vapour space volume, and (c) calculate the equivalent in TNT of the total volume, just as was done in the section on receptacle explosion. With this, the final expression is

$$W = 0.024 \frac{PV^*}{k-1}\left[1 - \left(\frac{1}{P}\right)^{\frac{k-1}{k}}\right]$$ (3.43)

where W is the equivalent weight of TNT (kg), and P is the pressure existing in the receptacle before rupture (bar). V^* is given as

$$V^* = V_v + V_1(f\,D_{Lo}/D_{Vs})$$ (3.44)

V_v and V_1 being the volumes of vapour and liquid in the vessel before the explosion (m³), D_{Lo} and D_{Vs} the densities of liquid and vapour at the pressure and temperature of the system before the explosion and f the fraction of liquid that flashes after depressurization. The methods for calculating f are detailed in the next chapter.

The other effects of a BLEVE, i.e. the formation of projectiles and the thermal radiation are generally more dangerous than the pressure wave produced by the explosion. As indicated in the previous section, the flying fragments produced in a BLEVE can be of considerable size and travel long distances. In fact they have caused a considerable number of deaths and, at times, have extended the range of the accident to other installations. In Figure 3.16 are shown the results of Holden and Reeves [30] as to the projection of fragments in BLEVEs of LPG. One can see that approximately 10% of the fragments reach distances of more than 400 m in explosions that involve relatively small quantities of liquefied gas. The fragments produced by the explosion are usually unequally distributed, usually finding more

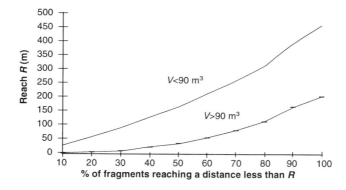

Figure 3.16   Projection of fragments in BLEVEs of LPG (adapted from Holden and Reeves [30])

fragments in the direction of the axle when cylindrical vessels explode. Based on the analysis of a small number of incidents Holden and Reeves [30] also give an approximate correlation of the number of fragments resulting from an explosion, for vessels with capacities between 700 and 2500 m^3:

$$\text{Number of fragments} = -3.77 + 0.0096 \; V \qquad (3.45)$$

where V is the vessel capacity in m^3. Birk [31] summarizes the hazards from propane BLEVEs (thermal radiation, blast overpressure, projectiles). Regarding the projectile range, results are quoted in which large projectiles reached over 14 fireball radii (the calculation of the fireball size is discussed below).

The thermal effects of BLEVEs are related to the formation of a fireball. This, which constitutes the major danger of a BLEVE, is produced when a mass of flammable expanding vapour catches fire. After ignition, the mass of hot gases is elevated into the air, which causes the characteristic radiant spheroid. Typical radiation values in fireballs associated with BLEVEs are quoted in the range 200–350 kW/m^2, much higher than those measured in pool fires, which is mainly due to the lower smoke content of the flame.

The effects of the fireball are usually accounted for through empirical equations related to the quantity of substance involved in the BLEVE. The following have been proposed ([14]:

$$\text{Maximum diameter of the fireball (m): } D_{max} = 6.48 \; M^{0.325} \qquad (3.46)$$

$$\text{Duration of the fireball (s): } t_{BLEVE} = 0.825 \; M^{0.26} \qquad (3.47)$$

$$\text{Height at centre of fireball (m): } H_{BLEVE} = 0.75 \; D_{max} \qquad (3.48)$$

Initial diameter of the hemisphere at ground level (m):

$$D_{init} = 1.3 \; D_{max} \qquad (3.49)$$

In the previous equations M is the initial mass of flammable liquid, in kg. Elia [20] gives similar equations for the calculation of the diameter and duration of the fireball associated with a BLEVE, and also proposes the estimation of the total radiated energy Q_R (J) from the fireball, as

$$Q_R = 0.27 \; M \; (-\Delta H_{comb}) \; P_o^{0.32} \qquad (3.50)$$

where P_o is the initial pressure at which the liquid is stored (MPa) and the rest of the parameters have already been discussed. The flow of radiation per unit of surface area and time is found by dividing Q_R in equation (3.50) by the external area of the sphere and the duration of the BLEVE. In a similar way, other equations [14] have been proposed in which the duration of the BLEVE is considered:

$$I = \frac{F_R(-\Delta H_{comb})M}{\pi \left(D_{MAX} \right)^2 t_{BLEVE}} \qquad (3.51)$$

where the value suggested for F_R is between 0.25 and 0.4. In the previous equation the radiation flow emitted is given in kW/m^2. From here, once the view factor and the atmospheric transmissivity have been estimated, equation (3.39) can be applied to calculate the intensity of radiation on the receiving surface. In this respect, when considering the vulnerability of people to the effects of a BLEVE, it is appropriate to use a geometric view factor corresponding to a surface perpendicular to a sphere

$$F_{vg} = \frac{D^2}{4X^2} \tag{3.52}$$

where the value of D is usually calculated with equation (3.46) and X is the distance to the sphere centre. However, if the requirement is to estimate the view factor from a sphere to a surface not perpendicular to it, values for F_{vg} as a function of the angle and the X/R ratio can be found in the literature [22].

3.6   Questions and problems

3.1   In a laboratory reactor the oxidative dehydrogenation of ethane to ethylene is carried out at 600°C and atmospheric pressure, co-feeding ethane and oxygen to the reactor. Estimate the flammability limits of the mixture under these conditions. What would be the amount of nitrogen needed to make the mixture inert?

3.2   The equation of Antoine for n-octane gives the variation of its vapour pressure with temperature according to [25]:

$$\log P_v = 6.9237 - \frac{1355.126}{209.517 + t}$$

where the vapour pressure is given in torr and the temperature t in degrees centigrade. Taking the lower flammability limit of octane in air as 1% by volume, calculate the flash point for n-octane and compare it with the value given by equation (3.1).

3.3   Perform an inspection of a laboratory in which flammable gases are used. It should preferably be a real laboratory, or in the case of an assumed one, its characteristics should be defined with great precision. Assume that a leak of a flammable gas occurs, and identify the possible sources of ignition, and areas where the confinement of the flammable vapour would be most likely. Propose solutions to avoid the ignition of the leaked material.

3.4   The reaction that took place at Flixborough consisted in the oxidation of cyclohexane at a temperature of 155–160°C, using air. Assume that a reactor contains 100 litres of liquid cyclohexane (the reactors at Flixborough contained 300 times this quantity). What would be the maximum pressure inside the reactor due to the explosion of its contents? Design a receptacle capable of containing it.

3.5   Too often, newspapers carry the notice of an explosion of propane or butane gas in a home. Calculate the total energy released in the combustion of a cylinder of gas for domestic use (about 12 kg of hydrocarbon). Assume that a cylinder of butane is kept in the open air on a terrace, and an important leak is produced that forms a cloud containing all of the gas in it. Make an estimate of the overpressure produced at a distance of 60 metres if ignition takes place before the leaked material has had time to disperse.

3.6   A leak happens in a natural gas distribution station. The resulting explosion totally destroys the buildings (this corresponds to an overpressure of some 70 kPa), in a radius of approximately 150 metres. Calculate the mass of combustible gas involved.

3.7   A 25 m diameter gas sphere, 50% full of liquid propane, suffers a BLEVE caused by an external fire. Initially the sphere is at 293 K and 7 atm-g. The relative humidity of the air is 35%. Estimate the radiation received from the resulting fireball by workers watching the fire at a distance of 400 m.

3.7   References

1. Planas Cored, G. La prevención de incendios y explosiones en las instalaciones industriales. (1989) *Ingeniería Química*, **246**, 141–55.
2. Bodhurtha, F. P. (1980) *Industrial Explosion Prevention and Protection*, McGraw-Hill, New York.
3. Crowl, D. A. and Louvar, J. F. (1990) *Chemical Process Safety, Fundamentals with Applications*, Prentice Hall, Englewood Cliffs.
4. Perry, R. H. and Green, D. (eds) (1984) *Perry's Chemical Engineer's Handbook*, 6th edn, McGraw-Hill, New York, Chapter 3.
5. Lees, F. P. (1980) *Loss Prevention in the Process Industries*, Butterworth-Heinemann, London.
6. Zabetakis, M. G. (1965) *Fire and Explosion Hazards at Temperature and Pressure Extremes*, AIChE-IChemE Symp. Ser. 2, 99–104.
7. King, R. (1990) *Safety in the Process Industries*. Butterworth-Heinemann, London.
8. U.S. Bureau of Mines (1952) Bulletin No 503, p. 6.
9. NFPA (National Fire Protection Association) (1990) Standard 704: *Identification of the Fire Hazards of Materials*, NFPA, Quincy, Massachussets; Standard 325M: *Fire Hazard Properties of Flammable Liquids, Gases and Volatile Solids*, NFPA, Quincy, Massachussets.
10. CCPS (Center for Chemical Process Safety) (1993) *Guidelines for Engineering Design for Process Safety*, American Institute of Chemical Engineers, New York.
11. Bartknecht, W. (1981) *Explosions*, Springer-Verlag, New York.
12. Wiekema, B. J. (1979) Vapour cloud explosion, in *Methods for the Calculation of the Physical Effects of the Escape of Dangerous Material (Liquid and Gases), (The Yellow Book)*, TNO, Directorate General of Labour, 2273 KH Vooburg, the Netherlands.
13. Sichel, M., Kauffman, C. W. and Li, Y. C. (1995) Transition from deflagration to detonation in layered dust explosions. *Process Safety Progress*, **14**(4), 257–65.
14. CCPS (Center for Chemical Process Safety) (1989) *Guidelines for Chemical Process Quantitative Risk Analysis*, American Institute of Chemical Engineers, New York.
15. Clancey, V. (1972) *Diagnosis Features of Explosive Damage*. Proceedings of the 6th International Meeting of Forensic Science, Edinburgh.
16. Koshy, A., Mallikarjunan, M. M. and Raghavan, K. V. (1995) Causative factors for vapour cloud explosions determined from post-accident analysis. *J. Loss Prev. Process Industries*, **8**(6), 355–8.
17. Moore, C.V. (1967) The design of barricades for hazardous pressure systems. *Nucl. Eng. Des.*, **5**, 1550–66.

18. Prugh, R. H. (1988) *Quantitative Evaluation of BLEVE Hazards*, AIChE Loss Prevention Symposium, AIChE Spring National Meeting, New Orleans.
19. Opschoor, G. (1979) Evaporation, in *Methods for the Calculation of the Physical Effects of the Escape of Dangerous Material (Liquid and Gases), (The Yellow Book)*, TNO, Directorate General of Labour, 2273 KH Vooburg, the Netherlands.
20. Elia, F. (1991) Explosion and fire analysis, in *Risk Assessment and Risk Management for the Chemical Process Industry* (eds H. R. Greenberg and J. J. Cramer), Van Nostrand Reinhold, New York.
21. Thomas, P. H. (1963) *The Size of Flames from Natural Fires.* Proceedings of the 9th International Symposium on Combustion, Academic Press, New York.
22. Hoftijzer, G. W. (1979) Heat radiation, in *Methods for the Calculation of the Physical Effects of the Escape of Dangerous Material (Liquid and Gases), (The Yellow Book)*, TNO, Directorate General of Labour, 2273 KH Vooburg, the Netherlands.
23. Sigales, B. (1985) Condicionantes técnicos del riesgo–II. *Ingeniería Química*, **199**, 67–77.
24. Pietersen, C. M. and Huerta, S. C. (1985) *Analysis of the LPG incident in San Juan de Ixhuatepec, Mexico City, 19-11-84*, TNO Report B4-0222, TNO, Directorate General of Labour, 2273 KH Vooburg, the Netherlands.
25. Foust, A. S., *et al.* (1980) *Principles of Unit Operations*, J. Wiley and Sons, New York.
26. Romano, A., Piccinini, N. and Bello, G. C. (1985) Evaluación de las consecuencias de incendios, explosiones y escapes de sustancias tóxicas en plantas industriales. *Ingeniería Química*, **200**, 271–8.
27. Raj, P. K. and Kalelkar, A. S. (1974) *Assessment Models in Support of the Hazard Assessment Handbook*, US Coast Guard Report CG-D-65-74.
28. Prugh, R.W. (1991) *Chem. Eng. Prog.*, **87**(2), 66–72.
29. Casal, J. (1990) Riscs potencials del foc en installacions industrials, in *Plasmes i Focs*, Institut Déstudis Catalans.
30. Holden, P. L. and Reeves, A. B. (1985) Fragment hazards from failure of pressurized liquefied gas vessels, in *Assessment and Control of Major Hazards*, IChemE Symp. Ser. 93, IChemE, Rugby.
31. Birk, A. M. (1996) *J. Loss Prev. Process Industries*, **9**(2), 173–81.

4 Consequence analysis: release of hazardous substances

The very air that you breathe will come from the depths of hell. It will be foul, poisoned and your faces will turn a ghastly hue... Flee! Flee rash people, it is your only chance of survival

Asterix and the Soothsayer
Goscinny–Uderzo

4.1 Introduction

In the previous chapter the methods of estimating the consequences of fires and explosions were reviewed. In all cases, knowledge of the characteristics of the accident scenario was required. Therefore, we need to know the quantity of flammable vapour in a cloud to estimate the effects of the explosion which takes place upon its ignition. On the other hand, when considering a toxic release, the concentration as a function of distance and time must be estimated to calculate the vulnerability of the exposed population. In this chapter we will be primarily concerned with the models used to estimate the relevant variables in the release of hazardous substances, such as flow, concentration, total quantity discharged, etc.

By way of example, let us consider the release of a flammable liquid that gives rise to an unconfined vapour cloud explosion (UVCE). Figure 4.1 outlines the calculation stages for the estimation of its effects:

1. If, as is assumed in the Figure, the cloud is formed from the evaporation of a liquid pool, which in turn was caused by an accidental spillage, the first step consists of estimating the size of the liquid leak (source strength). Given the height of the liquid above the orifice, the pressure in the vapour space of the deposit, and the characteristics of the liquid and the orifice, one can predict the evolution of the discharge flow until the leak ceases (e.g. the level of liquid reaches the height of the perforation), or is detected and detained.

2. On the other hand, the spilt liquid evaporates. The rate at which this happens depends on the characteristics of the liquid and the ambient conditions (atmospheric temperature, wind, solar radiation, type of terrain, etc.), as well as the surface of the exposed liquid, which could be approximately constant (e.g. if a containment dike around the tank exists), or variable. The evaporation allows us to estimate the intensity of vapour flow from the liquid surface as a function of time.

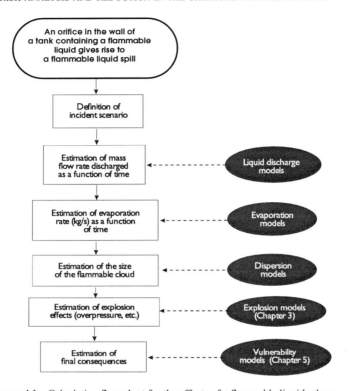

Figure 4.1   Calculation flow chart for the effects of a flammable liquid release.

3.   The vapour produced disperses in the surrounding air, and a concentration gradient can be calculated. The region containing vapour mixtures within the flammability limit (or the region in which concentrations dangerous to health are reached, in the case of a toxic release) can be estimated with the help of adequate dispersion models. Once the quantity of flammable vapour in the cloud at the moment of ignition is known the models discussed in Chapter 3 can be applied to estimate the effects of the UVCE.

4.   Once the overpressure–distance curves (or the concentration–distance–time curves in the case of a toxic emission) have been calculated, the vulnerability models given in the next chapter give us the probable final effects upon people and installations.

4.1.1   Sequences of events

Figure 4.2 illustrates sequences of events following the release of a hazardous substance (toxic or flammable). It is important to distinguish between an instantaneous escape (e.g. the collapse of a vessel), and a semi-continuous one (e.g. a perforation or fissure sufficiently small to cause a significant duration of

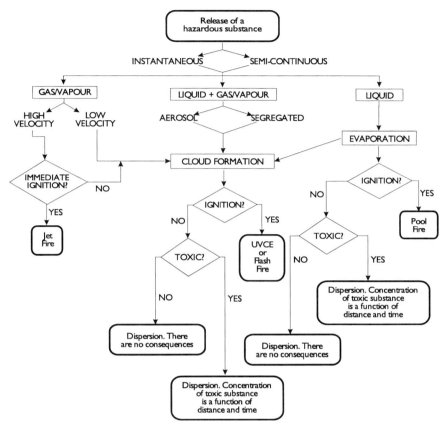

Figure 4.2   Possible sequences of events following the release of a hazardous substance.

the discharge). In the case of an instantaneous escape it can be assumed that all of the fluid is immediately available for dispersion in the atmosphere when gases are involved, or for spreading over the terrain and evaporation in the case of a liquid leak. In general, during a semi-continuous escape, the conditions in the vessel from which the escape is produced will change with time (lowering of the liquid level, or lower pressure, if the vessel contains a gas), although, if the leak is small, the pseudo-steady-state approximation (i.e. assuming constant conditions in the vessel) is acceptable for a limited period of time.

Once it is established whether an escape is instantaneous or semi-continuous, the physical state of the material released is the most important factor. In the case of two-phase discharge the relative proportions of gas/vapour and liquid are important in determining the friction loss and, therefore, the flow rate under a given set of conditions. However, it is not always possible to know beforehand

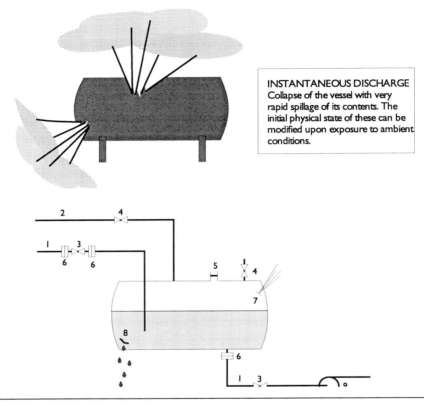

INSTANTANEOUS DISCHARGE
Collapse of the vessel with very rapid spillage of its contents. The initial physical state of these can be modified upon exposure to ambient conditions.

SEMI-CONTINUOUS ESCAPE: Loss of containment of limited magnitude.
Some possible locations:

1. Perforations/fissures in pipes carrying liquid.
2. Perforations/fissures in pipes carrying gas.
3. Leaking valves in the liquid lines.
4. Leaking valves in the gas lines.
5. Rupture disk (or safety valves).
6. Flanges, joints, etc.
7. Perforations/fissures in the gas space of the vessel.
8. Perforations/fissures in the liquid space of the vessel.
9. Pumps, etc.

Figure 4.3   Some possible causes of containment loss in a process.

the vapour fraction, which can also change during the discharge. Figure 4.3 gives examples of different ways in which the discharge of the contents of a deposit can be produced. It is clear that the location of the leak relative to the vapour–liquid interface is an important factor in determining the physical state of the discharge. However, the initial proportions of the phases can be modified. Thus, with an instantaneous discharge, part or all of the liquid content of the vessel, if suddenly exposed to ambient pressure and temperature, can experience a flash evaporation. In turn, the evaporation of an appreciable fraction of liquid could entrain liquid droplets into the vapour phase, giving rise to a mist cloud where the drops could either sediment or evaporate before reaching the ground, etc. The same can be said of semi-continuous escapes corresponding to points 1, 3, 6, 8 and 9 in Figure 4.3. On the other hand, escapes initially in the vapour phase may also undergo phase changes through partial condensation processes when exposed to cooler ambient conditions.

As shown in Figure 4.2, if the loss of containment gives rise to the spillage of a flammable liquid, the vapours on its surface can ignite, giving rise to a pool fire. In the case of discharge of a pressurized flammable gas, after an immediate ignition a jet fire occurs, while if the ignition is delayed the amount of gas released increases, giving rise to a UVCE. The characteristics of the mixture of toxic or flammable vapour/gas with the surrounding air determine the dangerousness of the cloud formed, together with other concomitant circumstances, such as the moment at which ignition occurs (in the case of flammable clouds), atmospheric conditions, proximity of uninhabited areas, topography of terrain, etc.

Figure 4.4 shows different patterns of evolution of the discharge flow over time. Curve 1 corresponds to an instantaneous discharge. In a short period of time (that taken for the vessel to physically disintegrate) maximum flow is reached, followed by a sudden decrease after collapse of the vessel. In curve 2 there is a period of increase until the maximum flow is established, and then a steady decrease as driving force for the discharge flow decreases. If for a certain period of time the material inside the vessel is replaced at the rate it is consumed, a period of approximately constant flow appears (curve 3). Lastly, curve 4 represents the case in which the opening which causes the leak gets larger with time. In any case the total quantity discharged may be less than, equal to or greater than the initial

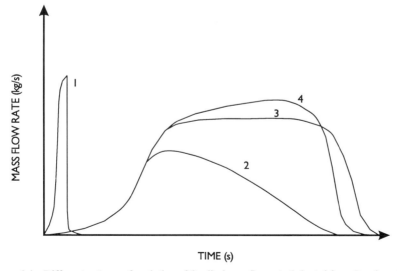

TIME (s)

Figure 4.4   Different patterns of evolution of the discharge flow rate (adapted from Opschoor [1]).

contents of the vessel, depending on the height of the perforation and the specific configuration of the pipes, isolation valves, etc. The magnitude and duration of the escape can be determined by the total amount of fluid that can be spilled (e.g. the liquid above the opening in the tank wall, all of the gas in a pressurized gas container, etc.), or by the response time for the incident (leak detection plus corrective action).

4.2   Accidental discharge of liquids

The escape of liquids from a vessel can be caused by a perforation in its wall or in the discharge pipes, valves, etc., as shown in Figure 4.3. The equations giving the flow of liquid for a given system configuration can be found in any conventional text of fluid mechanics. For the case in which the escape takes place through an orifice in the wall of a vessel, the discharge mass flow is given by

$$m^*(\text{kg/s}) = F_c A \sqrt{2\rho(P_1 - P_2)} \tag{4.1a}$$

or

$$m^*(\text{kg/s}) = F_c A \rho \sqrt{\frac{2(P - P_2)}{\rho} + 2gh} \tag{4.1b}$$

where F_c is the discharge coefficient for the orifice and A its area (m²), P_1 and P_2 are the pressures on either side of the orifice (Pa), P is the total pressure in the vapour space of the vessel (Pa), h is the height of the liquid above the orifice (m) and ρ is the liquid density (kg/m³). To estimate the values A and F_c the type of hole and its dimensions are required. Sometimes analysis of specific cases where the characteristics of the exit orifice are known can be made (e.g. study of the case where the exit pipe of the tank is torn apart at the weld due to the impact of a vehicle). In others, a set of suitable scenarios may be considered (e.g. irregular cracks in the wall of a tank, at different heights and with different equivalent areas). In both cases, the analysis is aimed at predicting the evolution of the discharge flow rate.

In equation (4.1) the pressure on the inside of the perforation, P_1, depends on the height of the liquid above the hole and the pressure in the vapour space of the tank. Because the height varies with time, the discharge flow also varies, and equation (4.1) predicts different values of m^* depending on P_1. The calculation of h (and, therefore, of P_1) as a function of discharge time is straightforward for vertical cylindrical tanks. For spherical tanks or horizontal cylindrical tanks the relationship between height and volume of liquid can be found in appropriate texts [2]. When the level of liquid reaches that of the perforation the spillage of liquid stops, to be replaced (if the pressure in the vapour space of the tank is above atmospheric), by the escape of gas, or by a two-phase discharge.

In the case of a vertical cylindrical tank, with a constant pressure P in the vapour space above the liquid level (as would occur in the case of tanks with a nitrogen padding system for inertization and pressure control, isothermal tanks containing a liquid in equilibrium with its vapour or tanks open to the atmosphere), the variation of flow with time is given by [3]:

$$m * \text{(kg/s)} = F_c A \rho \sqrt{\frac{2(P - P_2)}{\rho} + g h_0} - \frac{\rho g (F_c A)^2 t}{A_R} \qquad (4.2)$$

where P is the pressure in the vapour space of the tank (equal to P_2 if the tank is vented to the atmosphere), h_0 is the initial height above the orifice, t the time in seconds and A_R the cross-sectional area of the cylinder. The time taken for the deposit to empty to the orifice level t_F is obtained by making $m* = 0$ in equation (4.2):

$$t_F = \frac{A_R}{F_c A g} \sqrt{\frac{2(P - P_2)}{\rho} + g h_0} \qquad (4.3)$$

and the total quantity (kg) of spilt liquid up to this moment is given by

$$M_T = A_R h_0 \rho \qquad (4.4)$$

If the spillage of liquid is not from an orifice in the tank wall, but from a perforation in the exit pipe, an equation analogous to (4.1) can be used, in which P_1 is now the internal pressure in the pipe at the position where the perforation has occurred. The value of P_1 is calculated as a function of the flow rate in the tube (up to the perforation), the conditions in the deposit and the duct characteristics (length, roughness, accessories, etc.), upstream from the orifice. Other cases of liquid spillage considered in Figure 4.3 can be dealt with in a similar way.

4.3   Accidental discharge of gases and vapours

In the case of gas discharge through a perforation, a distinction must be made between a sonic and a subsonic discharge, depending on whether the quotient value between P_2 (generally equal to atmospheric pressure) and P_1 is less than the critical ratio, given by

$$r_{crit} = \left(\frac{P_2}{P_1} \right)_{crit} = \left(\frac{2}{k + 1} \right)^{\frac{k}{k-1}} \qquad (4.5)$$

Thus, in the case of moderately high initial values of P_1 (it is enough that P_1 is greater than 1.9 bars for the majority of diatomic gases), the discharge starts with a value of the pressure ratio lower than r_{crit}, and the flow is sonic. Eventually, if the gas discharged is not replaced by new material in the deposit, the pressure P_1

decreases sufficiently for the value P_2/P_1 to become greater than r_{crit}, and from this moment the flow is subsonic. Generally, the time required for the passage of gas through the orifice is so short that the exchange of heat with the tube wall and the exterior can be neglected, and the process is considered as an adiabatic expansion. The equations of fluid mechanics for subsonic flow of ideal gases give the mass flow through an orifice as

$$m^* = F_c A P_1 \sqrt{\frac{2M}{R_g T_1} \frac{k}{k-1} \left[\left(\frac{P_2}{P_1} \right)^{\frac{2}{k}} - \left(\frac{P_2}{P_1} \right)^{\frac{k+1}{k}} \right]} \qquad (4.6)$$

where P_1 and T_1 are the pressure and temperature on the deposit side, P_2 is the pressure on the discharge side of the orifice (usually atmospheric), M is the molecular weight of the gas and R_g is the universal gas law constant. k has a value equal to the ratio of heat capacities if the expansion is isentropic, and a lower value if it is not. In the case that (P_2/P_1) has a value lower than the critical ratio, the flow is sonic at the throat, and m^* has a value given by

$$m^* = (m^*)_{max} = F_c A \sqrt{P_1 \rho_1 \frac{2k}{k-1} \left[(r_{crit})^{\frac{2}{k}} - (r_{crit})^{\frac{k+1}{k}} \right]} \qquad (4.7a)$$

or

$$m^* = (m^*)_{max} = F_c A \sqrt{P_1 \rho_1 k \left(\frac{2}{k+1} \right)^{\frac{k+1}{k-1}}} \qquad (4.7b)$$

It should be noted that under sonic flow conditions the pressure P_2 on the discharge side of the orifice does not influence the mass flow obtained, as indicated in the previous equations, which do not depend on P_2. The variation of flow is given by equation (4.7b) up to the moment at which the flow becomes subsonic, and from then on the equation (4.6) gives the mass flow through the orifice.

A very important case of discharge of fluids through openings is that of pressure relief of a vessel through rupture discs or safety valves. This case is treated in detail in Chapter 7, where the possible existence of two-phase flow is also discussed.

Concerning the possibility of leaks in gas pipes, a wide variety of situations is possible, corresponding to the different configurations of piping and modes of failure (perforations in the pipe walls, defective valves, leaks in the joints, flanges, etc.), so that each case must be treated separately. In any case, a conventional treatment of gas flow in pipes enables the pressure at the point where the leak is produced to be determined. The two extreme cases correspond to isothermal flow (approached when long residence times and non-insulated pipes are used) and adiabatic flow (insulated pipes and high gas velocities, which give short residence times and little opportunity for heat interchange). In reality, the flow would be between both extremes, and normally the difference between the mass flow predicted by the isothermal and adiabatic flow equations is small, so that the equations for the adiabatic case can be used in most cases. In the case of a deposit connected to a discharge pipe, Levenspiel [4] gives charts that relate pressure drop to the flow rate obtained under adiabatic conditions. These relationships can be used to obtain P_{in}, the pressure inside the duct at the point of the leak. Once P_{in} is obtained, the treatment is analogous to that indicated previously for perforations in the wall of a deposit.

4.4 Two-phase discharge

There are many instances in which the fluid that escapes because of a perforation or through a pipe is a mixture of liquid and vapour, with a vapour/liquid ratio that may change as the discharge takes place. Two-phase flow can occur during the emptying of a vessel containing gas and liquid, especially if the liquid tends to form froth and the opening is found close to the inter-phase level. Alternatively, the second phase may originate from phase changes during the discharge. Thus, for example, liquids stored under pressure at temperatures higher than their normal boiling point undergo a sudden evaporation as they encounter lower pressures in their passage through the perforation or the discharge pipe.

It is important to know with some precision when two-phase flow can be produced and what are its characteristics, because the mass flow measured in a two-phase release can be considerably different from that which is obtained when only one of the phases is present. Thus, in one of the most important applications of two-phase discharge through an orifice, which is the emergency venting of vessels, it has been consistently shown [5, 6] that the area required in the presence of two-phase flow is larger than in the case of vapour-only discharge. Two-phase flow is a long way from being well understood, and there persists a considerable degree of empiricism in the proposed expressions. Among the problems which still remain is that of the degree of equilibrium that exists between liquid and vapour when the discharge of liquids is accompanied by a partial flash evaporation, and the determination of the conditions at which critical (choked) two-phase flow can be reached.

For two-phase flow in pipes empirical expressions like those given in the majority of texts on fluid mechanics can be applied. These expressions are often based on the well-known Baker-type diagrams and the Lockart–Martinelli correlation for the prediction of the type of flow and pressure drop, respectively [7].

For a two-phase discharge through an orifice, the CCPS [8] gives the following expression, proposed by Fauske et al. [9], for the estimation of the flow rate in a liquid discharge with simultaneous flash evapora% on:

$$G = F_c \sqrt{\left(G_{sub}\right)^2 + \frac{\left(G_e\right)^2}{N}} \qquad (4.8)$$

where G represents the mass flow density (kg/m²s). G_{sub} is the contribution of subcooling and is given by

$$G_{sub} = \sqrt{2(P - P_v)\rho_f} \tag{4.9}$$

P being the total pressure in the vessel, P_v the vapour pressure at storage temperature (Pa), and ρ_f the density of the liquid (kg/m³). In the case of saturated liquids it can be assumed that equilibrium is approached if the discharge duct is more than 0.1 m in length (or if the length of the discharge duct is greater than 10 diameters). In this case G_e is given by

$$G_e = \frac{\Delta h_v}{v_{fg}\sqrt{TC_p}} \tag{4.10}$$

where Δh_v is the latent heat of vaporization (kJ/kg), C_p is the specific heat (kJ/kg K), T is the temperature of the deposit (K) and v_{fg} is the change in specific volume when passing from liquid to vapour (m³/kg). In pipes less than 0.1 m long, the quantity of liquid that flashes during discharge rapidly decreases as the length of the pipe decreases, tending towards zero when $L = 0$. The correction factor N in equation (4.8) accounts for this effect:

$$N = \frac{(\Delta h_v)^2}{2\Delta P \rho_f v_{fg} TC_p (F_c)^2} + \frac{L}{0.1} \tag{4.11}$$

where ΔP is the pressure difference across the vessel wall. The previous equation is valid for lengths between 0 and 0.1 m. For $L = 0$, single-phase liquid discharge occurs, and the previous equations are reduced to equation (4.1), i.e. the discharge flow through a perforation in a wall of negligible thickness can be estimated as if there were no flash evaporation. In other words, flash evaporation will occur as soon as the liquid leaves the orifice, but since the evaporation produced during its passage through the wall can be neglected, it does not influence the discharge velocity.

The estimate given by equations (4.8) to (4.11) is only approximate. There are more sophisticated methods, which make intensive use of correlations for the calculation of physical properties of the fluids involved, plus mass and energy balances and semi-empirical flow equations based on an extensive range of experiments at different conditions. The results obtained by AIChE–DIERS (Design Institute for Emergency Relief Systems) are worth special attention, as this is the obligatory reference in these kinds of study. Revised in Chapter 7 are the methods for the design of pressure relief devices in process vessels and the equations giving the discharge flow rate through them, for single and two-phase flow.

Example 4.1

A fissure with a 2 mm equivalent diameter is produced in a deposit containing methane (k = 1.31) at ambient temperature (15°C). The discharge coefficient *for the orifice is estimated at 0.61. Calculate the initial discharge flow for inside pressures (P_1) of 1.5 and 15 atmospheres respectively.*

The vapour pressure of methane at 15°C is considerably greater than 15 atmospheres, so that the discharge is produced in gaseous phase in both cases. The fissure area is 3.14 mm², or 3.14×10^{-6} m².

The critical pressure ratio is obtained by making $k = 1.31$ in equation (4.5). This gives $r_{crit} = 0.54$. With this value, $P_{1,crit} = 1/0.54 = 1.85$ atm = 187400 Pa. In the first case, P_1 is less than this value and the flow is subsonic, in the second it is considerably higher, and the flow is sonic.

When $P_1 = 1.5$ atm, equation (4.6) gives a flow of 154.6 kg/m²s, i.e. 0.49 g/s for the fissure area considered. On the other hand, when $P_1 = 15$ atm (1.52×10^6 Pa), equation (4.7b) yields $G/A = 1603.4$ kg/m²s, or 5.0 g/s for the area considered.

4.5   Evaporation of a liquid release

In the previous sections we have examined the calculation methods for estimating the flow and the duration of the discharge when liquids, gases and gas/liquid mixtures escape through an orifice. The evolution of the discharged material is examined next. Figure 4.5 illustrates some of the possible interactions between the fluid released from a damaged vessel and its surroundings. Discussed below are three cases, corresponding to the evaporation of superheated liquids (usually originating from liquefied gases stored under pressure), boiling liquids, and non-boiling liquids.

4.5.1   Evaporation of superheated liquids

When a deposit contains a liquefied gas under pressure and a loss of containment occurs, the escaping liquid is subjected to a sudden decrease in pressure, down to atmospheric pressure. The liquid is initially at ambient temperature if the deposit was in thermal equilibrium with its surroundings, or lower if the liquid was kept under refrigeration. Whatever the case, when the liquid that escapes is superheated, i.e. with a temperature greater than its normal boiling point, a partial flash evaporation takes place upon exposure to ambient conditions.

During flashing, a considerable quantity of liquid can be dragged in the form of droplets, with a very large increase in the area of evaporation. Part of the liquid in drops may eventually fall again to the ground, while the rest (sometimes all of it) evaporates before sedimentation takes place, using heat from the ambient air. As to the liquid pool that remains on the ground, if the spill is small, the evaporation of superheated liquids is usually so fast that all of the liquid is consumed in a very short time and no pool is formed. In larger discharges, after the rapid initial evaporation, the liquid and the adjacent ground have cooled sufficiently for evaporation from the surface of the pool to proceed more gradually.

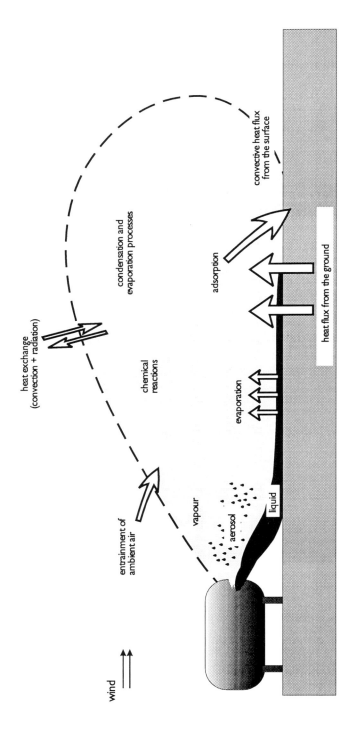

Figure 4.5 Some of the processes that can occur after an accidental fluid discharge (adapted from Hanna and Drivas [10]).

The initial flash evaporation is so fast that the process can be considered as approximately adiabatic. The excess energy contained in the superheated liquid would then be used as latent heat of vaporization, which allows the extent of flash to be calculated: The evaporation of a differential amount of liquid, dm, decreases the temperature of the remaining liquid by dT, according to

$$\Delta h_v dm = mC_p dT \tag{4.12}$$

The flash evaporation continues until the temperature of the liquid has decreased to its boiling point. After integration, the above equation gives the relationship between the remaining mass of liquid and its temperature. For only one component, the mass evaporated when the boiling temperature is reached is calculated as

$$m_v = m_0 \left(1 - \exp^{-\dfrac{C_p\left(T_0 - T_b\right)}{\Delta h_v}} \right) \tag{4.13}$$

where m_0 is the initial liquid mass, m_v the amount vaporized in the flash, T_0 and T_b the initial and boiling temperatures respectively, and C_p and Δh_v represent respectively the specific heat and the latent heat of vaporization, averaged between T_0 and T_b. In multicomponent mixtures manual calculation can be tedious, and it is appropriate to carry out the adiabatic flash calculations using any of the commercial process simulators (e.g. PROCESS™ or ASPEN PLUS™).

The value of the fraction vaporized, f_v found experimentally is usually much lower than that estimated as the ratio between m_v (from equation (4.13)), and m_0. This is principally due to the dragging of liquid droplets (aerosol formation) mentioned above. Aerosol formation not only increases the mass incorporated into the cloud, but also changes its characteristics, increasing its apparent density and producing a supplementary cooling (through evaporation of the liquid droplets), which can lead to the condensation of humidity in the atmospheric air. Reliable prediction of the aerosol content of the cloud is difficult. It requires detailed modelling which accounts for drop formation and acceleration, followed by sedimentation/evaporation processes. Kletz [11] proposes a simplified estimate, suggesting that the contribution coming from the aerosol equals that of the flash evaporation. Using this, the approximate value of the vaporized fraction is obtained as $2f_v$. However, this corrected value can still be considerably lower than experimental observations. In the CCPS revision [8] experimental work is cited in which liquid was not obtained after the initial flash, even though the fraction vaporized was estimated at only 20% (and in some cases 10%). Therefore, in the absence of more sophisticated calculations, it can be admitted as a conservative estimate that all of the spilt liquid is rapidly incorporated into the cloud, unless the calculated value of f_v is lower than 10–20%. In this case the correction suggested by Kletz can be applied.

If after the initial flash an appreciable mass of liquid still remains, a liquid pool at its boiling point results, and from this moment the evaporation can be calculated with the methods shown next.

4.5.2   Evaporation of boiling liquids

This heading covers the evaporation of gases liquefied by cooling (and that are at or near their boiling point), and also the evaporation of superheated liquids already discussed, once the initial flash has occurred and the remaining liquid is at boiling point.

When a liquid with the previous characteristics comes into contact with the ground, heat transfer from the ground to the liquid, which is at a lower temperature, takes place immediately. If the spill occurs in a dry medium and the liquid does not penetrate it significantly, the heat transfer process can be described by the Fourier equation for the cooling of a semi-infinite medium:

$$\rho_s C_{ps} \frac{dT_s}{dt} = K_s \frac{d^2 T_s}{dz^2} \tag{4.14}$$

where ρ_s, C_{ps}, K_s and T_s are, respectively, the density, specific heat, thermal conductivity and temperature of the substrate on which the spill has occurred, and the direction of the increase of the z co-ordinate is into the terrain. Equation (4.14) is solved with the following boundary conditions:

At $t = 0$, $T = T_s$, for any z

At $t > 0$, $T = T_b$ for $z = 0$

                                          $T = T_s$ for $z = \infty$ $\qquad$ (4.15)

The heat flux by conduction (W/m²) at the interface ($z = 0$) is calculated as

$$Q = K_s \frac{dT_s}{dz}\bigg|_{z=0} \tag{4.16}$$

The solution of equations (4.14) to (4.16) yields the heat flow transmitted per unit area from the ground to the liquid [1] :

$$Q = \frac{K_s (T_s - T_b)}{\sqrt{\pi \alpha_s t}} \tag{4.17}$$

where α_s is the thermal diffusivity of the substrate (m²/s). Table 4.1 gives some approximate values of conductivity and thermal diffusivity for various substrates. The rate of evaporation M_e (kg/m²s) is calculated by

$$M_e = \frac{K_s (T_s - T_b)}{\Delta h_v \sqrt{\pi \alpha_s t}} \tag{4.18}$$

Table 4.1 Some typical conductivity and thermal diffusivity values for use with equation (4.18) (from Opschoor [1]).

Material	K_s (W/mK)	$\alpha_s \times 10^7$ (m²/s)
Average ground (8% water)	0.9	4.3
Sand (dry)	0.3	2.3
Sandy ground (dry)	0.3	2.0
Sandy ground (8% water)	0.6	3.3
Wood	0.2	4.5
Gravel	2.5	11
Carbon steel	45	127
Concrete*	1.1	10

* Highly variable, depending on water content

where Δh_v is given in J/kg. As a conservative estimate, it is recommended [1] that the rate of evaporation predicted by equation (4.18) is compared with that estimated for non-boiling liquids (discussed in the next section), and take the greater of the two.

In equations (4.17) and (4.18) the square root of time appears in the denominator, which indicates that as time passes and the substrate cools, the rate of evaporation continuously decreases towards zero. This is due to the implicit assumption when obtaining the previous equations, that all the heat for evaporation comes from the cooling of the ground. However, when making calculations for extended periods of time, the Sun's radiation should be taken into account, and also it may be necessary to make corrections for heat transfer by convection from the surroundings. Solar radiation varies notably with time, the season of the year, ambient conditions and latitude. As an example, the maximum expected solar radiation values (at midday, on clear days), are quoted [12] between 425 (January) and 1070 W/m² (July), for latitude 45° north. The corresponding values for 35° north would be 650 and 1140 W/m² respectively.

The corrections due to solar radiation or heat transfer by convection could be dominant even over short periods of time if the spill occurs on thermally insulated surfaces. Opschoor [1] also suggests corrections for the case of liquids spilt on permeable substrates, such as gravel, in which case evaporation rates considerably higher than those calculated with equation (4.18) are possible, and also for the case of spills on substrates with a high humidity content. In this case higher evaporation rates are also possible, due to the additional heat released in freezing the water present.

Example 4.2

A refrigerated spherical deposit of 20 m diameter containing 2000 kg of liquid propane at –5°C collapses, with the instantaneous release of its contents. The liquid spreads over a surrounding area of dry sandy ground at 20°C, which causes the sudden evaporation of most of the spillage. The

remaining liquid forms a pool with an exposed area of 300 m². Calculate
the variation of the evaporation rate with time and estimate at what moment
the solar radiation (on this day 400 W/m²) becomes a significant contribution.

$T_o = 268$ K, $T_s = 293$ K and $T_b = 231$ K. From Table 4.1, $K_s = 0.3$ W/mK,
and $\alpha_s = 2 \times 10^{-7}$ m²/s. The specific heat of propane at 260 K is 2.64 kJ/kgK,
and this value can be used as representative of the temperature interval
considered (a rigorous calculation would take into account the variation of
specific heat with temperature). The heat of vaporization of propane at its
normal boiling point is 430 kJ/kg.

Substituting values in equation (4.18), $M_e = 0.055t^{-0.5}$ kg/m²s, i.e. the
evaporation rate for the area considered is $16.5t^{-0.5}$ kg/s. The previous
expression implies a strong variation over time: After one second the
evaporation rate is 16.5 kg/s, after one minute it is 2.12 kg/s, and after 5
minutes only 0.95 kg/s.

The heat flux from the ground is given by equation (4.17) as $Q = 23470t^{-0.5}$
kg/m²s. If one considers a 'significant contribution' for the solar radiation
received (400 W/m²) to be around 10%, then Q must decrease to approximately
4000 W/m². This takes place after $t = (23470/4000)^2 = 34.4$ seconds.

It must be noted however, that the scenario for this case is not very
realistic. Bearing in mind the initial temperature of the propane and its
physical properties, equation (4.13) predicts a value for the fraction vaporized
$m_v/m_o = 0.203$, i.e. 20.3%. According to the above discussion, for a value of
this magnitude the formation of aerosol in the initial flash is generally
sufficient to vaporize all the liquid, thus the formation of a pool is highly
unlikely, even less so on a penetrable substrate such as sandy ground.

4.5.3   Evaporation of non-boiling liquids

Let us consider the discharge of a liquid with a boiling point T_b, greater than the
ambient temperature. If the temperature at the moment of discharge is lower than
T_b, the result is a non-boiling pool of liquid from which evaporation occurs. In
this case evaporation is not the result of cooling of the substrate; instead, the main
driving force for evaporation is $(P_s - P_{amb})$, the difference between the vapour
pressure of the evaporating substance at the liquid surface and that in the
surroundings. This means that the process is controlled by mass transfer rather
than by heat transfer. As a consequence, the evaporation rate of non-boiling liquids
strongly depends on the existence or not of air streams over the pool, which affect
the mass transfer coefficient.

The rate of evaporation per unit of surface area can be written as [1]

$$M_e = \frac{kM_w}{RT}\left(P_s - P_{amb}\right) + \frac{P_s M_e}{P} \qquad (4.19)$$

where k is the mass transfer coefficient, P is the total pressure, M_w is the molecular weight of the evaporating component and R is the gas constant. The first term of the equation represents the mass transport due to the difference in vapour pressures, while the second corresponds to convection flow. In view of its magnitude in normal cases of evaporation of non-boiling liquids, the second term can almost always be ignored when compared to the first. Thus, the evaporation flow can be expressed as

$$M_e = \frac{kM_w}{RT}\left(P_s - P_{amb}\right) \qquad (4.20)$$

This equation can be applied for the case of moderate rates of mass transfer (values of P_s less than 2×10^4 Pa). For greater values, the convection flow term can no longer be neglected. Opschoor [1] proposes a more general equation, based on the two-film theory. The final expression obtained is

$$M_e = \frac{kM_w P}{RT}\ln\left(1 + \frac{P_s - P_{amb}}{P - P_s}\right) \qquad (4.21)$$

In the previous equations (4.20) and (4.21) P_s is calculated as the vapour pressure at the liquid surface temperature, T. Therefore, P_s is constant once the steady-state temperature has been reached. This results from the balance between the different heat flows existing: from the surrounding air, from the substrate, heat exchange by radiation and heat losses due to evaporation of the liquid. In most cases, until the steady state is reached, a period of decreasing temperature exists, so that the evaporation rate values given by the previous equations correspond to the initial moments, and can be considered as a conservative estimate regarding evaporation intensity (however, they will give rise to shorter emissions, which in some cases will no longer be conservative).

For estimating the mass transfer coefficient, the formula given by Sutton [13] can be used, using the parameter values recommended by Opschoor [1] for neutral stability conditions (discussed in the next section). In this case, for the evaporation rate from a circular liquid surface one obtains:

$$M_e = 2 \times 10^{-3}\left(U_{w,10}\right)^{0.78} r^{-0.11} \frac{M_w P}{RT}\ln\left(1 + \frac{P_s - P_{amb}}{P - P_s}\right) \qquad (4.22)$$

In the above equation $U_{w,10}$ represents the wind speed (m/s), measured 10 m above the ground (reference height), and r the radius of the circular pool (m). The evaporation velocity, M_e, is given in kg/m²s. In the case of evaporation from pools of rectangular surface, the length L is used instead of the radius r in the previous equation. For cases where the atmospheric conditions are not neutral, the constant and the exponents (wind velocity and pool radius) in the previous equation should be corrected.

Another empirical equation for the estimation of evaporation rates of non-boiling liquids is the following [14]:

$$M_e = 4.66 \times 10^{-6}\left(U_{w,10}\right)^{0.75} C_T \frac{P_s M_w}{P_{ref}} \qquad (4.23)$$

where M_e is given in lb/ft²min and $U_{w,10}$ in mph. P_{ref} is the vapour pressure of the reference compound, hydrazine in this case, which is calculated (in atmospheres) from

$$\ln\left(P_{ref}\right) = 65.3319 - \frac{7245.2}{T} - 8.22 \ln T + 6.1557 \times 10^{-3} T \qquad (4.24)$$

T being the absolute temperature (K). C_T is a correction factor for the temperature of the spillage. C_T is equal to unity if the temperature is less than 0°C, otherwise it is given by

$$C_T = 1 + (4.3 \times 10^{-3} t^2) \qquad (4.25)$$

where t is the pool temperature in degrees centigrade.

The review by Hanna and Drivas[10] compares different models for the prediction of evaporation rates from pools of spilled liquid. The comparison between the predictions of several models is interesting, because it shows that the differences can be very significant. Similarly, comparison between the predicted rates of evaporation and the relatively few existing experimental observations shows that, in general, the models have a tendency to overestimate the intensity of evaporation.

Example 4.3

A spillage of benzene spreads over an irregular terrain, with a total pool area of 80 m². Calculate the rate of evaporation at 26°C, with a wind speed of 2 m/s.

At 26°C the vapour pressure of benzene is 100 mm Hg, and the liquid is obviously not boiling. The vapour pressure of benzene in the atmosphere can be neglected (i.e. $P_{amb} = 0$).

A circular area equivalent to 80 m² has a radius of 10.1 m (if a different geometric shape had been assumed, e.g. a square, the results would be similar, because of the weak influence of the corresponding term in equation (4.22)). After substitution in equation (4.22) $M_e = 0.0012$ kg/m²s is obtained, i.e. 96 g/s are evaporated from the 80 m² liquid pool.

To use equation (4.23) previous calculation of the hydrazine vapour pressure is required. Equation (4.24) gives a vapour pressure of 0.0199 atm (15.13 mm Hg), and from equation (4.25) C_T is equal to 3.9. Substitution of these values in equation (4.23) gives $M_e = 0.0288$ lb/ft²min = 0.0023 kg/m²s. This is equivalent to an evaporation velocity of some 184 g/s, approximately twice the estimate obtained with equation (4.22).

4.6   Dispersion of gases and vapours in the atmosphere

Gaseous emissions as such, or the vapours produced from the discharge of volatile liquids, disperse in the atmosphere. This lowers the concentration of the emitted substance, and at the same time increases the area affected by the release. Dispersion models can be used to predict the airborne concentrations of the substance released for a given time and location. This requires previous knowledge of the characteristics of the emission (e.g. the source strength, calculated as described in previous sections), the atmospheric conditions and the topography of the surrounding terrain.

4.6.1   Types of emission

From the point of view of the nature of the emission and its continuity in time the following classification can be used:

Nature of the emission (relative density)	Neutral Positive flotation Negative flotation
Continuity of the discharge	Instantaneous Continuous Emission variable with time

With respect to the continuity of the discharge, the type of emission depends on the time-scale involved. **Instantaneous emissions** (also called 'puffs') as such do not exist. However, the term is applied to those emissions in which the time necessary to reach a receptor (e.g. a person) at a given distance from the origin is much greater than the time required for total discharge of the material. The classic example is the explosion of a vessel containing gas under pressure. As can be seen, the definition is relative, and thus an emission of specific duration may or may not be considered instantaneous, depending on the location of the receptor and the atmospheric conditions. In a **continuous emission** ('plume'), the duration of the emission is long compared to the time necessary to reach the receptor. Under these conditions, for a release of constant characteristics, the dispersion model can be solved to obtain the steady-state value of concentration expected for a given location. The typical example of this type of situation is the emission from a stack.

In practice, the majority of accidental emissions are found in an intermediate position between instantaneous and continuous, having in addition a certain degree of variation in the emission characteristics. The most complex case is a release of variable characteristics, continuous or intermittent, but with important changes in concentration, flow rate, etc. of the substance emitted, or with significant variations in atmospheric conditions during the time duration of the release. In spite of this, both extreme situations, instantaneous and continuous, are useful for modelling purposes, and in many cases give a sufficiently approximate description of accidental releases. In what follows, unless otherwise indicated, it will be assumed that in whatever type of emission the atmospheric conditions remain constant, and for continuous releases it will also be assumed that the characteristics of the emission (flow rate, concentration, etc.) remain constant with time.

As dispersion takes place, the centre of the cloud formed by an instantaneous emission will move away from the origin, due to the carrying effect of the wind. This process is depicted in Figure 4.6 as the displacement along the x axis of an approximately spherical cloud, that is expanding. The case of a continuous emission is shown in Figure 4.7. As time passes, the real $(0,0,h)$ and virtual $(0,0,H)$ origins of the emission remain fixed in space. As the substances released are transported downwind (x direction), atmospheric air enters the plume aided by ambient turbulence, which provokes the dilution of the plume as the distance to the emission point increases. For modelling purposes, the contours of the plume in a continuous emission (just as occurs with the contours of the cloud in the case of an instantaneous emission) are defined by a predetermined concentration, fixed arbitrarily (e.g. a concentration which is a thousandth of the value at the origin of the release, a certain concentration in parts per million, etc.).

As to the nature of the emission, the classification given above refers to the relative densities of the emitted material and air. That an emission is positively or negatively buoyant with respect to atmospheric air depends not only on the substance involved, but also on the temperature of the emission and of the surrounding air, the relative humidity and the possible formation of aerosols. Thus, for example, a release of sulphur dioxide ($M_w = 64$) would be denser than air, whether occurring at ambient temperature or at lower temperatures. On the contrary, methane ($M_w = 16$) is much lighter than air at ambient temperature, but if the emission takes place at its normal boiling temperature ($-162°C$), as would occur if the source were a pool of liquid methane, the density at the point of emission

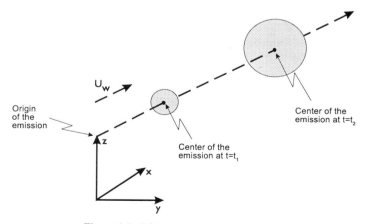

Figure 4.6 Dispersion of an instantaneous emission.

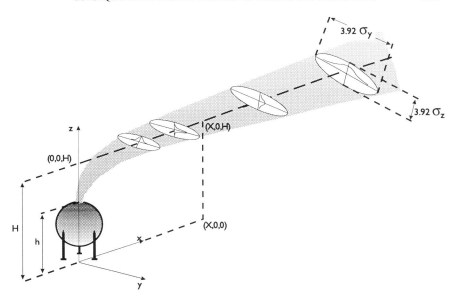

Figure 4.7 Dispersion of a continuous emission with positive buoyancy (adapted from Finch and Serth [15]).

would be greater, and therefore dense cloud behaviour (negative flotation) could take place initially. After dilution and exchange of heat with the surrounding air the characteristics of a dense emission can change rapidly, approximating neutral behaviour.

4.6.2   Atmospheric stability

In practice it is difficult to take into account all the meteorological conditions that affect the dispersion of a substance in the atmosphere. Thus, a parameter called **stability class** has been developed, which enables the calculation of dispersion coefficients. Stability in this context refers to the vertical mixing of the different layers of air. It depends not only on the wind speed, but also on the atmospheric temperature gradients. The scale of stability most frequently used is that of Pasquill, with six classes of stability from A (high instability, intense mixing) to F (the most stable, with little mixing). The classes of stability are shown in Table 4.2 for different wind speeds and degrees of heating due to solar radiation, which are in turn related to the time of day (day or night) and the cloudiness [10]. A more complete table is found in the work of Van Buijtenen [16], adapted to the conditions in the Netherlands.

During daytime, the temperature of atmospheric air generally decreases as the height above ground increases. If a mass of air is displaced from a height h_1 to a

Table 4.2(a) Meteorological conditions for the definition of Pasquill stability classes [10]

Surface wind speed (m/s), measured at 10 m height	Daytime insolation			Night-time conditions	
	Strong	Moderate	Slight	Thin overcast or at least 4/8 cloudiness*	Less than 3/8 cloudiness*
< 2	A	A–B	B		
2–3	A–B	B	C	E	F
3–4	B	B–C	C	D	E
4–6	C	C–D	D	D	D
>6	C	D	D	D	D

A: Extremely unstable; B: Moderately unstable; C: Slightly unstable; D: Neutral; E: Slightly stable; F: Moderately stable.
*The degree of cloudiness is defined as that fraction of the sky above the local apparent horizon that is covered by clouds.

Table 4.2(b) Determination of the appropriate insolation category

Sky cover	Solar elevation angle > 60°	Solar elevation angle between 35 and 60°	Solar elevation angle between 15 and 35°
4/8 or less, or any amount of high thin clouds	Strong	Moderate	Slight
5/8 to 7/8 middle clouds (2100 to 4900 m)	Moderate	Slight	Slight
5/8 to 7/8, low clouds (less than 2100 m)	Slight	Slight	Slight

higher one, h_2, where the pressure is $P_2 < P_1$, the air expands and cools. The associated cooling process can be quantified for a specific set of conditions. Thus, if we consider the process to occur adiabatically, the rate of decrease of temperature with height, known as the **lapse rate**, would be about 0.01°C/m for dry air, i.e. dT/dh is approximately –1°C each 100 metres. Therefore, in the absence of other effects, ascending into the atmosphere means finding air at ever-decreasing temperatures. These differences in temperature cause differences in density, so that the hot air (lower layers) tends to ascend and the cold air to sink, generating mixing in the vertical direction.

The conditions of maximum instability (classes A and B) are favoured with light winds and high or moderate insolation. Therefore, on a clear day with strong insolation intense mixing, i.e. high instability, can be expected. Slight instability (C) occurs with the same insolation conditions, but with higher winds, or with light winds if the solar heating is light. A typical case when these conditions may occur is, for example, a sunny autumn evening. Class D (neutral) prevails with strong winds, and under slight insolation or night-time conditions. The stable classes E and F occur at night-time, with moderate winds and clear or slightly cloudy skies. Under these conditions, the heat loss by radiation rapidly lowers the temperature of the ground, which in turn cools the lower layers of air. This places the coolest (and densest) layers of air at the lower levels, which inhibits natural convection and increases stability.

The phenomenon of increasing temperatures when ascending in the atmosphere is called an **inversion condition**, and is very important in the dispersion of substances in the atmosphere, as it hinders vertical mixing. Thermal inversion occurs frequently at night, as shown in the previous paragraph. Shown in Figure 4.8 is a typical example of variation of the temperature gradient with the time of day, for open country. It can be seen that the absolute value of the negative temperature gradient $(dT/dt < 0)$ increases during the central hours of the day, to be inverted during the night. On the coast, the heat sink effect due to the sea decreases the variations observed, and sometimes inhibits thermal inversion. Other causes of thermal inversion have been cited [17], e.g. subsidence of air from greater heights which results in compression and heating up of lower layers, or the presence of sea breezes (cold, due to the evaporation of sea water) that can introduce a layer of cold air at the lowest levels of the atmosphere.

4.6.3 Wind

It has already been shown that the wind speed has a strong influence on the stability class. Moreover, the wind is obviously important when determining the behaviour of an atmospheric release. The three most important characteristics from this viewpoint are:

1. *Direction*   This determines the main direction of propagation of the emission and, therefore, the areas most likely to suffer its consequences. It is usually necessary to consider more than one wind direction in risk analysis, even in areas where the predominant winds have a defined orientation.
2. *Speed*   As the wind speed increases, the substances released arrive sooner at the receiver situated in the wind's direction. However, the emission also experiences a greater dilution due to the increased turbulence generated by stronger winds. The wind speed varies with height, there being a boundary layer whose thickness depends on the roughness of the terrain. Figure 4.9 represents in a schematic way

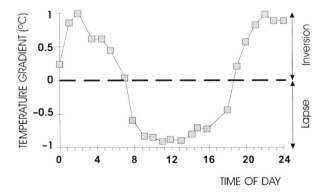

Figure 4.8   Typical example of variation of vertical temperature gradient with the time of day (data read from Lees [17]).

the effects of terrain roughness on the variation of wind speed with height. It can be seen that the extension of the boundary layer is greatest (some 500 m) in the case of tall buildings, such as skyscrapers or high industrial installations, reaching an extent of 400 m over terrain with low buildings (low houses, trees) and some 250 m in very open and flat terrain, such as the surface of the sea. Above the boundary layer we have gradient wind, thus called because its velocity can be determined with enough accuracy by considering only atmospheric pressure gradients, i.e. friction effects from the ground can be neglected.

The wind speed varies in the boundary layer region due to friction from the ground. The variation is usually expressed as an exponential relationship [18]:

$$U_{w,z} = U_{w,10} \left(\frac{z}{10} \right)^n \tag{4.26}$$

where $U_{w,z}$ is the wind speed at a height z above ground, $U_{w,10}$ is the wind speed at a height of 10 m, usually taken as a reference. Table 4.3 gives typical values for the exponent n, which is a function of the type of terrain and the class of stability. Since we are mainly concerned with the effects of a hazardous emission at ground level, the previous equation can be used to correct the values of wind speed, especially when considering dense emissions, that tend to remain at low levels for longer periods of time.

In addition the wind direction can be subjected to strong local influences that combine with the effects of terrain and temperature [17]. Thus, for example, it is typical in valleys that a descending night breeze exists, caused by the terrain being cooled by radiation, which in turn cools the adjacent air layers. Similarly, on the coast, the ground is usually warmer during the day, and the sea during the night, which frequently gives rise to breezes moving inland during the day (the

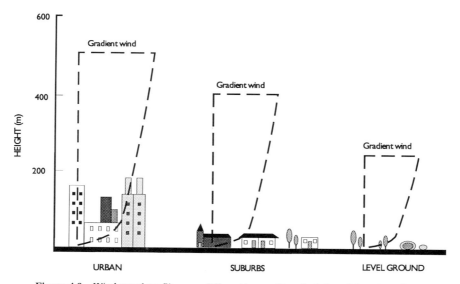

Figure 4.9   Wind speed profiles over different types of terrain (adapted from Lees [17]).

Table 4.3   Values of exponent n for use with equation (4.26)

Stability class	Urban terrain	Rural terrain
A	0.15	0.07
B	0.15	0.07
C	0.20	0.10
D	0.25	0.15
E	0.40	0.35
F	0.60	0.55

air over the ground, warmer and less dense, rises, being substituted by sea air) and the reverse occurs at night.

3. *Frequency and persistence*   The frequency of wind in a given direction indicates the percentage of time in which the wind has this orientation. Persistence is the quality which refers to the constancy with which the wind blows in a certain direction. The analysis of risks is considerably complicated if the changes in wind direction are frequent, so that the data on persistence must be known. The wind persistence cannot be obtained from frequency data because the same frequency can be obtained with very different values of persistence.

The data for wind direction and speed for a specific locality are usually presented as the classic wind rose diagram, with the meteorological station placed in the centre of a circle and radial arms in which the wind speed in a given direction and its frequency are represented. It is necessary that the values of wind direction, speed, frequency and persistence (together with other relevant data such as insolation, temperature and humidity of the air, precipitation, etc.) are collected over prolonged periods of time, thus taking into account seasonal variations, and also to obtain a sufficient amount of data to statistically assess the likelihood of a given set of meteorological conditions when making a risk analysis.

4.6.4   Dispersion models

The development of new dispersion models and the extension of existing ones constitutes a research area of intense activity, so that only a superficial view of the topic is given here. The discussion is concentrated on simple models, which are, however, frequently applied. For a critical revision of the main existing models the references specific to this topic can be consulted [10, 19].

The behaviour that can be expected from the release of a positive buoyancy vapour was illustrated in Figure 4.7. Shown in Figure 4.10 is a similar diagram for an emission initially denser than air, in which different stages can be observed. During the initial acceleration and dilution phase the initial momentum of the material released predominates, and the discharge can take place in any direction. During this phase a rapid expansion occurs that can dilute the concentration of the emission by one or two orders of magnitude [10]. After the initial phase, a

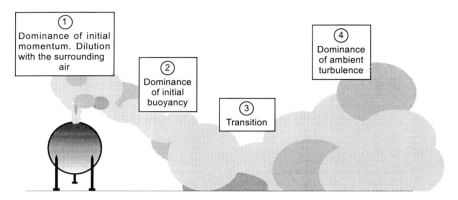

Figure 4.10 Different stages in the release of dense gases (adapted from Hanna and Drivas [10]).

second stage is entered in which negative flotation still exists, so that the cloud tends to sink towards the ground. Due to the progressive dilution of the cloud with the surrounding air, a moment is reached at which the flotation effects become less important than atmospheric turbulence, and transition between stages 2 and 4 occurs. The moment of transition is difficult to determine, and constitutes one of the issues that differentiate the models proposed for release of dense gases. From step 4 the cloud can be considered as an emission of neutral flotation.

Emissions of neutral or positive flotation:

Gaussian dispersion models

In the absence of important density effects (negative flotation), dispersion in the atmosphere is a process governed by atmospheric turbulence, which facilitates the mixing of the substance emitted with surrounding air.

Let us assume an **instantaneous emission** of Q kg of gas, at a time when the wind blows at a speed U (m/s), in the direction of the x-axis, with $x = 0$ corresponding to the point of origin of the emission (Figure 4.7). Assuming a turbulent diffusion coefficient for the emitted substance in air equal to D (m²/s), constant in all directions (isotropic diffusion), the dispersion in the atmosphere is given, in rectangular co-ordinates, by

$$\frac{\partial C}{\partial t} + U \frac{\partial C}{\partial x} = D\left(\frac{\partial^2 C}{\partial x^2} + \frac{\partial^2 C}{\partial y^2} + \frac{\partial^2 C}{\partial z^2} \right) \tag{4.27}$$

where C is the concentration of the emitted substance (kg/m³) for a given time and location. With no wind, equation (4.27) becomes

$$\frac{\partial C}{\partial t} = D\nabla^2 C \tag{4.28}$$

An instantaneous emission, in the absence of other effects (terrain, etc.), adopts approximately the form of a sphere, so that spherical co-ordinates can be conveniently used to describe the system. The previous equation can be solved with the following boundary conditions:

$$C = 0 \text{ when } t = 0, \text{ for } r > 0$$

$$C = 0 \text{ when } t = \infty, \text{ for any value of } r \qquad (4.29)$$

In addition, the mass conservation condition must be satisfied:

$$\int_V C dV = Q \qquad (4.30)$$

V being the volume of the cloud. The solution to equations (4.28) to (4.30) (instantaneous release, no wind) is [3]:

$$C = \frac{Q}{8(\pi Dt)^{3/2}} \exp\left(- \frac{r^2}{4Dt}\right) \qquad (4.31)$$

The solution in rectangular co-ordinates is obtained by simply replacing r^2 by $x^2 + y^2 + z^2$. When an anisotropic dispersion occurs the turbulent diffusivity depends on the direction considered, and equation (4.31) becomes

$$C = \frac{Q}{8(\pi t)^{3/2}\left(D_x D_y D_z\right)^{1/2}} \exp\left[- \frac{1}{4t}\left(\frac{x^2}{D_x} + \frac{y^2}{D_y} + \frac{z^2}{D_z}\right)\right] \qquad (4.32a)$$

Where wind is present, the same equations are used, with the exception that the centre of the instantaneous emission is displaced along the x-axis, at a distance $x = Ut$. The equation to use is

$$C = \frac{Q}{8(\pi t)^{3/2}\left(D_x D_y D_z\right)^{1/2}} \exp\left[- \frac{1}{4t}\left(\frac{(x - Ut)^2}{D_x} + \frac{y^2}{D_y} + \frac{z^2}{D_z}\right)\right] \qquad (4.32b)$$

As for **continuous emissions**, it has already been stated that considered as such are those in which the duration of the release is considerably greater than the ideal arrival time to a receptor, expressed as x/U. If the condition of steady state ($\frac{\partial C}{\partial t} = 0$), is used in equation (4.27), for a constant release rate of Q^*(kg/s), a similar treatment leads to [3]:

$$C = \frac{Q^*}{4\pi D\sqrt{x^2 + y^2 + z^2}} \exp\left[- \frac{U}{2D}\left(\sqrt{x^2 + y^2 + z^2} - x\right)\right] \qquad (4.33)$$

In the majority of cases the aim is to calculate the concentration relatively close to the x-axis (direction of propagation of the release), so that $(y^2 + z^2)/x^2 \ll 1$. Using $(1 + \delta)^{1/2} \approx 1 + \delta/2$, valid for small values of δ, the previous equation can be simplified to

$$C = \frac{Q^*}{4\pi Dx} \exp\left[- \frac{U}{4Dx}(y^2 + z^2)\right] \qquad (4.34)$$

It is often of interest to know the maximum concentration that can be reached at a given distance x from the origin. This is found along the x-axis ($y = z = 0$), and is given by

$$C = Q^* / 4\pi Dx \qquad (4.35)$$

Where there is anisotropy, equation (4.34) is transformed into

$$C = \frac{Q^*}{4\pi x \sqrt{D_y D_z}} \exp\left[-\frac{U}{4x}\left(\frac{y^2}{D_y} + \frac{z^2}{D_z} \right) \right] \qquad (4.36a)$$

The previous discussion has been concerned with elevated sources, i.e. the influence of the terrain was not considered. If a release, instantaneous or continuous, occurs at ground level and the terrain is impermeable to the substance emitted, the real concentration will be twice the concentration predicted in the previous equations, due to the reflection effect produced by the terrain. Thus, for a continuous emission at ground level the concentration distribution will be

$$C = \frac{Q^*}{2\pi x \sqrt{D_y D_z}} \exp\left[-\frac{U}{4x}\left(\frac{y^2}{D_y} + \frac{z^2}{D_z} \right) \right] \qquad (4.36b)$$

An intermediate situation exists for an elevated source, with an effective height H above the ground, as shown in Figure 4.7. In this case the reflection on the terrain does not occur immediately, but at a certain distance from the source, x_G, when the emission has already reached the ground. The equations to be used in this case (for continuous emission) are [17]:

$$C = \frac{Q^*}{4\pi Dx} \exp\left[-\frac{U y^2}{4Dx} \right]\left\{ \exp\left[-\frac{U (z - H)^2}{4Dx} \right] + \exp\left[-\frac{U (z + H)^2}{4Dx} \right] \right\} \qquad (4.37)$$

In the case of an emission at ground level ($H = 0$), the previous equation is equivalent to equation (4.36b), considering isotropy for the diffusion coefficients. The term containing ($z + H$) in equation (4.37) represents what is known as 'emission by reflection', and corresponds to the part of the plume from an elevated source that would be below the ground surface ($z = 0$), after the emission reaches the ground, at $x > x_G$, and that is 'reflected' [15].

A characteristic of equations like (4.36b) is that, for a given distance x, the variation in concentrations predicted in a horizontal direction (y), or vertical (z), corresponds to a normal or Gaussian distribution. Experimental observations of dispersion of releases of neutral or positive buoyancy in the atmosphere have corroborated this prediction. The parameters of the distribution are shown in Figure 4.11(a). The effect of progressive dilution of the emission by the surrounding air can be taken into account by using an ever higher value of standard deviation as the distance from the point of emission increases. This produces a curve which becomes increasingly flatter as shown in Figure 4.11(b). Following this approach, different authors have developed methods for obtaining the distribution parameters.

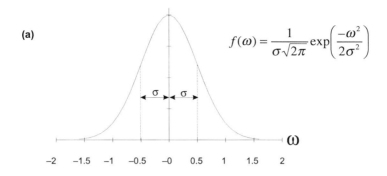

$$f(\omega) = \frac{1}{\sigma\sqrt{2\pi}}\exp\left(\frac{-\omega^2}{2\sigma^2}\right)$$

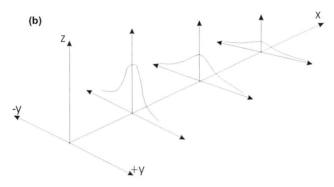

Figure 4.11 (a) Normal distribution curve for a standard deviation equal to 0.5; (b) variation of the concentration profile as the distance to the source increases.

Sutton [13] found experimentally that, although the distribution of concentrations at a given distance from the source could be reasonably represented by a normal distribution, the variation of concentrations along the x-axis could not be predicted by the above equations. Thus, the concentration on the emission axis ($y = z = 0$) for continuous emissions at ground level was proportional to $x^{-1.76}$ when, according to equation (4.36b), it should have been proportional to $1/x$. This and other irregularities led Sutton to propose a group of modified equations. An empirical expression was used to calculate the standard deviation for the distribution of concentrations:

$$\sigma_i^2 = 0.5\,C_i^2\,x^{2-n} \tag{4.38}$$

where σ_i is the standard deviation in direction i, and C_i and n are empirical constants. It can be seen that σ_i increases with distance x. Taking the Sutton model as a basis, a new set of equations was proposed, and new empirical coefficients were calculated. The results are known as the Pasquill–Gifford model [3, 8, 17], or Gaussian plume model [16].

In this model, the variation of concentrations with distance and time is given as

$$C = \frac{Q}{(2\pi)^{3/2}\sigma_x\sigma_y\sigma_z}\exp\left[-\frac{(x-Ut)^2}{2(\sigma_x)^2} - \frac{y^2}{2(\sigma_y)^2}\right]\left\{\exp\left[-\frac{(z-H)^2}{2(\sigma_z)^2}\right] + \exp\left[-\frac{(z+H)^2}{2(\sigma_z)^2}\right]\right\}$$

(4.39)

where H is the effective height of the emission ($H = 0$ for an emission at ground level), and the source is located at $x = 0$ (Figure 4.6). For a continuous emission,

$$C = \frac{Q*}{2\pi U\sigma_y\sigma_z}\exp\left[-\frac{y^2}{2(\sigma_y)^2}\right]\left\{\exp\left[-\frac{(z-H)^2}{2(\sigma_z)^2}\right] + \exp\left[-\frac{(z+H)^2}{2(\sigma_z)^2}\right]\right\}$$ (4.40)

In the derivation of equation (4.40) the following assumptions are implied [15]:

- The concentration profiles in the z and y directions can be adequately described as a Gaussian distribution, whose parameters depend on the distance to the source, x.
- The wind speed U, and the emission flow rate Q, are constant.
- The diffusive flow in the x direction can be neglected when compared to the mass transport by convection (wind).
- The product emitted is a gas, vapour or stable aerosol, that does not react or sediment.

In the case of emissions from a stack, or in situations in which the discharged material has sufficient momentum to reach heights significantly above the emission point it is necessary to calculate the effective source height, H, which as shown in Figure 4.7, may be different from the height of the discharge point. There are various expressions to calculate H [20], the Briggs method [21, 22] being widely used.

Application of the Gaussian model to atmospheric dispersion problems requires a knowledge of the variation of the standard deviation σ_i with distance. Different sets of equations [10, 16, 17, 23] have been proposed to express this dependence, according to the prevailing stability class. The values of σ_i estimated for different methods differ slightly. Presented here is the development of Van Buijtenen [16], based on the equations shown below, adjusted for distances between 100 m and 10 km.

$$\sigma_y = ax^b$$ (4.41a)

$$\sigma_z = cx^d$$ (4.41b)

where the values of the four coefficients a, b, c and d for the case of a continuous emission are obtained from the table below:

Stability class	a	b	c	d
A: Very unstable	0.527	0.865	0.28	0.90
B: Unstable	0.371	0.866	0.23	0.85
C: Slightly unstable	0.209	0.897	0.22	0.80
D: Neutral	0.128	0.905	0.20	0.76
E: Stable	0.098	0.902	0.15	0.73
F: Very stable	0.065	0.902	0.12	0.67

In the previous equations, x, σ_y and σ_z are given in metres. The standard deviation values in a horizontal direction (σ_y) should be interpreted as averaged over a period of 10 minutes, and the values of σ_z are for heights lower than 20 m and a value of the terrain roughness length z_0 (defined below) equal to 0.1 metres.

The values for standard deviations calculated with equation (4.1) and the previous coefficients are valid for continuous sources. In the case of instantaneous emissions, in view of the limited data available, the following expressions are recommended [16]:

$$\sigma_x = 0.13\ x \text{ for any stability class} \tag{4.41c}$$

$$\sigma_y = 0.5\ \sigma_{yc} \tag{4.41d}$$

$$\sigma_z = \sigma_{zc} \tag{4.41e}$$

where σ_{zc} and σ_{yc} are the values for continuous emissions in the z and y directions respectively, calculated with equations (4.41a) and (4.41b). Other authors [8] suggest taking a value of σ_x equal to that calculated for σ_y with equation (4.41d).

Corrections to the Gaussian dispersion model

1. Correction for the roughness length

The terrain roughness length parameter z_0 is introduced in order to quantify the effects of a non-uniform terrain surface on the vertical dispersion of the emission. The values of z_0 for different types of ground can be estimated as follows:

$z_0 = 0.03$ Flat terrain, with few trees
$z_0 = 0.10$ Open terrain (flat with abundant trees, arable land, etc.).
$z_0 = 0.30$ Cultivated land (greenhouses, isolated houses, etc.)
$z_0 = 1.0$ Residential area, with dense but low construction, industrial site with low structures, etc.
$z_0 = 3.0$ Urban area, high buildings, industrial buildings with large structures

The values of σ_z obtained with the coefficients c and d given previously can be used as such for a value of z_0 equal to 0.1. For other values it is necessary to introduce a correction according to

$$\sigma_z = cx^d (10\ z_0)^m \tag{4.42a}$$

where

$$m = 0.53\ x^{-0.22} \tag{4.42b}$$

2. Correction for the duration of exposure

Continuous emission with constant characteristics over a period of time is an idealization. In reality, as expected, the emission boundaries at a certain distance from the origin experience erratic variations of limited amplitude, which must be accounted for. The situation can be visualized by imagining the emission of smoke from a stack. If at a given moment a snapshot is taken of the emission, the cloud leaving the stack extends over a certain region (area corresponding to $t = t_1$ in Figure 4.12). A second photograph, a few seconds later ($t = t_2$), will show a different area. Overlapping the contours obtained by a long series of pictures, the region affected by the emission would be obtained, which corresponds to the envelope of the different boundaries. The area of this region is greater than that of the plume at a given moment, whilst its average concentration is lower.

The fluctuations can be accounted for approximately by the introduction of a correction factor for σ_y:

$$C_t = (t/600)^{0.2} \tag{4.43}$$

$$\sigma_y = C_t \sigma_{y,10} \tag{4.44}$$

where t is the time a receptor is exposed (in seconds) and $\sigma_{y,10}$ represents the standard deviation for an exposure of 10 minutes, which is obtained using equation (4.41a).

3. Correction for the source dimensions

In the previous discussion it was implicitly assumed that the release was produced by a point source, or that the estimates of concentration were made at a sufficiently large distance that the physical dimensions of the source were unimportant. However, it can happen that the source has considerable dimensions which must

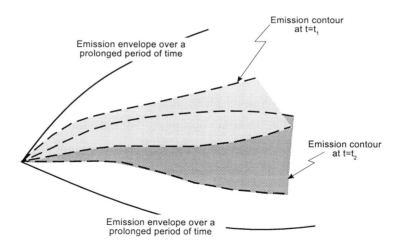

Figure 4.12   Fluctuations in the contour of an emission.

be taken into account (consider, for example, the spillage of a volatile liquid and the ensuing evaporation from the surface of the pool formed).

A method frequently used is that of a virtual point source, illustrated in Figure 4.13 for a source with rectangular dimensions $(2L_y) \times (2L_z)$. This procedure implies the selection of an imaginary point source A, situated at a distance x_v from the real source, in such a way that, after covering the distance x_v, the plume originated from the virtual point source would have the same dimensions as the real source. Van Buijtenen [16] suggests the following expression for the standard deviation in the y-axis:

$$\sigma_y = L_y / 2.15 \tag{4.45}$$

The previous equation implies that the limits of the real source along the y-axis are taken at the points where the concentration has fallen to 10% of its maximum value. If the concentration over the source does not follow a Gaussian distribution, but is approximately homogeneous, instead of the previous equation one can use [16]

$$\sigma_y = L_y / 1.25 \tag{4.46}$$

Once the standard deviation value is estimated (using the previous equations or others considered more suitable for a specific source), the virtual distance for the y direction can be calculated from equation (4.41a). Combining equations (4.45) and (4.41a) we can write

$$x_{vy} = \left(\frac{L_y}{2.15a} \right)^{\frac{1}{b}} \tag{4.47}$$

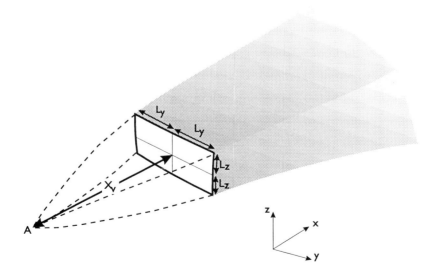

Figure 4.13   Virtual point source for a rectangular $(2L_y) \times (2L_z)$ emission.

Analogously, for the virtual distance corresponding to the z direction,

$$x_{vz} = \left(\frac{L_z}{2.15c}\right)^{\frac{1}{d}} \tag{4.48}$$

If necessary, the values σ_y and σ_z used in equations (4.47) and (4.48) can be modified with the corresponding corrections for roughness length or the emission duration.

4. Correction for transmission over heterogeneous terrain

It can happen that the type of terrain changes over the distance at which the dispersion calculations are made. For example, an emission coming from an area where there are industries with tall buildings can be carried over open country of low roughness length to arrive finally at a residential zone. If the calculation of the possible exposure of a receptor in the residential zone to an emission from the industrial area is required, it will be necessary to take into account the changes in the terrain's characteristics. The calculations in this case can be made with the aid of virtual sources, in a similar way to that explained in the previous section for sources with finite dimensions.

Thus, if an emission crosses a terrain of extension x_A with a roughness length equal to z_{oA}, followed by an extension x_B of roughness z_{oB}, the method of calculation consists of using the roughness z_{oA} to determine the standard deviation σ_A after the emission has travelled a distance x_A. Once σ_A is known, the problem can be treated as if the emission crossed only terrain of roughness length z_{oB}. To this end, a virtual distance x_{vB} is calculated which produces this same value of the standard deviation (σ_A) with the roughness parameters of zone B. In this way, the heterogeneous terrain of extension $x_A + x_B$ is equivalent to a homogeneous terrain of extension $x_{vB} + x_B$ and roughness z_{oB}. A calculation of this type is shown in Example 4.5.

4.6.5   Concentration contours

From the dispersion equations given above, the extension of the regions subjected to a particular concentration can also be calculated. This is important when making vulnerability estimations (Chapter 5), in which, for example, the extent of terrain affected by a certain concentration of a toxic emission, or the area within the flammability range in the case of an emission of flammable vapour, must be known.

In the case of a plume, we are interested in calculating the lateral distance (y direction) on both sides of the emission axis that corresponds to a certain concentration of the emitted substance. For a point situated at xm from the source of a continuous emission at ground level, the lateral distance at which a concentration C_i is reached is given by [16]:

$$y(x) = \sigma_{y(x+x_v)} \sqrt{\frac{2\ln C_{max}(x)}{C_i}} \tag{4.49}$$

where C_{max} is the concentration along the axis of the emission $(x,0,0)$ at a distance x from the source. The term $\sigma_{y(x+x_v)}$ indicates that, although the lateral distance (y) in the above equation is calculated at xm from the real source, in the case where the characteristics of the emission make it necessary to use a virtual source, the distance $x + x_v$ should be used for the calculation of σ_y.

4.6.6   Dispersion of dense emissions

As has already been mentioned, the dispersion of emissions denser than air (whether due to the molecular weight of the emitted substance, the temperature of emission or the presence of aerosols) after the initial acceleration and dilution, is followed by a period in which negative buoyancy predominates and finally, after further dilution and/or heat exchange, by transition to a behaviour dominated by ambient turbulence (neutral flotation). The previous stages are not well-defined; there is overlap between the different phases in the evolution of a dense emission, and uncertainty regarding their duration for different scenarios.

The dispersion of dense gases presents distinct characteristics with respect to the neutral or positive flotation emissions, which makes the use of models derived for these inappropriate, whilst a density greater than air persists. In particular, as shown in Figure 4.10, the emission extends under gravitational influence close to the source, with a behaviour resembling that of a liquid spillage. In these conditions the gas cloud adopts a relatively flat shape, flows over obstacles in the terrain and can partially spread upwind. Moreover, the difference in densities slows down the process of mixing with air. As mixing takes place, the emission exhibits a density gradient similar to that described for a thermal inversion situation (i.e. denser strata at the bottom of the cloud), which also reduces vertical mixing. All these effects reduce dispersion, and tend to increase concentrations at a specified distance from the source with respect to those obtained in an emission of neutral or positive flotation.

During the negative flotation stage, the relative influences of gravity and ambient turbulence are indicated by the Richardson number R_i, defined as the ratio of potential energy of the cloud to the turbulent energy of the environment. One of the expressions proposed to calculate R_i is [10]:

$$R_i = g\frac{\rho_p - \rho_a}{\rho_a}\frac{(V_o)^{1/3}}{(u*)^2} \quad \text{(for instantaneous emissions)} \qquad (4.50)$$

$$R_i = g\frac{\rho_p - \rho_a}{\rho_a}\frac{V'_o}{(u*)^2 ud} \quad \text{(for continuous emissions)} \qquad (4.51)$$

where ρ_p and ρ_a are, respectively, the densities of the dense emission and the surrounding air, V_0 is the initial volume of the instantaneous cloud (m^3), V'_0 the initial flow rate of a continuous emission (m^3/s), d the orifice diameter, u the wind speed and $u*$ the friction velocity, which is calculated as the square root of the ratio of surface momentum to cloud density [10]. The relationship between the values of u and u^* depends principally on the roughness of the terrain. Typical values quoted for u/u^* range between 10 and 15.

When the cloud contains aerosol (which can cause a dense behaviour in releases of vapours lighter than air), the value of cloud density (assuming that the aerosol occupies a small fraction of the total volume) can be approximated by:

$$\rho_p = (P/R'T) + \rho_{aer} \qquad (4.52)$$

where ρ_{aer} is the aerosol mass per unit volume of cloud and R' is given by

$$R' = \Sigma (m_j/m)(R_g/M_j) \qquad (4.52b)$$

where the summation is extended to all the components of the mixture, m_j is the mass of component j in the cloud, m the total cloud mass, M_j the molecular weight of component j and R_g the universal gas constant.

Equations (4.50) and (4.51) are useful for predicting the prevailing dispersion regime for a dense cloud. Although considerable uncertainty exists as to the critical value of the Richardson number which separates the stages dominated by gravitational effects and ambient turbulence, a value of the order of 10 is usually accepted. A higher value indicates the necessity of using dense gas models.

Initial extension of a dense cloud

A realistic model for dense emissions would require, in addition to the intrinsic dense cloud dispersion dynamics, consideration of the influence of the phenomena shown in Figure 4.5, and more specifically, at least the following: evaporation and condensation of drops in the cloud, heat (and on occasion mass) exchange with the ground, chemical reactions and heat exchange by radiation. Present-day models are still a long way from considering these phenomena in their full complexity, and in general contemplate only simplified cases. However, there are interesting theoretical/empirical approaches, and the reader is directed to the specialized bibliography or to specific commercial simulators for a more rigorous treatment of dense cloud dispersion. The very simplified treatment shown below is centred on the dispersion of a dense cloud for the case in which the influence of the above-mentioned phenomena can be neglected.

A dense gas cloud has well-defined limits compared to one of neutral flotation. Van Ulden [24] proposed describing the spread of a dense cloud with an equation similar to that describing the movement of a water front after the rupture of a containment dike. Assuming a dense cloud of vertical cylindrical shape with a radius R and height h_F, the horizontal velocity of the cloud's edge is given by [10]:

$$\frac{dR}{dt} = c\sqrt{\frac{gV^{1/3}(\rho_p - \rho_a)}{\rho_a}} \qquad (4.53)$$

V being the cloud volume at a given moment. It has been found experimentally that the proportionality constant c in the previous equation is close to unity. In the case of an instantaneous emission without significant heat transfer between the cloud and its surroundings, from the previous equation one can obtain [10]:

$$R^2 = \left(R_0\right)^2 + 2t\sqrt{\frac{g\left(V_0\right)^{1/3}\left(\rho_p - \rho_a\right)_0}{\rho_a}} \tag{4.54}$$

The above equation demonstrates a satisfactory agreement with experimental data after the initial moments of front formation [25].

As the dense emission expands, its concentration decreases. This dilution is determined by the rate at which air enters the cloud. Various semi-empirical models exist that predict the dilution rate and allow the calculation of cloud height and volume, but their discussion is outside the scope of this text. Below is shown only the result obtained from the combined use of dimensional analysis and experimental observations made with instantaneous Freon releases [10], giving the change in the cloud dilution as it travels:

$$\frac{V}{V_0} = \left[\frac{x}{V_0^{1/3}}\right]^{3/2} \tag{4.55}$$

where x is the distance in the direction of the wind, measured from the point where the instantaneous emission takes place. The above expression is valid only for dense emissions at distances greater than $(V_0)^{1/3}$. In spite of its simplicity, equation (4.55) adequately predicts the results of a number of experimental observations made within the stage in which negative cloud flotation predominates over ambient turbulence.

Transition to neutral or positive flotation

After a certain time (instantaneous emissions), or at a certain distance from the discharge point (continuous emissions), the dilution of the emission by ambient air is such that the effects of density cease to be significant and the cloud behaves as a neutral emission, so that the equations shown for this case can be used.

The criteria for determining the moment of transition are still the object of considerable discussion. For example, some models consider transition taking place when the normalized density difference $(\rho_p - \rho_a)/\rho_a$ falls below an arbitrary value, such as 0.01 or 0.001; in other cases the criteria for transition requires that the Richardson number decreases below a critical value between 1 and 10. There are also models which compare the dense front velocity and the friction velocity, and consider that transition to neutral flotation takes place when both become equal.

Whatever the criteria chosen, Gaussian dispersion models can be used to describe the evolution of the release after transition. The Gaussian model is applied using a virtual source that replaces the cloud after transition by an equivalent Gaussian emission. The dispersion parameters at the transition stage are estimated as [10]:

$$\sigma_z = 0.707\, h_F \qquad (4.56a)$$

$$\sigma_y = 0.707\, R \qquad (4.56b)$$

In the case of instantaneous emissions, the standard deviation in the direction of the wind σ_x can also be obtained from equation (4.56b). Once the standard deviations at the transition point are known, a virtual distance is calculated as shown previously, which allows the application of the Gaussian model equations over the corrected distance.

Example 4.4

A perforation in a pipe of a chemical plant releases 0.2 kg/s of ammonia in the vapour phase, with an effective height of 15 m above the surrounding terrain, which is open country. The wind speed is 7 m/s and the leak happens at night, with a temperature of 15°C. (a) Calculate the maximum concentration at a point situated 0.5 km from the source. (b) Calculate the concentration at a point at ground level, with co-ordinates (500, 50, 0).

(a) The emission is produced in the gas phase, and on account of the atmospheric conditions no condensation can be expected (normal boiling point is −34°C). In the proposed case $y = 0$, and $z = H = 15$, since the maximum concentration occurs along the emission axis. The stability class is D (neutral), whatever the cloudiness (Table 4.2), because the emission occurs at night and the wind speed is greater than 6 m/s. Under these conditions, $a = 0.128$, $b = 0.905$, $c = 0.20$, $d = 0.76$. It is not necessary to correct for the terrain roughness length, as this corresponds to $z_0 = 0.1$. Further, the case proposed (perforation of a pipe), corresponds well to a point source emission. At the distance considered, the standard deviations are calculated with equation (4.41) as:

$$\sigma_y = 0.128(500)^{0.905} = 35.5\,\text{m}; \; \sigma_z = 0.20(500)^{0.76} = 22.5\,\text{m}$$

With $H = 15$ m, $Q^* = 0.2$ kg/s and $U = 7$ m/s, equation (4.40) gives a concentration of 8.0×10^{-6} kg/m^3, i.e. approximately 11 ppm.

Presented in Figure 4.14 are the results obtained from the application of a relatively simple commercial software (CHEMS-PLUS 2.0 from Arthur D. Little Inc.), in conditions equivalent to those proposed in the example. The graph shows the variation of the concentration with distance along the emission axis. The model predicts that within a radius of 265 m concentrations higher than the TLV value for ammonia are reached (Chapter 5). For a concentration of 11 ppm the distance calculated for the model is 505 m, in very good agreement with the previous estimate.

(b) $z = 0$, $x = 500$, $y = 50$. With these values equation (4.40) predicts a concentration of 3.38×10^{-6} kg/m^3.

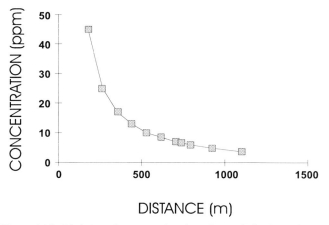

Figure 4.14   Variation of concentration along the x-axis for Example 4.4.

Example 4.5

Repeat the calculations of case (a) in Example 4.4 for a heterogeneous terrain, with the roughness length over the 500 m distance distributed as follows: 200 m with $z_0 = 1.0$, and 300 m with $z_0 = 0.1$.

The virtual source method is used. For the first 200 m roughness $z_0 = 1.0$ is used. With this, the standard deviation at $x = 200$ m can be obtained, after correcting for roughness length different from 0.1, according to equation (4.42):

$$m = 0.53(200)^{-0.22} = 0.165; \quad \sigma_{z,200} = 0.20(200)^{0.76}(10)^{0.165} = 16.4 \text{ m}$$

Now the virtual distance x_v for a standard deviation of 16.4 m at a distance of 200 m can be calculated, using the parameters corresponding to a roughness length of 0.1:

16.4 = $0.20(x_v)^{0.76}$, and therefore $x_v = 330$ m with $z_0 = 0.1$.

Thus, for the first 200 m a virtual source has been used corresponding to a terrain of the same characteristics as the next 300 m section. This implies that the standard deviation at the end of the real heterogeneous 500 m would be the same as that at the end of 300 + 330 m of a homogeneous terrain with roughness length 0.1. In this way,

$\sigma_{z,500} = 0.20(630)^{0.76} = 26.8$ m

Using this value in equation (4.40) together with $y = 0$ and $z = H = 15$, a concentration $C = 7.3 \times 10^{-6}$ kg/m³ is obtained, lower than that calculated in the previous case, due to the additional dispersion caused by the greater roughness length of the first 200 m.

Example 4.6

A gas container ruptures, instantly releasing at ground level 40 kg of ethane at ambient temperature (20°C). Calculate the distance at which the explosion of the cloud formed could occur, knowing that the wind speed is 4 m/s and atmospheric conditions correspond to stability class E.

Because the emission occurs at ambient temperature and ethane has a molecular weight close to that of air, significant density effects are not expected. The cloud will adopt an approximately hemispherical shape, with the centre of its base situated at $x = Ut$.

The explosion of the cloud will not occur from the moment in which dilution by air causes the concentration at any point in the cloud to drop below the lower flammability limit. For ethane this corresponds to a concentration in air of 3% by volume (Table 3.2), or a concentration of 0.0375 kg/m³ taking a density of 1.25 kg/m³ for ethane at 20°C and atmospheric pressure.

The maximum concentration is found at the centre of the cloud: therefore, it is sufficient that the concentration at this point has fallen below the lower flammability limit for no explosion to occur. Equation (4.39) can be used to calculate the distance at which the centre of the cloud, with co-ordinates $(Ut,0,0)$ reaches a concentration of 0.0375 kg/m³. Equation (4.39) becomes

$$C = \frac{Q}{(2\pi)^{3/2} \sigma_x \sigma_y \sigma_z} \{2\}$$

The standard deviation values are calculated with equations (4.41) for the applicable stability class. In this way,

$$\sigma_x = 0.13\, x;\ \sigma_y = (0.5)0.098\, x^{0.902};\ \sigma_z = 0.15\, x^{0.73}$$

Substituting these values together with $C = 0.0375$ kg/m³ and $Q = 40$ kg in the previous equation yields $x = 90.6$ m. Therefore, the explosion could occur at distances less than 90.6 m. Another way of visualizing this is to consider that the cloud's centre travels the 90.6 m in 90.6/4 = 22.65 s. The explosion will not occur if during approximately 23 seconds the cloud does not find an ignition source in its path.

4.7   Questions and problems

4.1   The rupture of a pipe gives rise to a flammable liquid pool which ignites. The radiation from the fire is not too serious, but the gases originating from the combustion are highly toxic. Discuss the data necessary to estimate the concentration of toxic substances over time at a given distance from the fire.

4.2   A tank contains liquid dimethylamine in equilibrium with its vapour, at 25°C. A fissure in the tank wall occurs, with an effective area of 0.7 cm². Calculate the physical state of the discharge and its flow over time if the fissure is: (a) 20 cm above and (b) 20 cm below the initial liquid level. Assume a circular hole, with sharp edges.

4.3   A perforation occurs in a deposit containing liquid propane. Discuss the calculation algorithm to obtain the discharge flow over time, depending on the position of the perforation and the storage temperature. Consider (a) that the deposit can be considered isothermal during discharge or (b) that it is well-insulated, and the evaporation of the propane produces a cooling of the remaining liquid mass.

4.4   The CCPS [26] proposes the following problem: a gas cylinder contains compressed air for use in the FID detector of a gas chromatograph. On changing the cylinder, the worker removes its safety cover and starts to adjust the pressure regulator when the cylinder slips from his hand and falls to the floor. Whilst falling, the valve knocks against the side of the table and is broken, exposing a 5/8" aperture in the upper part of the cylinder. The stored air is initially at 75°F and 2250 psi. Calculate the initial flow rate through the opening. To what force is the cylinder subjected as a consequence of the flow? What is the acceleration? Calculate the speed and kinetic energy when the cylinder reaches the laboratory wall, 20 m distant. The cylinder weighs 140 lbs and has an internal volume of 1.5 cubic feet. For simplicity, assume that the thrust is maintained constant whilst the critical pressure ratio is exceeded and the air behaves as an ideal gas. Assume a nozzle discharge with a discharge coefficient equal to 0.9.

4.5   As a consequence of the collapse of a deposit containing isobutane (normal boiling point –11.7°C) an instantaneous spillage of its contents occurs (80 m³), which were stored under pressure at ambient temperature (20°C). Calculate the fraction of isobutane that flashes. Calculate the evaporation rate as a function of time if the spill occurs on a cement floor.

4.6   A leak in a pipe 5 m above the ground releases 40 kg/min of chlorine. Assuming there are no appreciable density effects, calculate the concentration at 20, 80 and 200 m from the point of emission in the wind's direction. Consider each of the different stability classes that can arise during the day with a wind speed of 3.2 m/s. Discuss the hypothesis of negligible density effects. Estimate the size of the region where negative flotation predominates.

4.7   The instantaneous emission of 12 kg of gaseous HCl takes place at ground level. The wind speed is 3 m/s and insolation is moderate. Represent the contour of the area within which the TLV-C value of 5 ppm can be exceeded (Chapter 5).

4.8   Represent the variation with time of the concentration of HCl to which a person of height 170 cm situated at $x = 100$, $y = 10$ is exposed, under the conditions of the previous problem.

4.9   In this chapter simplified models of dispersion have been presented, which do not take into account aspects which occur frequently in reality. Discuss how the concentrations could be modified in a continuous emission: (a) by the topography of the terrain (consider both mountainous elevations as well as depressions, in relation to positive and negative flotation emissions); (b) in the presence of large masses of water (lakes); (c) by vegetation; (d) by rain or snow.

4.10   A road tanker carrying methanol suffers an accident and as a consequence its cargo is spilt, extending over an approximately circular area, with a 55 m diameter. The ambient temperature is 25°C and wind speed is 2.7 m/s, with stability class C. Use the virtual source method to calculate the concentration of methanol at a point in the wind's direction 250 m from the accident.

4.8   References

1. Opschoor, G. (1979) Evaporation, in *Methods for the Calculation of the Physical Effects of the Escape of Dangerous Material (Liquid and Gases), (The Yellow Book)*, TNO, Directorate General of Labour, 2273 KH Vooburg, the Netherlands.
2. Perry, R. H. and Green, D. (eds) (1984) *Perry´s Chemical Engineer´s Handbook*, 6th edn, McGraw-Hill, New York, Chapter 6.
3. Crowl, D. A. and Louvar, J. F. (1990) *Chemical Process Safety, Fundamentals with Applications*, Prentice Hall, Englewood Cliffs.
4. Levenspiel, O. (1984) *Engineering Flow and Heat Exchange*, Plenum Press, New York.
5. Leung, J. C. (1986) Simplified vent-sizing equations for emergency relief requirements in reactors and storage vessels. *AIChE J.*, **32**, 1622–34.
6. Fauske, H. K., Epstein, M., Grolmes, M. A. and Leung, J. C. (1986) Emergency relief vent sizing for fire emergencies involving liquid filled atmospheric storage vessels. *Plant Operation/ Progress*, **5**, 205–8.
7. Coker, A. K. (1990) Understand two-phase flow in process piping. *Chem. Eng. Prog.*, **86**(11), 60–5.
8. CCPS (Center for Chemical Process Safety) (1989) *Guidelines for Chemical Process Quantitative Risk Analysis*, American Institute of Chemical Engineers, New York.
9. Fauske, H. K., Epstein, M. and Grolmes, M. A. (1987) *Source Term Considerations in Connection with Chemical Accidents and Vapor Cloud Modeling*. Proceedings of the International Conference on Vapor Cloud Modeling, November 2–4, 1987, Cambridge, MA, AIChE/CCPS, New York, pp. 251–73.
10. Hanna, S. R. and Drivas, P. J. (1987) *Guidelines for Use of Vapor Cloud Dispersion Models*. AIChE/CCPS, New York.
11. Kletz, T. (1977) Unconfined vapor explosions, in *Loss Prevention 11*, Chemical Engineering Progress Technical Manual, American Institute of Chemical Engineers, New York.
12. Perry, R. H. and Green, D. (eds) (1984) *Perry´s Chemical Engineer´s Handbook*, 6th edn, McGraw-Hill, New York, Chapter 12.
13. Sutton, O. G. (1953) *Micro meteorology*, McGraw-Hill, New York.
14. CHEMS-PLUS 2.0 Reference Manual (1991) A. D. Little Inc., Cambridge, Massachusetts.
15. Finch, R. N. and Serth, R. W. (1990) Model air emissions better. *Hydrocarbon Processing*, **69**(1), 75–80.
16. Van Buijtenen, C. J. P. (1979) Dispersion, in *Methods for the Calculation of the Physical Effects of the Escape of Dangerous Material (Liquid and Gases), (The Yellow Book)*, TNO, Directorate General of Labour, 2273 KH Vooburg, the Netherlands.
17. Lees, F. P. (1980) *Loss Prevention in the Process Industries*, Butterworth-Heinemann, London.
18. Hanna, S. R., Briggs, S. A. and Hosker, R. P. Jr (1982) *Handbook on Atmospheric Diffusion*, US Department of Energy, Technical Information Center, Oak Ridge.
19. AIChE/CCPS (1989) *Workbook of Test Cases for Vapor Cloud Source Dispersion Models*. American Institute of Chemical Engineers, New York.

20. Ministerio de Industria y Energía (1987) *Manual de Cálculo de Altura de Chimeneas Industriales*, Centro de Publicaciones del Ministerio de Industria y Energía, Madrid.
21. Briggs, G. A. (1969) *Plume Rise*. AEC Critical Review Series, US Atomic Energy Commission.
22. Briggs, G. A. (1984) Plume rise buoyancy effects, in *Atmospheric Science and Power Production*, US Department of Energy, DOE/TIC-27601.
23. Turner, D. B. (1970) *Workbook of Atmospheric Dispersion Estimates*, US Department of Health, Education and Welfare, Cincinnati, Ohio.
24. Van Ulden, P. A. (1974) *On the Spreading of a Heavy Gas Released Near the Ground.* Proceedings of the International Loss Prevention and Safety Promotion in the Process Industries Symposium, The Hague/Delft, The Netherlands, pp. 221–6, Elsevier.
25. Brighton, P. W., Prince, A. J. and Webber, D. M. (1985) *J. Hazardous Materials*, **11**, 155–78.
26. AIChE/CCPS (1990) *Safety, Health and Loss Prevention in Chemical Processes. Problems for Undergraduate Engineering Curricula*, American Institute of Chemical Engineers, New York.

5 Vulnerability of persons and installations

To be immortal is trivial; except for man, all creatures are, they ignore death.

Jorge Luis Borges, The Immortal, Chapter IV

'The one of Fierabras is a balsam' replies Don Quixote, 'the recipe for which lies in my memory. With it there is no need to fear death nor so much as to think of dying of any wound. So, when I have made some and given it to you, if you ever see me cut through the middle in some battle - as very often happens - you have only to take the part of my body that has fallen to the ground and place it neatly and cunningly, before the blood congeals, on to the half that is still in the saddle, taking especial care to fit them exactly. Then you must give me just two drops of balsam to drink, and, you will see, I shall be as sound as an apple'

Miguel de Cervantes Saavedra,
The Adventures of Don Quixote, Chapter XVII

5.1 Introduction

Shown in Chapters 2 and 4 are the methods for identification of significant hazards in an industrial installation, and the procedures for estimating their foreseeable consequences. Until now, the effects produced by a particular accident have been expressed in terms of the physical variables which indicate the intensity of the phenomenon, e.g. the thermal radiation per unit of area received by a surface 150 m from a fire, the overpressure and the duration of the positive phase of the pressure wave at a point 0.5 km distant from the centre of an explosion, the map of concentrations five minutes after the instantaneous release of a given amount of a toxic substance, etc. The next step consists of the estimation of the vulnerability of persons and installations to the physical effects of a determined magnitude, which can be calculated with the procedures described in previous chapters and in the related references. One seeks, therefore, the quantification, even though it may be only approximate, of the final effects of a hypothetical accident whose characteristics are assumed to be known.

The vulnerability of persons is expressed as the number of individuals that can possibly be affected by a certain level of injury because of an accident. The level of injury predicted should be defined in the analysis, and can vary from slight pain and light injuries, up to the death of the exposed persons. On the other hand, the vulnerability of installations is quantified ultimately using economic magnitudes, although in this chapter only direct physical damage is dealt with (partial destruction of buildings, breakage of glass, fire in buildings, etc.). Logically, it is necessary to bear in mind in this case the vulnerability of persons in the interior of buildings and installations affected, who can suffer the consequences of an accident, although not exposed in a direct manner. Also important in certain cases are non-fatal effects, such as dizziness, temporary paralysis, disorientation, etc., that can diminish the capacity to react, cause errors and ultimately impede the evacuation or flight, causing death as a result.

The reactions of living organisms to adverse external agents (e.g. a toxic substance dispersed in the atmosphere) are many and various. Many possible biological effects caused by adverse agents exist, including sensitization of different organs, inflammation, narcosis, irritation, symptoms of asphyxia, necrosis of organic tissue, partial destruction of the immunity system, neoplasia, mutagenesis, teratogenesis, etc. Any of these may or may not cause death, depending on the severity of the affliction. In almost all biological populations there exist sensitive individuals who, confronting the same adverse factor, react severely, whilst the majority, apparently, are well able to tolerate the same level of intensity. Following the example of inhalation of a toxic substance, for a specific exposure to a fixed concentration, some individuals may show an intense reaction (e.g. asthmatic crisis), while others hardly suffer irritation. The diversity of reactions is accentuated when there are differences in age, sex, state of health, etc. within the exposed population.

A possible method of estimating vulnerability consists of directly relating the dose received with the effect considered. This can be achieved from empirical evidence showing that individuals who have been subjected to a certain dose of the injurious agent (e.g. a certain radiation intensity level during a given time) have suffered a particular effect (e.g. death from burns). As shown previously, on very few occasions is a certainty of this kind achieved, and the uncertainties are sufficiently important such that establishing the level of intensity that causes a determined injury, is generally of little practical use. Thus, let us assume that, following the study of an accident resulting in death through inhalation of a toxic substance, the concentration and the duration of exposure are determined. The intensity–effect relationship thus determined should not be used directly: a lower intensity (dose) could have caused the same effect, the victim could have been especially sensitive, or on the contrary have developed a certain tolerance to the poison, etc.

Therefore the methods that relate causes directly with effects are hardly used, and the approximations to the problem of estimation of vulnerability generally follow a probabilistic approach. In this chapter the Probit ('Probability Unit') method is shown, which provides simple relationships for predicting the adverse effects of different variables, provided that these can be described by transformations of the normal probability distribution.

5.2 Probit methodology

The Probit scale is a way of dealing with probabilities. The connection between Probit units (Y) and probability (P) is given by the following equation [1]:

$$P = \frac{1}{\sqrt{2\pi}} \int_{-\alpha}^{\gamma-5} \exp(-u^2/2)\,du \tag{5.1}$$

Equation (5.1) establishes a relationship between probability and Probit units. The result is the Probit distribution, with mean 5 and variance 1. The curve relating percentages and Probit units is shown in Figure 5.1.

As discussed above, it is often found that in a biological population there exist individuals capable of withstanding high levels of a harmful agent without manifesting significant adverse effects. Because of this, the cumulative percentage curves of number of affected individuals versus the intensity of the causative factor are significantly skewed and do not follow a normal distribution. However, it is also found experimentally that if the percentage of population affected is represented not against the intensity of the causative factor, but against its logarithm, the results very often follow a normal distribution. In this case, given the characteristics of the Probit variable, the following relationship can be written:

$$Y = 5 + \frac{\ln x - \mu}{\sigma} \tag{5.2}$$

where μ and σ are, respectively, the mean and the standard deviation of the normal distribution, and x is related to the intensity of the causative factor. The previous equation can be rewritten in the way usually used in vulnerability analysis as

$$Y = k_1 + k_2 \ln V \tag{5.3}$$

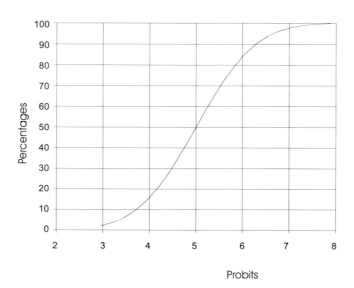

Figure 5.1   Relationship between percentages and Probit units (adapted from data given in [2])

In this case k_1 and k_2 are empirical constants, and V measures the intensity of the damage causative factor. The way in which V is expressed depends on the type of effect studied, as will be explained later.

The Probit methodology is widely used because the transformation of probabilities to Probit units according to equation (5.1) converts the typical sigmoid curve of the normal distribution (cumulative probability) into a straight line, which facilitates data fitting to obtain k_1 and k_2. This transformation is illustrated in Example 5.1. The use of a linear relationship is especially useful when one wishes to obtain vulnerability correlations from the analysis of previous accidents, which rarely yields sufficient reliable data that simultaneously relate the damage produced and the intensity of the causative factor.

Example 5.1

(a) From the data of Eisenberg and coworkers [3] on percentages of deaths from lung haemorrhage in explosions, presented in the table below, obtain the corresponding Probit equation, knowing that for this case the variable V (equation (5.3) can be expressed as overpressure P^o (Pa). (b) Estimate the percentage of deaths from lung haemorrhage at a distance of 50 m from an explosion equivalent to 4600 kg of TNT.

Overpressure $\times 10^{-5}$ (Pa)	1.00	1.20	1.41	1.76	2.00
% of deaths in the exposed population	1	10	50	90	99

(a) Figure 5.2 illustrates data representing the percentage of people affected against ln P^o. It can be seen that the results obtained follow approximately the typical sigmoid curve of the cumulative normal distribution.

The equivalent representation in Probit units can be seen in Figure 5.3, where the percentages 1, 10, 50, 90 and 99 have been replaced by 2.67, 3.72, 5.0, 6.28 and 7.33 Probit units, obtained from Figure 5.1. The representation is now a straight line ($r = 0.9985$). The regression parameters, with their standard errors, are respectively 6.693 ± 0.21 and -74.44 ± 2.54, slightly different to those given later (equation (5.10)), for the same case. The Probit equation would therefore read as

$$Y = -74.44 + 6.693 \ln P^o$$

(b) At 50 m from an explosion equivalent to 4600 kg of TNT, the normalized distance is $z = 50/(4600)^{1/3} = 3.0$ m/kg$^{1/3}$. With this value of normalized distance one obtains an overpressure of some 95 000 Pa (Figure 3.9). Substitution of this value into the previous equation yields $Y = 2.273$, which is lower than the threshold of 1% of people affected (because 1% corresponds to 2.67 Probit units).

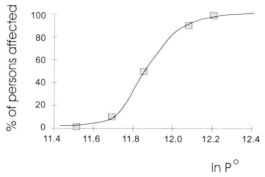

Figure 5.2   Percentage of deaths from lung haemorrhage as a function of ln P^o

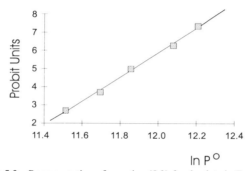

Figure 5.3   Representation of equation (5.3) for the data in Example 5.1.

5.3   Effects of toxic emissions

A discussion, even one that avoids going in depth, of the toxicological aspects that can occur with substances used by the chemical industry clearly falls outside the scope of this book. In what follows is given a summary of some of the fundamental concepts necessary for the development of risk analysis, and it is left to the reader who desires more information to look to more appropriate texts [e.g. 4–8].

The prediction of vulnerability to toxic substances constitutes a complex problem because it depends on many factors of a very diverse nature. Among these factors are:

- The entry route to the organism
- The intrinsic toxicity of the material considered
- The way in which the toxic material acts (e.g. as to the reversibility of its effects)
- The dose received, expressed as a function of the concentration of the toxic agent and the time of exposure
- The type of exposure (sudden emission of large quantities, or prolonged exposure at low concentrations, etc.)

- The variation of the individual biological reactions to the same dose of toxic agent.

With respect to the first point, the main entry routes are:

1. *Ingestion*: The human digestive system provides numerous possibilities for the action of toxic agents in the mouth, throat, oesophagus, stomach and intestinal tract. This route of entry is relatively easy to prevent, by modifying, if necessary, the conduct of workers, principally where personal hygiene, food and drink are concerned, as well as dangerous habits, such as sucking the ends of flexible tubes to start flow of liquid by the siphon effect.
2. *Epidermis*: The skin provides another important point of access for toxic agents used industrially, with effects that include chronic or acute inflammation, allergic reactions, neoplasia, and sometimes death in a short period of time. The entry of a toxic agent can occur by absorption through intact skin, or through skin that has lost some of its protective properties, through injuries caused mechanically or by corrosive chemical agents, etc. Other related entry routes, e.g. capillary follicles and sweat glands, are usually less important. The skin of the palm of the hand, in spite of being thicker than in other parts, has a greater porosity, which can lead to more intense absorption [9]. It is clear that entrance through the skin can be prevented or minimized through the use of adequate personal protection equipment.
3. *Eye contact*: The eyes can suffer adverse effects from contact with solid, liquid or gaseous chemical products. These range from temporary irritation to permanent damage resulting in blindness. From the toxicological point of view, the greatest danger comes from the accumulation of toxic material through absorption by the peripheral blood vessels.
4. *Inhalation*: A person at rest breathes some 8 litres of air each minute. Together with the air breathed, the toxic substances which it may contain have a rapid entrance path, which carries them directly to very sensitive areas of the organism. Gases and vapours can rapidly reach the pulmonary alveolus, responsible for the exchange of oxygen with carbon dioxide necessary to replenish the oxygen level in the blood. The effects of injurious substances in the alveolus can be either their physical blockage (insoluble powders) or a reaction in the wall of the alveolus of toxic or corrosive substances. The extent of the penetration and the effects caused by toxic vapours also depend on their solubility in water. Thus, due to their high solubility, both ammonia and hydrogen chloride are usually absorbed before reaching the alveolus. In this case, the main effects are found in the upper respiratory tract. Regarding solid particles, it is usually accepted [10] that particles larger than 40–50 microns do not enter the respiratory system, those larger than 10 microns are deposited in the upper paths, those between 2 and 10 microns reach successively the trachea, bronchi and bronchioles and only those smaller than 2 microns reach the alveolus. As to fibres, those with diameters up to 3 microns and lengths up to 50 microns can penetrate the respiratory system, the most dangerous being those less than 8 microns long and 1.5 microns in diameter.

The human body has different mechanisms to protect against the entry of a toxic substance that, in many instances, are effective in eliminating or at least reducing the levels of concentration reached. Thus, the toxic product can be expelled by the liver (excretion of substances in the spleen), lungs (respiration), kidneys (urine), and to a lesser degree via the skin (perspiration), hair and nails. A second path of elimination is the metabolic transformation into less toxic substances. This is the case with conversion of cyanides to less toxic thiocyanates, by way of an enzymatic route. However, on occasions the metabolic route can cause an increase in the toxicity level, transforming compounds into other more dangerous compounds, as occurs with carbon tetrachloride, which is converted into derivatives more toxic to the liver. Lastly, toxic substances can also be eliminated from the bloodstream via storage, fundamentally in fatty tissues, which can lead to a new release in the future, if the organism uses accumulated fat.

5.3.1   Toxicity indicators

As already shown, considerable difficulty exists when making precise estimates of the effects that a toxic substance can have on living organisms. In spite of this, or perhaps because of it, numerous toxicity indexes have been published, which give an indication of the health hazard posed by a toxic substance.

A classic work in this field is that of Hodge and Sterner [11], who established the following relative toxicity scale, based on oral administration:

Toxicity class	Grade of toxicity	LD_{50} value
6	Super-toxic	< 5 mg/kg
5	Extremely toxic	5–50 mg/kg
4	Very toxic	50–500 mg/kg
3	Moderately toxic	500–5000 mg/kg
2	Slightly toxic	5–15 g/kg
1	Practically non-toxic	> 15g/kg

According to the previous scale, with less than 7 drops of a class 6 product the death of a person weighing 70 kg is probable, whilst the equivalent in classes 4 and 1 would be a full tablespoon and more than 1 litre, respectively.

The LD_{50} values in the previous table are the **lethal doses** determined in laboratory animals (rats), using standardized experimental conditions and an adequate population size for statistical analysis of the results. The usual definition [10] of the LD_{50} value is expressed as the dose that, administered orally or by skin absorption, will cause the death of 50% of the test group within a 14-day observation period. Analogously, for airborne toxic products whose entry is via inhalation, the LC_{50} (**lethal concentration**) values are defined as the concentration of airborne material that, when inhaled over a period of 4 hours, results in the death of 50% of the test group within a 14-day observation period.

Many countries have established similar classifications of toxicity. Thus for instance, in Spain, Annex IV of the Royal Decree 886/1988 on the *prevention of major accidents* establishes orientative criteria for the classification of dangerous substances. Thus, substances with an LD_{50} value ≤5 mg/kg (determined for oral ingestion in rats), are classified as very toxic, as well as those with an LD_{50} value

≤10 mg/kg administered dermally in rats or rabbits, and those with LC_{50} values ≤0.1 mg/l in rats. Other substances with higher LC_{50} and LD_{50} values may represent a similar hazard, depending on the rest of their physical and chemical properties.

IDLH (Immediately Dangerous to Life or Health) values, are published by NIOSH (National Institute for Occupational Safety and Health), to indicate the maximum airborne concentration of a substance to which a healthy male worker can be exposed for 30 minutes without developing symptoms (such as severe eye or respiratory irritation), that could impair his capacity to escape, and without suffering loss of life or irreversible organ system damage. The IDLH values given in Table 5.1 are not up to date, and therefore should be complemented by more recent toxicological data. In addition it is necessary, when employing this or another similar indicator, to bear in mind that their values refer to an average worker, and that considerably greater individual sensitivities are possible.

For substances without a published IDLH value, the EPA [15] has defined the Emergency Airborne Exposure Concentration Levels (EAECL). The EAECL values can be estimated depending on available data as $0.1 \times LC_{50}$, $0.01 \times LD_{50}$, LC_{lo} or $0.1 \times LD_{lo}$, where the meaning of LC_{50} and LD_{50} has already been explained, while LC_{lo} and LD_{lo} respectively refer to the minimum concentration and the minimum dose considered capable of causing death in an individual of the population exposed (derived from lowest reported lethal concentration or dose respectively). If an EAECL value is obtained from LD_{50} or LD_{lo} values, which represent specific doses (mg/kg), it is necessary to convert these to airborne concentration. For a standard exposure (30 minutes) and for a person of 70 kg who inhales approximately 400 litres of air over this period we have [2]:

$$EAECL(mg/m^3) = \frac{EAECL(mg/kg)(70\,kg)}{0.4\,m^3} \qquad (5.4)$$

As for the everyday existence of toxic substances in the working environment, the ACGIH (American Conference of Governmental Industrial Hygienists) regularly publishes the TLVs (*Threshold Limit Values*) for a series of substances, based on experimental data with laboratory animals and with humans, as well as on industrial experience. The TLVs refer to concentrations of a particular substance in air, and correspond to conditions to which it is believed that normally all workers could be exposed to over a working lifetime, without suffering adverse effects. There are various kinds of TLV values: The time-averaged TLV (TLV-TWA) is the time-averaged concentration for an 8-hour working day (40-hour working week), to which the majority of workers may be repeatedly exposed, day after day, without adverse effects. The TLV for short time exposures (TLV-STEL) is the maximum concentration to which the majority of workers could be exposed for a continuous period of up to 15 minutes without suffering: (a) intolerable irritation, (b) chronic or irreversible tissue change or (c) narcosis sufficient to reduce work efficiency, impair self-rescue or increase accident proneness. The exposure should not exceed 15 minutes, should not be repeated more than 4 times

Table 5.1   TLV-TWA, TLV-STEL and ILDH values for selected substances

Substance	TLV-TWA (ppm)*	TLV-STEL (ppm)*	IDLH (ppm)†
Acetic acid	10	15	1000
Acetone	750	1000	20 000
Ammonia	25	35	500
Benzene(s)	10(0.1‡)		
Bromine	0.1	0.3	10
1-Butanol(s)	50(C)		8000
s-Butyl Alcohol	100		
t-Butyl alcohol	100	150	
n-Butylamine(s)	5(C)		2000
Carbon dioxide	5000	30 000	50 000
Carbon monoxide	50(25‡)	400	1500
Carbon tetrachloride(s)	5		
Chlorine	0.5	1	30
Chloroform	10		
Crotonaldehyde	2		400
Cumene(s)	50		8000
Cyclohexane	300		10 000
Cyclohexanol(s)	50		3500
Cyclohexanone(s)	25		5000
Ethyl alcohol	1000		
Ethylbenzene	100	125	2000
Ethylenediamine	10		2000
Ethylene glycol	50(C)		
Ethlene oxide	1		
Fluorine	1	2	25
Furfural(s)	2		250
Hydrazine(s)	0.1(0.01‡)		
Hydrogen chloride	5(C)		100
Hydrogen sulphide	10	15	300
Methyl alcohol(s)	200	250	25 000
Methyl isocyanate(s)	0.02		20
Nitric acid	2	4	100
Nitric oxide	25		100
Nitrobenzene(s)	1		200
Nitrogen dioxide	3	5	50
Nitrotoluene(s)	2		200
Phenol(s)	5		250
Phosgene	0.1		2
n-Propyl alcohol (s)	200	250	4000
Pyridine	5		3600
Styrene (monomer)(s)	50	100	5000
Sulphur dioxide	2	5	100
Toluene	100(50‡)	150	2000
Toluene-2,4-diisocyanate (TDI)	0.005	0.02	10
Xylene	100	150	10 000

Correct use of TLV values requires application of ACGIH guidelines.
To convert TLVs in parts per million (ppm) to mg/m³ (at a pressure of one atmosphere and 298K) the following expression can be used:

$$TLV(mg/m^3) = \frac{MW(g/mole) \times TLV(ppm)}{24.45}$$

(C) indicates ceiling value
(s) indicates absorption through the skin
* Taken from [12,13]
† Taken from [14]
‡ indicates value included in the list of intended changes for 1991–92

per day, and there should be at least 1 hour between exposure periods, all these without exceeding the TLV-TWA value for the working day. Lastly, a third TLV is the TLV-C, ('Ceiling'), which is the concentration that should not be exceeded, even instantaneously.

The use of TLVs requires accounting for certain considerations that are emphasized in the guidelines for the use of TLV in the ACGIH publications. Among these are the following: The different types of TLV should not be considered as a sharp dividing line between safe and hazardous concentrations, nor as a relative toxicity index of different substances. TLVs are intended for use in industrial hygiene and their interpretation and application to particular cases is a specialist matter. They should not be applied to continuous exposure over time. In the case of exposures different from 8 hours per day or 40 hours per week, or when various toxic substances interact, the published TLVs should be appropriately corrected. Finally, the fact that a substance is not found in the TLV list is not necessarily an indication of low toxicity.

A similar scale to the TLV is that of the PEL (*Permissible Exposure Level*) values, also based on exposures of 8 hours and published by OSHA (*Occupational Safety and Health Administration*). TLVs tend to be somewhat more conservative than the corresponding PELs, although they coincide in many cases. There are many other categories of toxicity indicators (ERPG, TXDS, EEGL, SPEGL, etc.) whose description would be too long for inclusion in this book, and the reader is referred to the texts on industrial hygiene for definitions and methods of use, as well as in-depth study on the above. Table 5.1 gives TLV and IDLH values for some selected substances.

5.3.2 Probit equations for toxic exposure

When applying Probit expressions such as equation (5.3) to the evaluation of vulnerability to toxic substances, the intensity of the causative factor V accounts for both the concentration and the exposure time, $(C^n t)$. Equation (5.3) can thus be written as

$$Y = k_1 + k_2 \ln [C^n t] \tag{5.5a}$$

where the exponent n (dimensionless) is determined empirically, taking values between 0.6 and 3, the concentration C is expressed in parts per million, and the time t in minutes. The previous equation, after substitution of the appropriate parameter values, gives the value of Y directly for the case where the exposure is of constant characteristics (continuous emission, invariable concentration with time for a given location). In the case of emissions with characteristics variable with time (e.g. instantaneous emissions) the concentration measured at a given position varies with time, so that the term $C^n t$ in the previous equation must be replaced by the corresponding integrated expression for the time duration of the emission as perceived by the receiver (t_{exp}):

$$Y = k_1 + k_2 \ln \left(\int_0^{t \, exp} C^n dt \right) \qquad (5.5b)$$

As indicated, the values k_1, k_2 and n are determined empirically. The main source of data is experiments with laboratory animals, although some data also exist from past accidents in which both the accident characteristics and the effects produced are known with some precision. In this case, given the accident scenario (quantity or flow involved, type of emission, meteorological conditions, etc.), the use of similar models to those described in the previous chapter permit the estimation of the relevant concentration–time data for a person at a particular location, and relate these to the effects suffered.

Whatever the source of the data used, to obtain the parameter values for equation (5.5) a statistical data fitting is required, as seen in Example 5.1. In view of the scarcity of data, the parameters obtained often have high associated uncertainty. Also, it is possible that different sets of parameters give a similar quality of fit for the same group of data. Table 5.2 shows the parameter values for Probit equations for various toxic substances.

In the cases of chlorine and ammonia two groups of parameters from different sources appear. The estimates of percentage of population affected using one or the other group of values can differ considerably, which indicates the difficulty of obtaining precise estimates of vulnerability. For this reason, when quantifying

Table 5.2 Values of k_1 and k_2 (equation (5.5)) for death from exposure to different toxic substances (selected from [2] and [16])

Substance	k_1	k_2	n
Acrolein	−9.931	2.049	1
Acrylonitrile	−29.42	3.008	1.43
Ammonia	−35.9	1.85	2
	−30.57	1.385	2.75
Benzene	−109.78	5.3	2
Bromine	−9.04	0.92	2
Carbon monoxide	−37.98	3.7	1
Chlorine	−8.29	0.92	2
	−17.1	1.69	2.75
Formaldehyde	−12.24	1.3	2
Hydrogen chloride	−16.85	2.00	1.00
Hydrogen cyanide	−29.42	3.008	1.43
Hydrogen fluoride	−35.87	3.354	1.00
Hydrogen sulphide	−31.42	3.008	1.43
Methyl isocianate	−5.642	1.637	0.653
Nitrogen dioxide	−13.79	1.4	2
Phosgene	−19.27	3.686	1
Sulphur dioxide	−15.67	2.10	1.00
Toluene	−6.794	0.408	2.50

vulnerability in the analysis of a hypothetical scenario the use of Probit equations sometimes is avoided, and the study is limited to the calculation of the concentration–time values for the individuals exposed. In this way, the concentrations experienced by these individuals can be related to some of the reference values previously indicated, so that the analyst can consider the probability that values such as TLV-C, IDLH, LC_{lo}, etc. are exceeded.

5.4 Vulnerability to thermal effects

Thermal radiation from a fire can cause adverse effects to both people and installations. With directly exposed subjects burns of variable severity are possible, with death resulting from certain levels of radiation received and exposure time. On the other hand, thermal effects can affect buildings and installations, weakening their structures and totally or partially destroying them, which can also give rise to death or injury to persons not directly exposed to the radiation. Another case that frequently occurs is that of a fire in a particular installation extending to others not originally affected (domino effect), which can carry with it consequences far exceeding those originally expected, when considering only the installation where the accident originated.

Ordinary fires can occur in the chemical industry that are not specifically related to the industrial activity being performed. Thus, an electrical installation in poor condition can cause a fire in an office block within a chemical company, in the same way that this occurs outside of industrial installations. Such fires are common to any other field of activity, and are not considered in this text, unless they can spread and cause fires or explosions in industrial installations. Instead, we are concerned with the effects of industrial fires, such as the burning of a liquid pool, a flash fire or a fireball from a BLEVE.

The estimation of vulnerability to thermal radiation within this context implies the application of models described in Chapter 3. Once the type and characteristics of a fire are known, these models allow the estimation of its duration and of the intensity emitted as a function of time, which provides the necessary basis for making vulnerability calculations.

5.4.1 Effects of thermal radiation on people

The effects of thermal radiation on people are strongly dependent on the type of accident involved. Thus, when confronted with a pool fire, the persons exposed to dangerous levels of radiation generally react in time, seeking refuge or escaping. In this case, as the potential victims move away from the focus, the radiation received decreases, so that in a vulnerability analysis a function that considers the decrease in radiation with time would have to be introduced. On the contrary, with a flash fire the possibility of protection through individual reaction diminishes, due to the short time available for response.

Obviously the variation in personal behaviour introduces an additional uncertainty in the estimation of human vulnerability for a given scenario. King [10] cites a flash fire at the Abadan refinery caused by a hydrocarbon cloud from a flammable release. Some of the workers, on seeing the danger, ran in the correct direction, while others selected an escape route in the wind's direction and were enveloped by the cloud, which ignited, causing the death of the majority. Another factor of great influence on the final effects of a specific level of radiation is the amount of skin covered, which varies considerably with the season of the year. However, in a conservative risk analysis the skin would be considered to be uncovered because it is always highly probable that there will be uncovered parts of the body (face, neck, hands, etc.) directly exposed to radiation.

The severity of a burn depends mainly on the amount of tissue destroyed and the extent of corporeal surface affected. Other factors (age, location of the burns, severity of associated injuries) also affect the capacity for recuperation after the burns. The skin consists of two layers, the external layer or epidermis, which forms a barrier against micro-organisms and other adverse agents, and the internal layer or dermis, where the majority of live cells are found, consisting of interconnected fibrous tissues, which are those that prevent the evaporation of corporeal fluids. Embedded in the dermis and forming a passage to the surface of the skin are the sweat glands, responsible for the transpiration that helps to regulate the body temperature, as well as blood vessels and nerves, including the nerve terminations that transmit tactile sensations, pressure, heat, cold and pain. The skin is also responsible for the formation of vitamin D. The destruction of skin through burns can alter or interrupt all the previous functions, with resultant complications.

Burns are classified into four groups according to the depth of the damage caused to the skin. First-degree burns only affect the epidermis, producing redness and some pain. There are no blisters and oedema is minimum. First-degree burns heal without scars in two or three days. Moderate sunburn is an example. On the contrary, second-degree burns cross the epidermis and part of the dermis. Blisters are caused, persisting longer according to depth, and can be very painful. If there is no infection, a superficial second-degree burn can heal in about two weeks. However, a deep second-degree burn causes important losses of corporeal fluid and metabolic alterations, and may require between one and three months for recuperation.

With third-degree burns (and some severe second-degree burns, which are difficult to distinguish) the entire dermis is affected. There is usually no sensation of pain because the nerves that transmit it have been destroyed, along with the blood vessels, sweat glands, capillary follicles, etc. These burns produce great loss of bodily fluid and grave metabolic alterations. Finally, fourth-degree burns are those which reach beyond the dermis, affecting muscles and bones. These can occur if part of the body is trapped in the flames and is not removed immediately, and also occur frequently with electrical burns. With these deep burns toxic materials can enter directly into the bloodstream. They often involve amputation of limbs, and in all cases represent a very grave situation for the individual affected.

Often, the degree of destruction of the dermis is determined by the speed of the protective reaction produced when the radiation received exceeds a certain level. For this reaction to take place, a sufficient flow of heat must be transmitted through the skin, until reaching the adequate nerve tips. If the thermal dose received (function of the radiation intensity and time of exposure) is sufficiently high, the pain threshold is reached, which activates the protective mechanism, causing automatic reactions, as well as conscious ones. Experimental evidence exists that places the pain threshold at the moment a temperature of 45°C is reached at a depth of 0.1 mm below the surface of the skin [2]. After this point, the formation of blisters is very rapid, when the skin surface temperature reaches 55°C. The time taken to reach the pain threshold diminishes as the intensity of radiation increases, as shown in Figure 5.4. Thus, for a radiation of 500 Btu/h ft² (1.74 kW/m²), the pain threshold is reached in approximately one minute, while on tripling this intensity, the time necessary is reduced to 16 seconds. By way of comparison, the radiation received from the Sun can be estimated at roughly 1 kW/m², and radiation levels equivalent to 1.6 kW/m² can be tolerated without sensation of discomfort for relatively prolonged periods of time [2].

As indicated above, burns appear rapidly upon reaching the pain threshold. For example, at a radiation level of 9.5 kW/m² the pain threshold is reached in approximately 8 seconds, and if the exposure continues at the same intensity, second-degree burns will be produced in some 20 seconds [2]. Similarly, small differences in radiation intensity or exposure time cause large variations in the final effects. Thus, from the data of Eisenberg *et al.* [3], it follows that second-degree burns result from an exposure of 1.43 seconds duration to a radiation intensity of 131 kW/m². If the radiation intensity increases to 146 kW/m² (an 11%

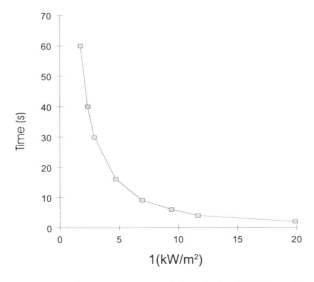

Figure 5.4 Exposure time necessary to reach the pain threshold (adapted from data in [2]).

variation), the mortality threshold is reached. On the other hand, a lower radiation level, of 128 kW/m^2, would cause 100% mortality if the exposure time was prolonged to 10.1 seconds.

The Probit expressions for the prediction of mortality due to thermal radiation from pool and flash fires use a causative factor V, proportional to the product $tI^{4/3}$. The expression usually used is that proposed by Eisenberg and coworkers [3]:

$$Y = -14.9 + 2.56 \ln(10^{-4}I^{4/3}t) \tag{5.6}$$

where the intensity of radiation received, I, is given in W/m^2 and the exposure time t in seconds. The same authors established criteria for non-lethal burns. Thus, first-degree burns can be expected if the product $tI^{1.15}$ exceeds a certain value. The expression to apply is

$$tI^{1.15} > 550\,000 \tag{5.7}$$

Example 5.2

Considering the characteristics of the pentane pool fire of 20 m diameter described in Example 3.8, estimate the time at which 1% (mortality threshold) and 50% human mortality respectively would be produced.

In Example 3.8 the intensity of radiation on a vertical surface at a distant of 40 m from the centre of a circular liquid pool (30 m from the surface of the flames) was calculated as 17.6 kW/m^2. Assuming a person situated at this distance, the vulnerability is given by equation [5.6]:

$$Y = -14.9 + 2.56 \ln(10^{-4}(17600)^{4/3}t)$$

When 1% of the population is affected, $Y = 2.67$, so that from the previous equation $t = 20.9$ seconds is obtained. Similarly, for 50% mortality, $Y = 5.0$ and $t = 51.9$ seconds. It must be taken into account that in this case the times involved are relatively long, so that it is not realistic to assume that those affected would continue to receive the same radiation level throughout the exposure time. On the contrary, one must count on auto-protection reactions (movement away from the fire, seeking shelter, coverage of the most sensitive areas), that diminish significantly the level of radiation received and its effects. However, there could be persons with physical disability of movement (fainting/unconscious, injured etc.), or trapped without possibility of escaping, who would be subject to the scenario as described.

5.4.2   *Effects of thermal radiation on buildings and structures*

The effects of thermal radiation are obviously different on structures of different types. The incidence of thermal radiation on combustible structures can cause their ignition and combustion. On the contrary, with non-combustible materials the most dangerous effect of thermal radiation is the weakening of the material's resistance, with the consequent risk of collapse. In this respect, direct flame impingement is more dangerous than thermal radiation from fires not in contact with the structure. Structural fireproofing (e.g. insulation of steel structures using special cement), is in fact customary, especially for the protection of supporting members.

When evaluating the damage to buildings and structures, it is very important to determine if the material considered can ignite. In this respect, one should note that the surface treatment of different materials can strongly modify its ignition characteristics. Wood is considered to be the solid combustible of reference in studies of ignition by thermal radiation. Lees [16] cites the work of Lawson and Simms [17], in which these authors investigate the ignition of wood with and without a pilot flame (at $\frac{1}{2}$" from the wood's surface). They obtained the following ignition criteria: In order for ignition to occur without the presence of a pilot flame, i.e. ignition caused by radiation proceeding from a source not contiguous with the material studied, the intensity of radiation received must exceed a value of 25 400 W/m², and must also comply with

$$(I - 25\,400)\ t^{4/5} > 6730 \tag{5.8}$$

where the intensity is expressed in W/m² and the time in seconds. For a value of radiation intensity higher than the critical value, equation (5.8) allows calculation of the minimum time required to produce ignition. This expression can be used, for example, to assess the likelihood of ignition in the case of radiation proceeding from a pool fire. For the case of piloted ignition, the threshold intensity is lower (13 400 W/m²), and the corresponding expression is

$$(I - 13\,400)\ t^{2/3} > 8050 \tag{5.9}$$

It is appropriate to use the formula for piloted ignition in the case of a flash fire [3]. For solids other than wood similar expressions can be obtained, where a critical intensity and a minimum time for ignition are stated. In the case where data are not available a heat balance can be carried out, estimating the amount of energy received by radiation and the dissipation of energy by convection and conduction. When the characteristics of the receiving body are known, the temperature evolution with time can be calculated, and compared with the values compatible with physical integrity of the material. On other occasions a threshold value is established at which considerable damage in particular structures can arise. Thus, 37 500 W/m² has been cited as a value of radiation intensity sufficient to cause damage to process equipment [2].

5.5 Vulnerability to explosions

The methods explained in Chapter 3 allow estimation of the direct effects of an explosion, such as the overpressure attained, the duration of the positive phase and the impulse as a function of distance. Likewise, the vulnerability models relate the magnitude of the effects produced with the final damage caused to people or installations. Thus, as a consequence of explosions, people can suffer directly diverse injuries that range from eardrum rupture to death from lung haemorrhage or projection of the body. Indirect injuries include those from glass fragments and flying debris, as well as death caused by the collapse of buildings or structures. On the other hand, damage to buildings and installations can vary widely, from breakage of windows to the complete destruction of structures and equipment and formation of craters.

Table 5.3 shows levels of damage that can be expected for different values of overpressure. It can be seen that with relatively low overpressures (0.35 bar or 5 psi) significant damage to buildings (sometimes their complete destruction) is possible. At even lower overpressures (2 psi) partial destruction of walls and roofing may occur, which often results in human fatalities. In comparison, the resistance of an organism to the direct effects of explosion is surprisingly high. Thus, partial destruction of walls and roofs occurs at an overpressure which is lower than the threshold for eardrum rupture (1% probability of eardrum rupture occurs at 2.4 psi), and much lower than the threshold of mortality from lung haemorrhage (14.5 psi). The conclusion is, therefore, that human fatalities from an explosion are more probable inside a building (indirect effects) than outside, from the direct effects of overpressure.

The Probit equations to use in evaluating the direct effects of an explosion are as follows [3]: for death from lung haemorrhage

$$Y = -77.1 + 6.91 \ln P^{\circ} \qquad (5.10a)$$

where P° is the overpressure (Pa). Similarly, for eardrum rupture the corresponding equation is

$$Y = -15.6 + 1.93 \ln P^{\circ} \qquad (5.10b)$$

In addition to the above effects are the injuries caused by high velocity fragments generated in an explosion, those due to impact by displacement or projection of the body and those that are related to other effects of the explosion, such as the collapse of buildings, generation of fires and release of toxic gases. To estimate the effects of the projectiles generated by an explosion, as well as the effects due to whole body translation, the intensity of the causative factor to use in the Probit expressions is the impulse I_p (Ns/m^2), defined (equation (3.19)), as

$$I_p = \int_0^{t+} P^{\circ}(t)\mathrm{d}t = \qquad (5.11)$$

where t_+ is the duration of the positive phase. Equation (3.20) gives the impulse as a function of the parameters of the modified Friedlander equation. Once the impulse has been estimated, the corresponding Probit equations are: For grave injuries caused by flying debris, particularly fragments of glass,

$$Y = -27.1 + 4.26 \ln I_p \qquad (5.12)$$

Table 5.3 Damage caused by explosions as a function of overpressure (data selected from [16, 18, 19]

Overpressure (psi)	Type of damage
0.03	Occasional breakage of large windows already under strain
0.04	Loud noise. Breakage of windows due to sound waves
0.1	Breakage of small panes of glass already under strain
0.3	Projectile limit. 95% probability of avoiding significant damage. 10% of windows broken
0.5–1.0	Destruction of windows with damage to frames
0.7	Minor structural damage to houses
1.0	Partial demolition of houses, which become uninhabitable
1–2	Failure of wooden and aluminium panels, etc.
2	Partial collapse of house roofs and walls
2–3	Destruction of cement walls of 20 – 30 cm width
2.4	Threshold (1%) of eardrum rupture
2.5	Destruction of 50% of brickwork of houses. Distortion of steel frame buildings
3–4	Rupture of storage tanks
5–7	Almost total destruction of houses
7	Loaded train wagons turned over
7–8	Breakage of brick walls of 20 – 30 cm width
10	Probable total destruction of buildings. Machines weighing 3500 kg displaced and highly damaged
12.2	90% probability of eardrum rupture
14.5	Threshold (1%) of death from lung haemorrhage
25.5	90% probability of death from lung haemorrhage
280	Crater formation

The shock wave can also cause displacement of the whole body, often to considerable distances, possibly causing cranial traumas. The Probit equation for death due to whole body translation is

$$Y = -46.1 + 4.82 \ln I_p \tag{5.13}$$

and that for grave injuries is

$$Y = -39.1 + 4.45 \ln I_p \tag{5.14}$$

As for the effects on structures of the projectiles formed in the explosion, in Chapter 3 some consideration was given to estimation of the velocity and range of the fragments produced in the explosion of a vessel. Lees [16] revises some empirical equations for the penetration of fragments in different materials. These equations give an estimate of the depth of penetration according to the type of material receiving the impact. Another associated effect of explosions is the production of a shock wave transmitted by the ground, whose amplitude can be sufficient to damage some structures. Lees [16] also cites the following equation to calculate the amplitude:

$$A = 0.001 \, K \frac{\sqrt{E}}{d} \tag{5.15}$$

where the amplitude of perturbation A is given in inches, E is the explosive mass in pounds and d the distance in feet. The empirical constant K depends on the type of terrain, estimated as 100 for rocky ground and 300 for clay. The damage threshold would correspond to an amplitude of 0.008" for structures of medium resistance, or 0.003" for weak structures. As an element of comparison, 0.003" corresponds to the order of magnitude for vibrations produced by heavy traffic.

Example 5.3

Estimate the vulnerability to the effects of an explosion equivalent to 1000 kg of TNT at distances of 60 and 120 m from its origin.

To calculate the overpressure the normalised distance is obtained, according to equation (3.17), $z = 60/(1000)^{1/3} = 6\,m/kg^{-1/3}$. With this value of z one obtains (Figure 3.19), an overpressure $P^o = 25\,kPa = 3.6\,psig$. Similarly, for a distance of 120 m, $z = 12\,mkg^{-1/3}$, which gives an overpressure of 8 kPa = 1.2 psig. With these values of overpressure the effects of the explosion can be calculated. Other parameters of the shock wave are obtained from the table given in Chapter 3 where α, t_+, and t_a are given for the distances of 60 and 120 m in explosions of 1 tonne of TNT.

Distance	t_a	t_+	α
60 m	18	12	1.08
120m	59	18	0.93

The variation of overpressure with time for both distances is presented (equation (3.13)) in Figure 5.5 from the moment of arrival of the pressure

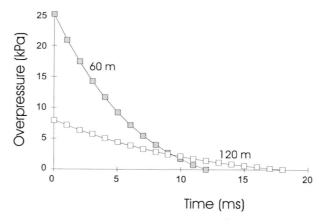

Figure 5.5   Variation of overpressure with time at 60 and 120 m from the centre of the explosion in Example 5.3.

wave at 60 and 120 m respectively. As expected, lower values of P^o and higher values of t_+ are observed as the distance increases.

(a) Effects at 60 m from the origin

Table 5.3 shows that for the overpressure existing at 60 m (3.6 psi), the damage to buildings and structures is very significant. Amongst other damage, rupture of some storage tanks and extensive demolition of buildings are likely.

As for the direct vulnerability of people, equation (5.10) for death by lung haemorrhage gives

$$Y = -77.1 + 6.91 \ln P^o = -7.1$$

Therefore, fatalities are not probable from this cause, which is in agreement with the fact that the threshold for death by lung haemorrhage occurs at an overpressure value of 14.5 psi (table 5.3). Similarly, for rupture of the eardrum, equation (5.10b) gives $Y = 3.94$. Transforming Probit units to probability percentages using Figure 5.1 gives a percentage of eardrum ruptures of approximately 15% of the population exposed.

To calculate the impulse it is necessary to account for the variation of overpressure with time, as indicated in equation (5.12). Substituting values in the integrated expression (3.20), $I_p = 107.92$ (kPa)(ms) = 107.92 N.s/m^2 is obtained. From equations (5.12) to (5.14) one can estimate that neither serious injuries nor death can be expected from flying debris or from whole body translation. Again, this is in agreement with the data from Eisenberg *et al.* [3], that situate the threshold of serious injuries at an impulse of 148 N.s/m^2, although the threshold for skin lacerations would be lower (74.1 N.s/m^2).

(b) Effects at 120 m from the origin.

With an overpressure of 8 kPa (1.2 psi) breakage of glass windows is to be expected, as well as damage to houses, some of which will become uninhabitable (Table 5.3). The application of the previous equations shows that eardrum rupture is not expected, and even less so death by lung haemorrhage, whole body translation, etc. which could be disregarded even at 60 m.

At both 60 and 120 m the most probable causes of death to affected people are the indirect ones (people inside buildings demolished by the explosion), rather than direct causes.

5.6   Factors that modify the vulnerability of people and installations

After an accident, if the models described previously are applied to the circumstances in which the accident took place, it is often found that the actual consequences, expressed in terms of human fatalities/injuries or damage to property, are less than those expected. On the contrary, examples exist of accidents where the final effects have exceeded those reasonably foreseen. In those cases where the final effects are of less importance than predicted, a possible cause may be the conservative character of the assumptions made when applying some models, as well as the empirical correlations used. In other cases chance intervenes, with effects that can be either beneficial or prejudicial. Finally, there are cases in which the emergency services (evacuation, fire-fighting, first aid, etc.) function with great efficiency and the consequences of an accident can be considerably mitigated.

In all cases, the scenario used when carrying out a risk analysis should be clearly defined (i.e. the accident characteristics and concomitant circumstances). Because human and material resources, as well as time available for analysis are limited, it is essential to restrict the number of cases to be analysed, concentrating on those of greater interest, whether because of their greater probability or the magnitude of the consequences, etc. In the selection of a representative set of circumstances for the study a historical record analysis of accidents/incidents plays an important part in setting out the pattern and most frequent types of accidents as seen in Chapter 2. Thus, for example, when carrying out a risk analysis of transport by train of dangerous substances, a historic analysis shows [20] that the predominant scenarios of accidents are collisions and derailments. Other possible accidents (failure of containers due to material fatigue, external fire, explosion, etc.) constitute a relatively small fraction of accidents by train, and may be ignored in many cases of risk analysis.

For a given type of accident, its characteristics and the circumstances in which it takes place can cause changes of various orders of magnitude regarding the final effects on people and installations. Within the factors of major influence are:

- Location: This is very important when determining who are the potential victims (which people or installations) of an accident. Factors such as the site of the accident relative to other installations in the plant come into play, workers in the proximity, the closeness of external populations, or of external property such as buildings, farms, public roads, etc.
- Meteorological conditions: Temperature, humidity, wind speed and direction, rain, etc., that can considerably aggravate or attenuate the effects of the accident.
- Time: It is clear that the effects will be different between night and day, or in and out of working hours. The percentage of the population on the outside of buildings also varies considerably during the day.
- Season of the year is of great influence in determining the degrees to which the population is exposed. Thus, as previously shown, the covering of the skin varies considerably over the period of a year, and the same is true of the population density (e.g. large variations during the holiday season), the number of people inside buildings, etc.
- Sequence of events: e.g. the possibility that other units are affected, giving rise to a domino effect that aggravates the consequences of an accident.

- Other factors: special consideration must be given to the performance of the protection systems and the emergency services, communications, capacity for evacuation and the training for emergencies of the people that could be affected, etc. Evasive action (evacuation, escape, refuge) are of great importance in the reduction of the final effects in many types of accident and, therefore, they are treated in somewhat greater detail in the next section and in Chapter 9, which deals with planning for emergencies.

5.6.1 Evasive actions

In certain kinds of accidents, the total duration of the phenomenon is so short there is not time to react and, therefore, the term evasive action makes no sense. Such is the case with flash fires, or the majority of explosions (excluding those situations where it is known beforehand that a high probability of explosion exists, as would be the case of a fire in a flammable gas storage park, in which case there may be a sufficient period of time to start an evacuation). On the contrary, in other types of accident, such as toxic emissions, the phenomenon can be sufficiently prolonged to consider the possibility of evasive action. This may consist of disorganized flight, evacuation or refuge of the affected individuals, understanding refuge in its widest sense to include any type of on-site protection from the effects of the accident, from shelter behind a wall to protect from the thermal effects up to the use of protective equipment against a toxic emission.

An adequate place of refuge depends on the type of accident. Thus, personnel inside buildings generally are more protected from fires and toxic emissions (outside the building) than those found on the exterior, although it must be said that the use of air conditioning or forced ventilation can decrease or cancel the protection against a toxic emission. As for explosions, we have already seen that buildings are more vulnerable than the human body, so that in the case of explosions, the injuries within a building are frequently greater than those outside it, due to the possibility of structural collapse. In addition, although a building can protect against primary flying debris in an explosion, the people taking refuge inside it can suffer from the effects of secondary flying debris, generated by the very same building.

Very often, the models of vulnerability to toxic emissions include a correction for the evacuation or refuge of the people exposed. In this respect it is necessary to distinguish between different kinds of emission. In many cases, the physical characteristics of the emission (smell, colour, etc.) do not allow immediate detection, and the only warning of the release comes too late, when its effects are obvious. Logically, this reduces the time available for evacuation, and often makes it impossible. In other cases, although an emission is accompanied by a distinctive smell, its effects are toxic at very low concentrations, so that again the warning comes too late. Neither is the direction of propagation of the toxic cloud usually obvious, so that the escape route chosen by an isolated individual in a moment of panic can take him directly to a greater concentration. Lastly, although buildings in principle provide some protection from toxic emissions, diminishing the concentration in the interior, one must bear in mind that they also retain the toxic substance for a longer time after the external cloud has dispersed, so that the toxic dose received (concentration–time) can be substantial.

With the previous considerations, the evacuation models often permit correction of theoretical predictions, in a way that brings the simulations closer to reality. A simple evacuation model is [16]:

$$\phi(t) = 1 \qquad\qquad\qquad\qquad\qquad \text{for } t < t_D \qquad\qquad (5.16a)$$

$$\phi(t) = \phi(t_\infty) + [1 -\phi(t_\infty)][\exp{(-\lambda(t - t_D))}] \quad \text{for } t \geq t_D \qquad (5.16b)$$

where t_D is the time elapsed until the evacuation commences, $\phi(t)$ is the fraction of the population not evacuated at time t, and $\phi(t_\infty)$ is the fraction of population which cannot be evacuated. The constant λ relates to the evacuation half-life according to

$$\lambda = \frac{0.693}{t_{1/2}} \qquad\qquad\qquad\qquad (5.17)$$

The previous model does not take into account factors which are often critical, such as the area to be evacuated, and the population density. Prugh [21] proposed a model that takes these factors into account, by means of a chart that relates evacuation failure (percentage of population that could not be evacuated), to warning time, population density and area to be evacuated.

5.7   Questions and problems

5.1   Consider the domestic tasks undertaken at home. To which toxic substances could we be exposed? What are their points of entry to the organism?

5.2   Repeat the calculations from Example 5.2 considering that the people affected retreat from the fire at a speed of 2 m/s.

5.3   The population of a small island is affected for a period of two hours by an average concentration of 30 ppm of chlorine. Calculate the percentage of deaths in the population. Discuss under what circumstances (time, climate, season of the year, etc.), the number of people injured could increase or decrease considerably.

5.4   Modify the calculations of the previous problem if the island has a population of 500 and a boat is available capable of evacuating 100 persons on each trip, if the time taken to reach a secure island and return is 30 minutes. The time before the evacuation can commence (t_D) is 10 minutes.

5.5   Estimate the vulnerability to toxic effects of people 80 m away from an ammonia container that explodes, instantly releasing 50 kg into the vapour phase. Wind speed is 6 m/s and the escape occurs in conditions of slight instability.

5.6   What is the physical meaning of the intensity threshold that appears in equations (5.8) and (5.9)? How would the value of 25.4 kW/m² for non-piloted ignition be modified in the case where a strong wind blows over the exposed surface?

5.7   When discussing the beneficial effects of different evasive actions the CCPS [12] classifies evacuation after occurrence of a toxic release as 'action of uncertain benefit'. Under what conditions could it be of no benefit?

5.8   Estimate the mass of the cloud produced in an escape of propane knowing that the breakage of glass took place up to 1.5 km away. At what distance can the threshold of serious injuries caused by flying glass be expected?

5.8   References

1. Finney, D. J. (1974) *Probit Analysis*, Cambridge University Press, Cambridge.
2. CCPS (Center for Chemical Process Safety) (1989) *Guidelines for Chemical Process Quantitative Risk Analysis*, American Institute of Chemical Engineers, New York.
3. Eisenberg, N. A., Lynch, C. J. and Breeding, R. J. (1975) *Vulnerability Model. A Simulation System for Assessing Damage Resulting from Marine Spills*. National Technology Information Service Report AD-A015-245, Springfield, MA.
4. Clayton, G. D., Clayton, F. E., Cralley, L. V. and Cralley, L. J. (eds) (1985) *Patty's Industrial Hygiene and Toxicology*, Wiley-Interscience, New York.
5. Gutiérrez Marco, A.- Fundación Mapfre (1983) *Curso de Higiene Industrial*, Mapfre, Madrid.
6. Clayson, B. B. and Krewskli, J. (eds) (1985) *Toxicological Risk Assessment*, CRC Press, Boca Raton, Florida.
7. Instituto Nacional de Seguridad e Higiene en el Trabajo (1989) *Toxicología Laboral Básica*, INSHT, Madrid.
8. Instituto Nacional de Seguridad e Higiene en el Trabajo (1986) *Higiene Industrial Básica*, INSHT, Madrid.
9. Crowl, D. A. and Louvar, J. F. (1990) *Chemical Process Safety, Fundamentals with Applications*, Prentice Hall, Englewood Cliffs.
10. King, R. (1990) *Safety in the Process Industries,* Butterworth-Heinemann, London.
11. Hodge, H. C. and Sterner, J. H. (1949) Tabulation of toxicity classes, *Am. Indl. Hygiene Assoc. Quart.*, **10**, 93.
12. American Conference of Governmental Industrial Hygienists (1991) *Threshold Limit Values and Biological Exposure Indices for 1991-92*, ACGIH, Cincinnati, Ohio.
13. Instituto Nacional de Seguridad e Higiene en el Trabajo (1992) *Fichas Internacionales de Seguridad Química* (translation of the "International Safety Cards", published by the European Commission), INSHT, Madrid.
14. National Oceanic and Atmospheric Administration (1988) *Database of Program ALOHA 4.0*, NOAA, Seattle, Washington.
15. Environmental Protection Agency (1987) *Technical Guidance for Hazard Analysis: Emergency Planning for Extremely Hazardous Substances*, EPA Federal Emergency Management Agency, US Department of Transportation, Washington DC.
16. Lees, F. P. (1980) *Loss Prevention in the Process Industries*, Butterworth-Heinemann, London.
17. Lawson, D. I. and Simms, D. L. (1952) The ignition of wood by radiation. *Br. J. Appl. Phys.*, **3**, 288.
18. Clancey, V. (1972) *Diagnosis Features of Explosive Damage*. Proceedings of the 6th International Meeting of Forensic Science, Edinburgh.
19. Bodhurtha, F. P. (1980) *Industrial Explosion Prevention and Protection*, McGraw-Hill, New York.
20. Boykin, R. F. and Kazarians, M. (1987) *Quantitative Risk Assessment for Chemical Operations*. Proceedings of the International Symposium on Preventing Major Chemical Accidents (ed. J. L. Woodward), Feb. 3–5, 1987, Washington DC. American Institute of Chemical Engineers, New York.
21. Prugh, R. W. (1985) Mitigation of vapour cloud hazards. *Plant Operation /Progress*, **4**, 95.

6 Quantitative risk assessment

It was often impossible to see how solid state electronic components could fail; yet they did

Arthur C. Clarke, 2001 A Space Odyssey, Chapter XXII

Sir Knight, knights errant should engage in adventures which offer a prospect of success, and not in such as are altogether desperate. For valour which merges on temerity is more like madness than bravery

Miguel de Cervantes Saavedra, The Adventures of Don Quixote, Chapter XVII

6.1   Introduction

Treated briefly in this chapter are the methods that allow quantification of the probability of a particular type of accident occurring. It is not enough simply to identify all possible types of accidents, their causes and their evolution sequences. Neither is it sufficient to be able to predict the effects of an accident, assuming a specific group of circumstances. The risk analyst needs, in addition, the capacity to estimate the frequency predicted for the accident, or alternatively, the probability that the accident will occur within a specified period of time. This allows him to evaluate the statistically expected losses (SEL), as the product of the magnitude of the effects of a given accident and the probability of its taking place during the useful life of the installation. The SEL is therefore an index that allows the comparison of the risk levels between potential accidents.

Thus, let us assume that our risk analysis reveals that a specified accident (A) is capable of producing economic losses totalling some $100 million (considering direct material damage, losses due to interruption of production, indemnities, repair of environmental damage, etc.) with an estimated frequency of 10^{-4} years^{-1}, i.e. once every ten thousand years. A second accident (B), should it occur, would cause economic damage totalling $0.5 million, with an estimated frequency of 0.1 years^{-1}, i.e. once every ten years. Let us also assume that human injury is not expected in either case, and that the useful life of the plant is estimated at 25 years. In case A, the SEL value during the life of the plant is $0.25 million, while in case B it would be $1.25 million. That is, in spite of the economic damage expected for accident B being 200 times less than A, the SEL value is 5 times greater, and from this point of view accident B is more serious and should have a higher priority when implementing new safety measurements.

As will be shown in this chapter, the above is not strictly true, as additional corrections are required for the impact perceived for major accidents. Besides, it is not very realistic to assume that an accident capable of causing $100 million of economic losses does not present a direct danger to people. However the SEL indicator, with the corrections deemed necessary, is useful in establishing a hierarchy of hazards, and priorities for investment in safety. To be able to establish a value, even approximate, for the SEL of a particular accident, it is necessary to know the rudiments of quantitative probability evaluation. Throughout this chapter the basis for estimating SEL values will be illustrated. Table 6.1 gives the definitions of some commonly used terms [1–4].

Table 6.1 Definition of some terms commonly used in quantitative risk analysis

Availability (A): The fraction of time a system is fully operational.

Confidence interval: A range of values of a variable with a specific probability (e.g. 95%), that the value of the variable lies within the range.

Demand rate (D): The rate (occasions/year) at which a protective system is called on to act, e.g. the pressure rises to the relief valve set pressure, or a level rises to the set point of the high level trip.

Equipment reliability (R): The probability that, when operating under stated environment conditions, process equipment will perform its intended function adequately, for a specified exposure period.

Failure: The inability of a system, subsystem or component to perform its required function.

Failure mode: A symptom, condition or fashion in which hardware fails. A mode might be identified as a loss of function; premature function (function without demand); an out of tolerance condition; or a simple physical characteristic, such as a leak (incipient failure mode), observed during inspection.

Failure probability: The probability (a value from 0 to 1), that a piece of equipment will fail on demand, or will fail in a given time interval.

Failure rate (m): The rate (occasions/year) at which a protective system develops faults. It may also be expressed as failures per demand, i.e. the number of failure events in a given number of demands.

Fractional dead time (FDT): The fraction of time that a protective system is inactive, i.e. the non-availability or the probability that it will fail to operate when required. $FDT = 1 - A$.

Hazard rate (H): The rate (occasions/year) at which hazards occur; for example, the rate at which the pressure in a vessel exceeds the design pressure or the rate at which the level in a tank becomes too high.

Human error: Physical and cognitive actions by designers, operators or managers that may contribute to or result in undesired events.

Human reliability: The probability of a person successfully performing a task.

Likelihood (L): A measure of the expected occurrence of an event. This may be expressed as a frequency (e.g. events per year); a probability of occurrence during a time interval (e.g. annual probability); or a conditional probability (e.g. probability of occurrence given that a precursor event has occurred).

Mean time between failures (MTBF): The average time between successive failures.

Mean time to failure (MTF): The above defined *MTBF* only has meaning when applied to a population of repairable components. If no repairs are performed *MTF* is used instead, defined as the statistical mean of the distribution of times to failure.

Mean time to repair (MTR): The statistical mean of the distribution of times-to-repair. In other words, the summation of active repair times during a given period of time divided by the total number of malfunctions during the same interval.

Test interval (T): Protective systems should be tested at regular intervals, to see if they are inactive or 'dead'. *T* is the time between successive tests.

6.2   Equipment reliability

6.2.1   Probability distributions

The equipment used by any industry, and by the chemical industry in particular, consists of a series of components made available according to a previous design. The possibilities of failure in an installation are virtually infinite. Thus, the initial design could be flawed, or the equipment may be used under conditions it was not designed for, there could have been undetected defects during construction, or the installation could be used improperly or without the necessary maintenance. Failure from external causes can also arise (electrical supply failure, earthquake, rupture due to a vehicle impact, etc.), or some of the components may simply have reached their wear and tear limits, etc. Reliability engineering is the branch of engineering which deals with the relationship between the reliability of equipment and the correct working of its components. Its fundamentals were developed for military use at the end of the second world war, extending later to the aerospace industry and finally to the rest of industry [5].

A failure can on occasions be directly attributed to human error. Some have been quoted, such as incorrect design or lack of maintenance. In reality any failure is ultimately due to human failure, because all equipment has been conceived, installed and used by humans, but when we talk of human error as the cause of an accident it usually means that the failure is related directly to an erroneous human action. Human errors will be discussed later in this chapter and in Chapter 8.

Equipment failures occur as a result of a complex interaction of individual components and the operational circumstances. The prediction of failures in an equipment is usually empirical, involving the accumulation of working data from a representative number of units over a sufficient period of time, and statistically fitting the failures observed to a specific probability distribution. The probability that a component, working satisfactorily at $t = 0$, has a life equal to or less than t (i.e. fails in the interval between 0 and t), is given by the failure probability function, $P(t)$. The complement of this function is called reliability $R(t)$, so that

$$R(t) = 1 - P(t) \tag{6.1}$$

Also important is the failure density function, $f(t)$, defined as

$$f(t) = \frac{dP(t)}{dt} \tag{6.2}$$

The product $f(t)dt$ gives the probability of system failure between t and $t + dt$, assuming the system worked correctly to time t. Similarly, the probability of system failure between any two times t_1 and t_2 is given by

$$P(t_1,t_2) = \int_{t_1}^{t_2} f(t)dt \tag{6.3}$$

The instantaneous failure rate for a population of equipment or components is expressed as

$$\mu(t) = -\frac{1}{N}\frac{dN(t)}{dt} \qquad (6.4)$$

where N is the number of components that remain working at time t and $\mu(t)$ is the instantaneous failure rate at time t, expressed as number of failures per component per unit of time. The integration of equation (6.4) between 0 and t yields

$$N = N_0 \exp\left[-\int_0^t \mu(t)dt\right] \qquad (6.5)$$

where N_0 is the initial number of components working at time zero. Therefore the reliability is defined as

$$R(t) = \exp\left[-\int_0^t \mu(t)dt\right] \qquad (6.6)$$

In the case of a constant failure rate, equation (6.5) becomes

$$N = N_0 \exp(-\mu t) \qquad (6.7)$$

Equation (6.7) corresponds to a special type of probability distribution known as the **exponential distribution**, characterized by a constant value of μ. According to the above equations, the reliability, failure probability and failure density for an exponential distribution are given by

$$R(t) = e^{-\mu t} \qquad (6.8)$$

$$P(t) = 1 - e^{-\mu t} \qquad (6.9)$$

$$f(t) = \mu e^{-\mu t} \qquad (6.10)$$

From equation (6.10) the mean time to failure (MTF) is calculated as the first moment of the failure density function:

$$MTF = \int_0^\infty tf(t)dt = 1/\mu \qquad (6.11)$$

Example 6.1

A system has a constant instantaneous failure rate. Calculate the variation with time of the reliability, failure probability and failure density functions for failure rates of 0.01 and 0.05 failures per hour respectively.

The applicable equations are (6.8) and (6.10). With the above values of μ, the variation of $R(t)$, $P(t)$ and $f(t)$ with time is shown in Figure 6.1. The reliability (in this case, the probability that the equipment will not fail during the interval 0 to t) falls rapidly with time (the failure rates are quite high), and the decrease is faster the greater the value of μ. The failure probability function follows a complementary behaviour. Lastly, the area beneath the failure density function curve in both cases is close to 1 (it would be exactly 1 if the representation in Figure 6.1 was extended to infinity). The initial value of failure density is 5 times greater for $\mu = 0.05$, but the decrease is more subdued in the case of $\mu = 0.01$, which compensates this effect.

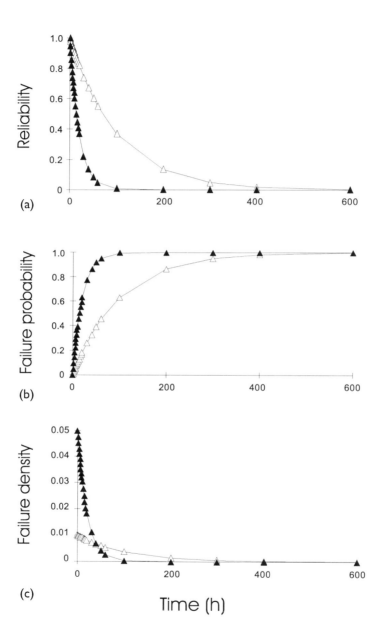

Figure 6.1   Variation of (a) reliability, (b) failure probability and (c) failure density with operating time for different values of the failure rate. The open and filled triangles correspond to 0.01 and 0.05 failures/hour, respectively.

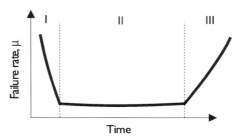

Figure 6.2 Typical failure rate versus time curve.

Up to now the case of an exponential distribution that corresponds to a constant failure rate has been discussed. However, the failure rate generally varies with time. The curve showing the variation of the failure rate with time typically has the characteristic shape of Figure 6.2, so that the failure rate curves are often called 'bathtub curves'. The shape of the curve comes from the three different stages in the useful life of equipment. In the initial stage, corresponding to stage I of the Figure, the early failures are produced, due to manufacturing defects, incorrect installation etc. Also, failures may occur because the operator is in the learning phase of equipment handling, and makes mistakes more often. With equipment that passes stage I there is a reasonable probability that it is well made and properly installed, and the operator is familiarized with its use. In stage II, the failure rate is practically constant, caused to a large extent by random fluctuations in the load supported by the equipment. Finally, the rapid increase in the failure rate observed in stage III is a consequence of equipment wear.

The curve in Figure 6.2 represents a typical case in which the failure rate is dependent on time. There are also elements where the failures depend on the demand to which they are subjected. The failure rate usually refers to demand when dealing with those elements that are normally inactive, and only work during a short period of time in which the demand occurs. This would be, for example, in the case of an electric switch, or a pressure relief valve.

The constant failure rate hypothesis implicit in the use of an exponential distribution is reasonable for many types of industrial equipment, as often a running-in period exists prior to normal use, and also it frequently occurs that equipment is replaced after a certain number of hours of operation. Therefore, although the application of this principle to mechanical equipment is not always possible [5], in many other cases it is reasonable to expect a more or less constant failure rate over the active life of an element. In fact, the exponential distribution has been used with success for studies of risk in nuclear power stations, and in numerous chemical installations [6].

There are many other distributions used in reliability studies. The **Poisson distribution** pertains to the group of discrete probability distributions, and has the following properties:

1. The number of events that occur in a given interval of time is independent of the number that occur in any other unconnected interval (i.e. Poisson-type processes have no memory).
2. The probability that a single event occurs in a very short interval of time is proportional to the length of the interval and does not depend on the number of events that occur outside this interval.
3. There is a very low probability that more than one event occurs in this very short period of time.

According to the above properties, the probability that a specified number of events (failures) x will occur in a time t when the average rate of events is μ is given by [7]

$$P(x, \mu t) = \frac{e^{-\mu}}{x!} (\mu t)^x \qquad (6.12)$$

Making $x = 0$ in the above equation gives the reliability, i.e. the probability of zero failures in the interval 0 to t. Under these conditions equation (6.12) coincides with equation (6.8).

When the failure rate is not constant the **Weibull distribution** is usually employed, introduced by the Swedish physicist of the same name in 1939. This distribution is useful as a generalization of the exponential distribution, because its flexibility allows the handling of failure rates which are variable with time. In the Weibull distribution the failure rate is given by [6]

$$\mu(t) = \nu\alpha(\nu t)^{\alpha-1} \qquad (6.13)$$

where ν and α are positive parameters, and $1/\nu$ is known as the characteristic life of the element while α is called the shape factor. When α is less than 1 a decreasing failure rate is obtained, corresponding to the running-in period shown in Figure 6.2. When $\alpha = 1$, the distribution has a constant failure rate with respect to time, and coincides with an exponential distribution. Finally if α is greater than 1 the behaviour shown in stage III of Figure 6.2 is obtained, corresponding to failure caused principally by ageing of components [5]. The reliability for the case of systems that follow the Weibull distribution is obtained by combining equations (6.6) and (6.13):

$$R(t) = \exp\left[-(\nu t)^\alpha\right] \qquad (6.14)$$

According to the above discussion, the mean time to failure for a system following the Weibull distribution will be [6]:

$$MTF = \int_0^\alpha \exp\left[-\nu t\right]^\alpha dt \qquad (6.15)$$

Among the distributions encountered, the exponential is usually applied if no other information is available. Since this is often the case in risk analysis, there are numerous analyses based on this distribution. The Poisson is a discrete distribution, the use of which is appropriate when an event can occur at any moment in time, i.e. the number of failures occurring in an arbitrary period does not indicate anything about the number of failures that could be produced in another distinct period. Finally the Weibull distribution is capable of providing great flexibility in the treatment of data, whenever enough experimental data exist to obtain its characteristic parameters with sufficient accuracy.

Apart from the above, there are many other failure probability distributions that may be used. Of course, the **normal distribution** is well-known and widely used, especially to describe the types of failure due to wear, variations in dimensions of parts made in automated processes, failures due to physical and natural phenomena, etc. The **log–normal distribution** implies that the logarithm of the values taken by a random variable follows a normal distribution. This distribution is preferred when the deviations with respect to the mean value are in proportions or percentages rather than in absolute values, and is used in applications such as metal fatigue studies, life of electrical insulators, data on repair times and numerous cases of failures in continuous processes [5]. Other distributions frequently used in reliability studies are the *binomial, multinomial* or *geometric* distributions (discrete), and the *Rayleigh, gamma, rectangular, Pareto* and *extreme values* distributions (continuous).

The identification of the distribution that best represents the observed data is not an obvious step in the majority of cases, and requires considerable experience on the part of the analyst. Currently, however, computer-based techniques are being developed, which are very useful for statistical data fitting, providing greater efficiency in the discrimination of models and the estimation of their parameters. Figure 6.3 shows schematically the procedure for selection of a distribution model and calculation of its characteristic parameters.

QUANTITATIVE ANALYSIS OF FAILURE DATA

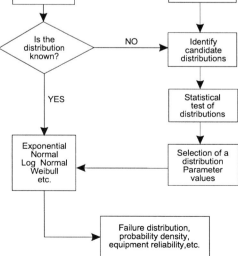

Figure 6.3 Schematic for the selection of a failure probability distribution model (adapted from [3]).

A first step in the determination of the most suitable failure distribution for a given equipment or component is the collection of data from a sufficient number of failures. The data available can be plant-specific or generic. The plant-specific data is obtained directly by the operators, noting failures for a particular equipment, and the conditions under which they are produced. In this respect, standardized failure report forms exist [4] in which are listed complete equipment identification data, date and time of failure, description of type of failure and the circumstances, method of failure detection, status of equipment (stopped, at full load, at 80%, etc.) and of the plant (normal operation, start-up, etc.) on production of the failure, effect on other equipment and repair details. All this information is very useful for constructing an adequate database. The problem of using plant-specific data is its scarcity. Fortunately, the equipment used in most large plants today has a high reliability. This means that seldom is enough failure data available for a specific piece of equipment to carry out a reliable statistical analysis.

When it is not possible to use specific data then generic data is used, which may include all the company's plants, or if insufficient, data from other process industry plants and bibliographic data. In this way a sufficient amount of data can be obtained to provide reliable statistical parameters, but specific information is lost as to the conditions under which the equipment failure is produced. The data from the bibliography usually presents ample intervals of variability, due to the disparity of the sources used. An important contribution of the observed variability corresponds to the use of a particular equipment in processes with different degrees of severity regarding operating or environmental conditions. Other sources of variability in failure data include the definition of the equipment's limits, the different standards of design and construction, as well as the production techniques and quality control used when the equipment was manufactured, the installation, maintenance, age and conditions of use.

Of the factors mentioned, the definition of the limits is especially relevant for constructing a database which will be useful in reliability studies. The following example [4] clarifies the importance of this point: An operator carries out a normal test of a pump. He tries to start it using a switch located on a panel in the control room but the pump does not respond. The operator completes a work order which states that the pump failed to start. The technician in charge of the repair discovers an electrical failure in the starting resistor of the pump controller (circuit breaker). This undoubtedly constitutes a catastrophic failure, because it prevents the pump from starting on demand. However, a problem persists in assigning the failure. If the resistor that produced the failure is not considered to be within the limits defined for the pump, but is included as a part of its controller, which includes the starting and protective systems, then it is the controller of the pump that has failed. In the CCPS taxonomy, the equipment boundary for a motor-operated pump includes the motor and the transmission, shaft, seals, casing, impeller and the circuit breaker. Therefore in this case the failure is attributed to the pump. It is important to define precisely the taxonomy used in a particular source, so as to be able to assign failures accurately and obtain significant reliability figures for a specific component.

Figure 6.4 shows various examples of failure data, taken from the literature [4] where, in addition to failure figures, the limits considered for each equipment are indicated. It can be observed in figure 6.4(a) that, for the same equipment (a motor-driven centrifugal pump), the failure frequency is considerably different depending on the mode of operation used. This is logical, because the failure modes and their causes are different. Thus, for example, in some operations it is common to find two pumps assigned to the same task, one of which works continuously, while the other (spare) is on stand-by. The probability that the working pump ceases to run in a certain time interval is different from the probability that the stand-by pump fails to start when required [8]. In the first case, the causes of failure may include factors such as over-heating which does not occur in the stand-by pump before being started, while the stand-by pump can have failures due to other reasons, such as a defect in the starter contact.

In Figure 6.4 the severity levels of the processes from which the data were taken are not shown, which, as previously indicated, is a consequence of using generic data. However, the process severity levels strongly influence the failure rate so that they should be taken into account whenever possible. The CCPS [4] gives lower and upper limits, although the intervals can vary widely. Thus, for centrifugal pumps in stand-by mode (Figure 6.4(a)), the average 18.6 failures per 1000 demands are bound by lower and upper limits corresponding to 1.9 and 59.9 failures respectively. Alternatively, the use of adjustment factors has been suggested. The failure rates would be multiplied by the adjustment factors to bring them into line with the most severe conditions. The CCPS [3] quotes du Pont's Process Safety Management Reference Manual (6th Edition, 1987), where increases of 7% are recommended (factor of 1.07) for instruments and control valves under the following conditions: operation under extreme temperatures, high humidity, dirty atmosphere, and inadequate location (exposure to possible damage by mechanical causes or inaccessible for inspection). For a corrosive atmosphere a factor of 1.21 is suggested. Suggested for other damaging agents are different factors, for instrumentation in general and control valves (indicated in parenthesis), thus for corrosion the factor suggested is 1.07 (1.14), for erosion 1.07 (1.28), for possibility of fouling/plugging 1.07 (1.14), for pulsating flow 1.14 (1.07) and for vibration 1.42 (1.21). Sometimes it is possible to apply more complete models, that take into account a greater number of factors. Thus for instance, models for the prediction of vessel failure rates (e.g. [9]), take into account the vessel age, the learning factor, vessel geometry and thickness, number of welds, and a quality factor that accounts for periodic inspections, including ultrasonic and radiographic tests for the welds and the vessel itself.

Example 6.2

A leading chemical company needs to carry out a reliability study on a particular type of motor-operated control valve, in widespread use in its

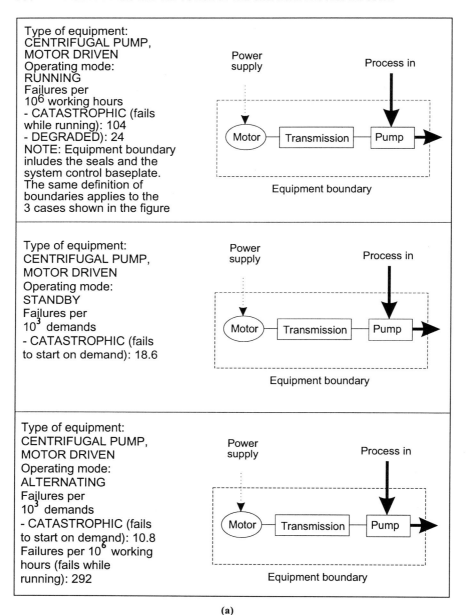

(a)

Figure 6.4   Examples of failure data for selected process equipment (from [4]).

Type of equipment:
METAL PIPE,
STRAIGHT SECTION
Failures per
10^6 mile-hours:
- CATASTROPHIC :0.0268

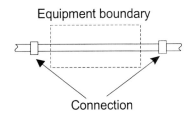

Type of equipment:
METAL PIPE,
CONNECTIONS
Failures per
10^6hours:
- CATASTROPHIC : 0.57

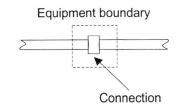

Type of equipment:
CONTROL VALVE,
PNEUMATIC
Failures per
10^6 hours:
– CATASTROPHIC
(Spurious Operation): 3.59
Failures per
10^3 demands:
– CATASTROPHIC (No
change of position on
demand): 2.2

Positioner

Actuator

Air supply

Signal

Process in

Equipment boundary

(b)

continuous process plants. Data collected over a period of 5 years on 158 active process valves is available, totalling 4.8 million working hours. Logged during this period as spurious operation while working is a total of 11 catastrophic failures (the system boundary in this case includes the motor as well as the pump). Estimate the probability of this kind of failure for the equipment under consideration.

Assuming a constant failure rate gives 2.3 failures for each million working hours (2.3×10^{-6} failures per valve-hour). Generic data for this equipment indicates [4] an average failure rate of 1.36 per million working hours, so that the value of 2.3 failures could be indicative of defects in maintenance, or of a process severity higher than average.

To estimate the confidence limits for the failure rate the following expressions are used [3]:

$$UL = \frac{(x+1)F_1}{(n-x)+(x+1)F_2} \tag{6.16}$$

$$LL = \frac{x}{(n-x+1)F_2 + x} \tag{6.17}$$

where UL and LL are, respectively, the upper and lower limits for the estimated confidence interval, x is the number of failures observed and n the sample size. In this case the sample size is 4.8 million valve-hours. F_1 and F_2 are the values of the F distribution for a specified level of probability, with degrees of freedom (f_1, f_2) and (f_3, f_4) respectively, which are defined as:

$$f_1 = 2(x + 1)$$

$$f_2 = 2(n - x) \tag{6.18}$$

$$f_3 = 2(n + x + 1)$$

$$f_4 = 2x$$

The values F_1 and F_2 are obtained from conventional tables for the F distribution, e.g. [7]. According to equations (6.18), the following values are obtained: $f_1 = 24, f_2 = f_3 = 9.6 \times 10^6, f_4 = 22$. Under these conditions, from the table for the F distribution, are obtained for a 95% confidence interval, values for F_1 and F_2 of 1.52 and 1.78 respectively. From equations (6.16) and (6.17), the UL and LL values are 3.8×10^{-6} and 1.3×10^{-6}, i.e. at a 95% confidence level the interval ranges from 1.3 (lower limit) to 3.8 (upper limit) failures per million working hours. For a 99% confidence level the interval would extend from 1 to 4.5 failures per million working hours.

6.3   Reliability and availability of protective systems

A protective system is a device which is installed to prevent the development of a dangerous situation. Examples of protective systems are a pressure relief device, a high temperature alarm or an emergency shut down system. Because protective systems are designed to reduce the probability of accidents, it is especially important to check their reliability. The reliability of complex systems requires detailed study which is not attempted here. In other references given at the end of this chapter [1, 3, 5, 6, 10, 11] this topic is developed further.

The majority part of protective systems only work periodically. This means it is possible that a protective system has failed (e.g. a relief valve is stuck in the closed position) and the failure will not show until a demand is made on the system. This type of failure is called a **hidden** or **unrevealed failure**, and is extremely important from the safety point of view. In other cases, the operator knows of the fault at the moment it occurs, and, therefore, can take corrective measures before the demand is produced. This type of failure is called a **revealed failure**; an example of this type would be the opening of a rupture disc at a lower pressure than the set pressure, due to chemical corrosion.

If a protective system is never tested, it eventually degrades and fails, with a probability of failure that increases with time. The only way to reduce the probability of unrevealed failures occurring is by frequent inspection of the protective systems. When a failure is discovered, a repair is required, and during this time the system is also unavailable. As shown in Table 6.1, the fractional dead time (*FDT*) is defined as the fraction of time that a system is not available, whether through failure or because of maintenance.

For a dangerous situation to arise a demand must be made when the system is unavailable, i.e. during the 'dead time'. For the case in which both the *FDT* and the demand rate D are low (the most frequent situation in protective systems), the hazard rate H defined in Table 6.1 can be calculated as the product of the demand rate and the probability that the system is unavailable, i.e.

$$H = D \times FDT \tag{6.19}$$

An equipment without a protective system is equivalent to an equipment with a protective system permanently unavailable (*FDT* = 1), in which case the hazard rate is equal to the demand rate. Let us assume that the time taken to repair a system is small compared to the time between checks, or that, immediately a failure is discovered during inspection, the damaged system is replaced by another while repair is taking place. If the interval between checks is T, the fractional dead time is given by

$$FDT = \frac{1}{T} \int_0^T P(t) \mathrm{d}t \tag{6.20}$$

Therefore, to find the fractional dead time the failure probability distribution must be known. In the case where the failure probability follows an exponential distribution, from equations (6.9) and (6.20) it follows that

$$FDT = 1 - \frac{1}{\mu T}\left(1 - e^{-\mu T}\right) \qquad (6.21)$$

Using a Taylor series for $e^{-\mu T}$ and neglecting the terms of the expansion from the third onwards gives

$$FDT \approx \frac{1}{2}\mu T \qquad (6.22)$$

which is a valid approximation when $\mu T < 0.1$. Equation (6.22) is almost an 'intuitive' expression for estimating the fractional dead time. Thus, in Figure 6.5, the protective system, initially active, becomes inactive upon failure. The system's condition only becomes manifest when a maintenance inspection is made, or when a demand is produced. In the example of the diagram, the revisions are made every three weeks ($T = 3$ weeks). The first failure occurs two weeks after installing the system, but as no demand has been made, the failure remains undiscovered until the first revision takes place, at three weeks (note that this condition is not realistic: a recently installed system, especially a protective system, should be checked more frequently, at least until after period I in Figure 6.2). After the first revision, the equipment is repaired, and during the second three-week period there are no failures, so that the second check shows that it is working correctly. The next failure occurs at close to eight weeks, and is not found until the nine-week check, and so on. Although some failures occur at an early stage in the period between checks and others at a later stage, on average a failure would remain undiscovered during half of the time interval between checks, T. In this case the fractional dead time can be estimated as the product of the failure rate μ and $T/2$ (equation (6.22)). Therefore, if the system fails on average once per year and the test interval is three weeks, the fractional dead time will be 1.5 weeks per year, i.e. 0.0288. From equation (6.20) it follows that equation (6.22) will be exact if the failure probability could be written as $P(t) = \mu t$, i.e. if the probability of failure was directly proportional to the transpired time (hence its 'intuitive' character). Equation (6.21), corresponding to the exponential distribution, is more general.

On the other hand, equation (6.19) is, as indicated, valid if both D and FDT are low. Otherwise, the following equation should be used [2]:

$$H = \mu(1 - e^{-DT/2}) \qquad (6.23)$$

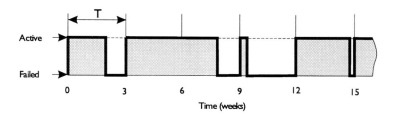

Figure 6.5   Status diagram for a component showing undiscovered failures.

The predictions of H obtained from equations (6.19) and (6.23) coincide for low values of the product DT (for $DT < 0.4$ the differences are less than 10%, but can increase notably for higher values). In fact, when DT reaches very high values the previous equation establishes that H tends towards μ, i.e. the hazard rate and the system failure rate are approximately equal, because at very high values of demand (or very high interval values of the test interval) it is probable that a demand will occur simultaneously with a failure of the protective system.

Equations (6.21) and (6.22) give estimates of FDT due to failed equipment. Obviously, these equations should be corrected to also take into account the down time due to revisions and repairs, as well as the probability of a defective repair [3]. Thus, the total value of the fractional dead time is calculated as the sum of dead times due to component failure discussed previously, plus the contribution to the FDT of to the duration of the revision and/or repair, the additional FDT caused by human error in the revision or repair of the system, and those due to common cause failures, which are discussed later.

Example 6.3

A system has a failure rate of 0.5 years⁻¹. Assuming that an exponential probability distribution can be applied, calculate the hazard rate if tests are made (a) quarterly, and (b) annually, knowing that on average there are 2 demands per year on the system. (c) What should the interval between tests be to obtain a hazard rate as low as 0.005 years⁻¹?

The product DT is greater than 0.4 in both parts (a) and (b) of the problem. Thus, the hazard rate is given by equation (6.23) rather than by (6.19), although in this case the error would be small. (a) With $D = 2$ years⁻¹, $\mu = 0.5$ years⁻¹ and $T = 0.25$ years, equation (6.23) gives $H = 0.11$ years⁻¹, or once every 9.1 years (equation (6.19) would have yielded $H = 0.12$ years⁻¹ a difference of about 10%).

(b) Similarly, the hazard rate is $H = 0.32$ years⁻¹ (once every 3.1 years, while equation (6.19) would have given a 30% higher value for H). One can see that lowering the frequency of checks from once per quarter to annually triples the hazard rate.

(c) The hazard rates obtained in the above sections are high. To reduce the hazard rate to a value of $H = 0.005$ years⁻¹ (once every 200 years), equation (6.23) (or (6.19)), since in this case DT is considerably lower) gives $T = 0.01$ years, i.e. checks are required approximately twice a week.

6.3.1 Associations of systems

In the previous example it was seen that it is possible to reduce the hazard rate by reducing the interval between checks. As we have seen, for low values of T, the

FDT varies almost proportionally to *T*. Obviously, a practical limit exists in the reduction of the fractional dead time that can be achieved by increasing the frequency of checks. On the other hand, even in the case of quick repairs, the assumption that the time for repair is very low compared to *T* is no longer true when inspections are carried out very frequently (low *T*).

Another way of reducing the hazard rate consists of doubling and at times multiplying essential equipment. This can be done in two ways: **redundancy** consists of the use of various identical units, each one of which is capable of carrying out the required function. The term **diversity** is applied to the use of various different units, each one of which is capable of performing the required function. Its objective is to diminish the probability of failure by common cause, which can sometimes affect redundant units.

The two most simple forms of protective equipment association are **series association** and **parallel association**, which are shown schematically in Figure 6.6. The series association implies that the system is protected only when all of the associated protective equipment works correctly, i.e. if any of the associated equipment fails the system fails, which in principle means a higher failure probability. Therefore, when association in series is used it is not because we find mathematically that the system reliability increases, but because it is advisable for physical reasons. A typical example is that of a pressure relief system consisting of a rupture disc in series with a safety valve. Pressure relief requires that both the rupture disc and the safety valve open at their respective set pressures. Therefore, mathematically, it follows that the probability of the association failing is greater than either of its components failing separately. In spite of this, the association is used, for example, when the process fluid can cause corrosion in the safety valve. In this case, the function of the rupture disc is to protect the safety valve, and this action considerably improves the individual reliability of the safety valve. As a consequence, the reliability of the rupture disc/safety valve system is greater than that of a safety valve exposed to corrosion by the process fluid.

The association of protective systems in parallel always increases safety, and is typical of redundant systems. A parallel system functions if any of its component systems is operative. Using the previous example, if the rupture disc fails, a protective system failure occurs. To prevent this, a second rupture disc/safety

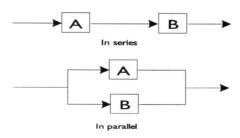

Figure 6.6   Association of protective systems.

valve system can be installed in parallel with the first or alternatively a second rupture disc with a slightly higher working pressure. For the protective system to fail now *both* rupture discs must fail. Because the probability of both failing (the product of the individual probabilities) is much lower, redundancy has greatly increased the system reliability.

The use of redundancy to increase the reliability of a protective system is, however, restricted. Consider a process in which an exothermic runaway reaction may take place. Let us assume that a temperature alarm has been installed that triggers an emergency shut-down of the system when the temperature goes beyond a certain value, T_T. Obviously, this emergency system relies on the reading of a temperature sensor (e.g. a thermocouple), that can be considered as a critical element for the protection of the reactor. In the majority of cases it would be considered unacceptable to place the action of a critical protective system on the measurement of only one element. In this case we can install a second sensor, so that emergency action begins if either of the two detects a temperature $T > T_T$. Thus, even though one of the sensors might work incorrectly and give temperature readings below the real value, it is probable that the second sensor gives correct readings and activates the emergency system.

Since thermocouples are relatively inexpensive one could implement a much higher level of redundancy with the installation of, say, a dozen sensors, so that when any one of them registers $T > T_T$ the emergency system is activated. This reduces to negligible values the probability that the emergency system is not activated because of a thermocouple failure. In exchange one would have a system with frequent unnecessary emergency stops due to **fail-safe** malfunctions of the temperature sensors. Each time one of them fails, giving a temperature in excess of T_T, the emergency shut-down system would activate, which is also unacceptable, not only for economic reasons (lost production), but also because it is very probable that frequent interruptions in steady-state processes would lead to unsafe operation. To prevent this type of situation **voting systems** are frequently used, where a minimum number of signals (e.g. 2 sensors out of 3, 2 out of 4, etc.) with values above the alarm activation level are required [2].

The calculation of the fractional dead time in simple systems where redundancy or diversity exists is carried out by following the general principles already discussed. For example, if a system is protected by two different elements (diversity) in parallel, whose failure rates are μ_1 and μ_2 (e.g. a rupture disc and a safety valve or two different rupture discs), the failure of the protective system would require both components to fail. Let us assume that both failures are independent and their probabilities are given by an exponential distribution. From equation (6.9) we can write for each one of the components,

$$P_i(t) = 1 - e^{-\mu_i t} \qquad (6.24)$$

For low values of the product $\mu_i t$, the failure probabilities can be approximated by $P_i(t) = \mu_i t$, and from equation (6.20), the protective system's fractional dead time is given by

$$FDT = \frac{1}{T}\int_0^T (\mu_1 t)(\mu_2 t)\mathrm{d}t = \mu_1 \mu_2 \frac{T^2}{3} \tag{6.25}$$

and is also equal to $(1/3)\mu^2 T^2$ in the case where redundancy rather than diversity exists, and systems 1 and 2 are identical. Generally, for the failure of n redundant protective systems (which are tested at the same time), the fractional dead time is given by [3]

$$FDT = \left[\frac{n!}{r!(n-r)!}\right]\frac{1}{(r+1)}\mu^r T^r \tag{6.26}$$

where n is the number of redundant units or components, $r = n-m + 1$, and m is the number of systems that must work correctly for the system to remain protected. Thus, in a system with three redundant components where one working is sufficient for the system to be protected (*1 out of 3 system*), the *FDT* is given by $^1/_4\mu^3 T^3$, whereas in a *2 out of 3 system* (voting system), the *FDT* would be $\mu^2 T^2$.

6.3.2   Common cause failures

It was considered in the previous discussion that the probability of failure of a given component is independent from that of other components. Thus, if two rupture discs in parallel are considered, one assumes that their probabilities of failure are not related. However, let us consider the following hypothetical case: Two rupture discs were made by the same company and approximately at the same time. An accidental defect in the composition of the raw material used, which was not detected by quality control, occurred in this period. As a consequence, the rupture discs will not open at their set pressures. If we consider in isolation the probability P of a manufacturing defect in a rupture disc which prevents it from opening at the set pressure, we find a very low value, because these elements undergo rigorous manufacturing controls. Therefore, with a redundant rupture disc we ought to be adequately protected, because the probability that both discs have suffered *independent* failures in the production process (P^2) is very low. However, in the case considered a common cause for both failures exists, which increases extraordinarily the probability of its occurrence.

Hauptmanns [6] quotes the following classification for the analysis of common cause failures:

A   Failure of two (or more) similar or identical components due to a common cause.

B   Failure of two (or more) similar or identical components as a consequence of a single initial failure (secondary failures).

C   Failure of two (or more) similar or identical components due to functional dependence, e.g. dependence on a common utility such as instrument air, process air, electricity, etc. Also included within this section would be a human error affecting various systems simultaneously. This type of failure is relatively easy to detect using FTA or HAZOP analysis (Chapter 2), while potential failures belonging to group A are frequently overlooked in preliminary analysis.

Common cause failures originate from a variety of sources, including design defects, improper operation (inappropriate operation/control parameters, operation under extreme conditions of temperature, humidity, vibration, etc.), external events (fire, sabotage, earthquake, etc.), maintenance defects (e.g. a periodical calibration of sensors incorrectly completed), etc.

6.4 Use of fault tree analysis in quantitative risk assessment

Fault tree analysis was introduced in Chapter 2 as a technique for the identification and quantitative evaluation of hazards, with an explanation of the basic symbols and some examples of its application to simple cases. Beyond simple examples, FTA can become quite complicated when applied to real systems in the chemical industry, so that its use requires previous experience. A detailed treatment of the construction of fault trees or the methods of resolution existing is outside the scope of this book. However, the importance of fault tree analysis for chemical process safety is considerable, and therefore a brief extension of the explanations given in Chapter 2 concerning the basics of FTA is included below. Those requiring a more in-depth account are referred to more specialized texts [6, 10–13].

A fault tree is a logical representation of the sequences of events that can lead to an event, arbitrarily designated as the **top event**. When all the reasonable sequences have been identified and the fault tree is correctly constructed, the FTA is possibly the most powerful tool for the quantifying of risks. On the other hand, the analysis is often applied to complex systems, and errors may occur in its construction and application. The most frequent analysis errors are of a qualitative nature, and usually arise from the following causes [3]:

- The system under analysis is not well understood by the analyst (comprehension of the physical/chemical nature of the system, its failure mechanisms, etc.). This frequently results in the omission of important sequences (it should always be remembered that no existing method guarantees that all of the reasonable cases have been taken into account), or inclusion of wrong sequences.
- Incorrect fault tree logic describing the system failures. This will result in incorrect quantitative evaluations.
- Lack of understanding or improper accounting for common cause failures.

A fault tree consists of various levels of events, connected by logic gates, normally AND or OR gates. When constructing a fault tree the events are usually identified by letters and/or numbers. The tree logic is dealt with using Boolean algebra (the rules most frequently applied to FTA are shown in Table 6.2). There are three classes of equipment failure that can be described using FTA [14]:

1. **Primary faults** are those which occur when operating under the conditions for which the equipment was designed. They are attributable to the equipment and not to external conditions. An example is the case of a vessel that ruptures at a pressure within the design limits of the vessel.

Table 6.2   Boolean rules frequently used in fault tree analysis

Commutative:	$A + B = B + A$ $AB = BA$	
Associative:	$A(BC) = (AB)C$ $A + (B + C) = (A + B) + C$	
Distributive:	$A + BC = (A + B)(A + C)$ $A(B + C) = AB + AC$	
Others:	$AA = A$ $A(A + B) = A$ $AA^* = 0$ $0A = 0$ $1A = A$ $(A^*)^* = A$	$A + A = A$ $A + AB = A$ $A + A^* = 1$ $0 + A = A$ $1 + A = 1$

AB corresponds to 'event A and event B', A + B corresponds to 'event A or event B'. For a given event A, A* (sometimes $\overline{A}$ or A′ is used instead) represents the complement of event A, i.e. the negation of A.

2. **Secondary faults** are those produced in an environment for which the equipment was not designed. For example, a vessel ruptures because some external event causes the internal pressure to exceed its design limits. The failure is not attributable to the equipment, but to external conditions.

3. **Command faults** are those in which the equipment performs correctly, but at the wrong moment, or in a place different from that intended. Again, the failure is not attributable to the equipment, but to the source of the signal received. For example, a high-temperature alarm may fail to warn of a temperature above the alarm level because a temperature sensor has previously failed (i.e. a primary fault in another element is responsible for the lack of action of the alarm).

In a fault tree the equipment primary faults are at the tips of the 'branches' or the 'roots', while the secondary and control failures are intermediate events, also connected to other events through logic gates. Other failures in principle not related to the equipment, such as external events and human errors, also normally appear in primary levels.

A fault tree can always be described by an equivalent Boolean algebra expression. Thus for a gate OR with two inputs A and B, the output is $A + B - AB$. As indicated in Chapter 2, when the frequencies or probabilities assigned to events A and B are low, it is common to neglect the product term, and express the output as the sum of the individual terms.

An important part of FTA is the identification of the groups of events that can give rise to the top event. These groups are called **cut sets**. Generally, the identified cut sets can be simplified to an equivalent series with a lower number of groups called **minimal cut sets**. A minimal cut set does not contain other groups and cannot be further simplified. Frequently the reduction is carried out using computer programs, but it can also be done manually in simple cases, as shown in the following example taken from the work of Schreiber [15].

Example 6.4

Following the study of a system the fault tree shown in Figure 6.7(a) was constructed. Obtain the Boolean expression that represents the fault tree, obtain the reduced expression and represent the equivalent fault tree.

In Figure 6.7(a) it can be seen that the same event (B) appears on different branches of the tree. When this occurs, it is often necessary to restructure the tree, so that the common failures can be adequately treated and errors are avoided. According to the rules already stated we can write:

$T = [(A + B) C][DB + (B + E))$

$T = [AC + BC] [B + E]$, since $DB + B = B$

$T = ABC + ACE + BCB + BCE$

$T = ACB + ACE + BC + BCE$, since $BC \cap B = BC$

Taking into account that $ACB \subset BC$ and $BCE \subset BC$, it follows that $T = ACE + BC = C [AE + B]$

The reduced fault tree that corresponds to this expression is shown in Figure 6.7(b), in which there no longer appears a common event on different branches. The minimum cut sets for this case are the groups of events (A,C,E) and (B,C), respectively.

6.4.1   Ranking of minimal cut sets

The ranking (hierarchical ordering) of the minimal cut sets identified is carried out after completing the FTA. For a qualitative ranking it is sufficient to consider two factors [14]:

1. Structural importance: Considers how many basic events are involved in each of the minimal cut sets. From this point of view, a unitary group (only one event) is more important than one containing two, and this is more important than one containing three, etc. The implicit assumption is that, other conditions being equal, a sequence to the top event involving only two events is more probable than another that involves three, and this more than another involving four, and so on.
2. Type of events: This factor considers the ranking within the groups of a particular size, taking into account the type of events involved. The rule in this case is: first, human errors, second, errors due to failure of active equipment (i.e. equipment that is actually working) and third, errors due to failures of passive equipment (static equipment, such as piping or a storage tank). Once again, this ranking is based on the consideration that a human error is more probable than one due to the failure of an active equipment, and that this is more probable than a fault affecting a passive equipment. Thus, within binary minimal cut sets (two events), one involving a human error and an equipment fault (whether active or passive),

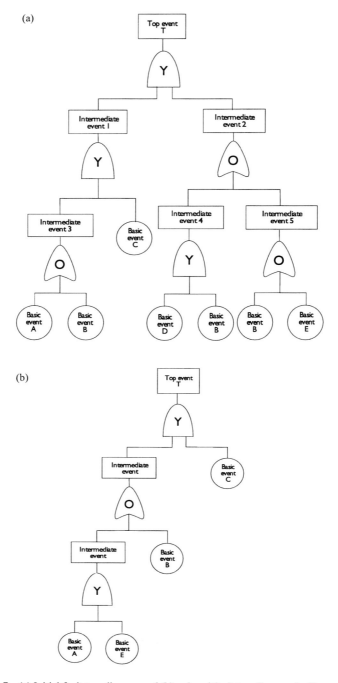

Figure 6.7   (a) Initial fault tree diagram and (b) reduced fault tree diagram for Example 6.4.

will be ranked higher than, for example, another group involving two active equipment failures.

The previous ranking only gives a general qualitative orientation and may have to be modified in specific cases, depending on the type and quality of equipment, operating conditions, operator training, maintenance policy, etc. In the end, the ranking of the most probable events can often be established from the experience of the personnel operating the plant. Moreover, one must take into account that the probability of failure for a given type of equipment (active or passive) may vary greatly with the operating environment, as was shown for the examples given in Figure 6.4. This means that a **quantitative** ranking of the minimal cut sets is also frequently required.

The following example, taken from the literature [14], shows the development of a fault tree, as well as the construction and ranking of the corresponding minimal cut sets, for a specific case: a continuous reactor with a significant runaway hazard.

Example 6.5

A highly unstable reaction is being carried out in the reactor shown in Figure 6.8. The system is sensitive to small increases in temperature and, therefore, a reaction quenching system has been set up to protect the system against an uncontrolled reaction. The reactor temperature is continually monitored by two sensors T1 and T2. The quench tank outlet valve, V-2, activates automatically when T1 detects a certain increase in temperature. Independently, T2 activates an alarm in the control room to alert the operator as to the possible loss of control of the reaction. When the alarm sounds, the operator should press a button that closes valve V-1, stopping the reactor feed. On hearing the alarm the operator is also instructed to press the button that opens the quench tank outlet valve in case sensor T1 failed to operate it. If either valve V-2 opens or V-1 closes, the reactor enters a stable emergency shut down, without damage to the system. Carry out a fault tree

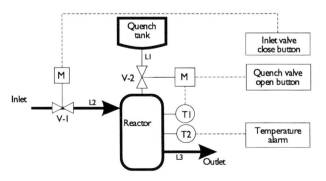

Figure 6.8 Diagram of the reactor used in Example 6.5.

analysis for this system, using as the top event 'reactor damage due to excessive temperature'.

Note: Prior to carrying out the FTA, the team of analysts decide not to take into account events such as electrical power failure, wiring or push button failures, which will be accounted for in later analyses. Also, the analysis is limited to the equipment included in Figure 6.8, i.e. process upstream or downstream from the reactor is not considered. The normal system state is with V-1 open and V-2 closed.

The fault tree (Figure 6.9) is developed in the normal way, seeking first the events that can directly generate the top event T. In this case the materialization of the top event requires that the extinction agent is not discharged and also that valve V-1 does not close in time, because in the

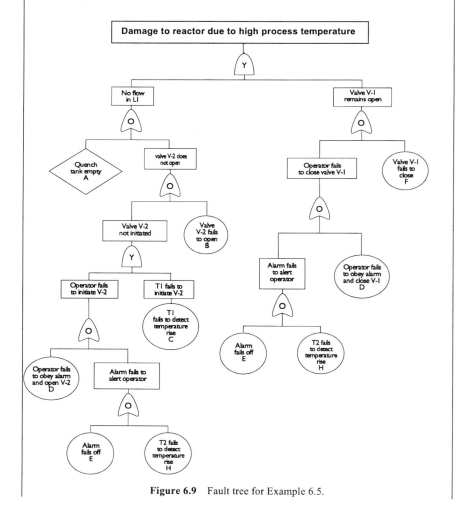

Figure 6.9   Fault tree for Example 6.5.

scenario assumed for the analysis it is assumed that either of the two actions would prevent the top event. From this point, each branch of the tree is developed, as indicated in the Figure. Note that event A is marked in the diagram with a rhomboid, indicating that the event is not developed further, although it could be done (evidently there may be a number of reasonable causes for the quench tank to be empty).

The expression representing the fault tree is:

$$T = (A + B + C[D + E + H])(F + G + E + H)$$

Taking into account that CHH = CH, CEE = CE, CDH $\subset$ CD, CEG $\subset$ CE, etc. we can write

$$T = AE + AF + AG + AH + BE + BF + BG + BH + CE + CH + CDF + CDG$$

Next, a qualitative ranking for the groups of events can be established. Regarding structural importance we distinguish between binary (AE, AF, AG, AH, BE, BF, BG, BH, CE, CH) and tertiary (CDF, CDG) groups. Within each group we can establish a hierarchy based on the types of failure. Thus, for the binary groups the order (assuming that the non-developed event A is due to a human error) is the following: AG (two human errors), AH (human error and failure of the temperature sensor, which is an active equipment), AF = AE = BG (the three cases combine a human error and an active equipment failure), CH = BH = CE = BE = BF (two active equipment failures). Note that if the non-developed event had been classified as a passive failure (e.g. if the cause of the tank being empty was a perforation in the tank, passive equipment), the ranking would change considerably.

For the tertiary groups: CDG (two human errors, one active equipment failure), CDF (one human error, two active equipment failures).

6.5   Human errors and reliability

In the previous section repeated reference was made to **human errors** without defining them and without consideration of their characteristics or the factors influencing the frequency with which they are produced. However, their importance from the point of view of safety is enormous, with estimates that they are directly responsible for more than half of industrial incidents and accidents registered [16]. Therefore, it is useful to consider, if only summarily, the nature of human errors and some of the methods used to estimate their frequency.

In a manner similar to a component failure, a human error can be defined as an action, performed by a person, whose consequences exceed the tolerable limits defined for a system. The action may be simple (pushing the start button of a pump) or complex (carrying out an analysis to predict the system response and taking a decision based on it).

As already stated, from a general point of view all accidents, including equipment failures, are ultimately a consequence of human error, because if an equipment fails it is due to a flawed design, a defect in manufacture, incorrect installation, etc. If we focus the analysis exclusively on actions taken directly by humans operating a given industrial process, we can use the classification of human errors given by Kletz [17]:

- Errors due to a momentary distraction. In this case the operator's intention is correct, but in spite of this the action is incorrect. This would be the case of an operator who knows his work and pays attention to completing it correctly, but, in spite of this, from time to time he mistakenly pushes a button or he does it too late, etc.
- Errors caused by insufficient training or deficient instructions. The operator does not know what to do, or even worse, thinks he knows but does not, leading him to important errors. This would correspond to the case of the operator of a reactor who has received insufficient instruction about the system's behaviour. One day, the reaction rate is too low and, on inspecting the equipment, he discovers that during the reactor start-up he forgot to activate the stirrer. He starts it immediately, even though the reaction temperature has already been reached. The sudden mixture of reactants provokes a rapid increase in the reaction rate, causing the reactor to explode.
- Errors due to the operator's physical or mental incapacity to confront a particular situation. Examples of this are a manual valve stuck in such a way the worker does not have sufficient strength to make it work, or the case of an operator in the control room overloaded by having to supervise too much equipment.
- Errors due to a lack of motivation, or to a deliberate decision not to follow certain instructions. Obviously, in some cases an act of this type could be classed as industrial sabotage, but in other cases the operator can sincerely believe it is better not to follow specific instructions under certain circumstances.
- Errors attributable to erroneous company management policies. Within this group are lax policies on work permits to speed up repairs, installation of less reliable equipment or reduction in safety measures to cut costs, work systems that impose excessive pressure on workers, accident/incident investigation procedures that places the emphasis on looking for culprits, etc.

A clear demarcation does not exist between the above groups, and very often more than one of the circumstances mentioned contributes to an accident. Even in the case of accidents due to distraction of the operator, very often the design of the equipment could have been improved to prevent such accidents taking place. The following examples, from the work of Kletz [17], illustrate this possibility.

Example 6.6

Many aviation accidents have been due to pilot error, by erroneous operation of a control. For example, modern aircraft are equipped with ground spoilers, flat metal plates fixed to the upper part of the wings, which are raised after *touch-down with the object of reducing lift. If by error they are raised before touch-down they produce a sharp descent of the aircraft. In a DC-8 the pilot has two alternatives: (a) lift a lever before touch-down to arm the spoilers so that, once armed, they lift automatically on landing, or (b) wait until after touch-down and pull the same lever.*

One day a pilot pulled the lever out before touch-down. The result was the deaths of 109 people. The reaction of the US FAA was to suggest placing a sign on the cockpit alongside the spoiler lever stating 'deployment in flight prohibited'. They might just as well have put up a notice saying 'Please do not crash the plane'.

The accident was not, ultimately, a pilot error, but the result of bad design. It was inevitable that sooner or later someone would move the lever the wrong way. The aeroplane manufacturer did not take any initiative. Only after two or three more planes suffered the same accident did they decide to install locks to prevent the raising of the ground spoilers before the aircraft had touched down.

Example 6.7

The batch reactor in Figure 6.10 was used for a reaction under pressure. Once the reaction was finished, the pressure was reduced and the product discharged into a product tank. To prevent the discharge valve opening before time, an interlock prevented its opening until the pressure had fallen below a gauge pressure of 0.3 bars.

A batch failed to react and it was decided to stop the reaction and vent off the gas. The remote actuator of the discharge valve had been left in the open position, and the drainage valve was also open, so that when the pressure dropped below 0.3 bars the interlock allowed the valve to open. The result was a discharge of the flammable product of the reactor into the working area.

Obviously in this case a human error by the operator was the direct cause of the discharge. However, the accident could have been prevented with a protective system of better design. In this case, a HAZOP analysis would have been very useful in highlighting the defects in the system.

The previous examples demonstrate the difficulty of placing an accident within the category of human errors, even in simple cases where an accident is apparently due to the momentary distraction of an operator. It is even more difficult to predict the likelihood of human error, i.e. the probability that an operator confronting a certain scenario takes the wrong decision, or, having taken the right one, executes it incorrectly. In spite of this, there is a considerable amount of experimental work that allows an approximate estimation of the probability of human error in cases corresponding to relatively simple situations, where the decision–action

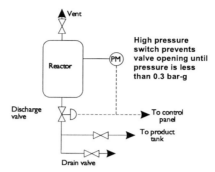

Figure 6.10 Arrangement of valves for the batch reactor of Example 6.7.

process takes place based on a limited number of alternatives. Presented in Table 6.3 are some examples taken from the literature.

On other occasions a more structured approximation has been followed, capable of differentiating to a certain extent the characteristics of each task. An example is the work of Bello and Columbiori [18], who proposed a method to take into account different factors contributing to human errors. The probability of error is

Table 6.3 Some estimates of human errors for simple tasks (selected from those quoted in Kletz [17]).

Probability	Operator action/omission
0.04	Fails to observe level indicator or take action
0.03	Fails to observe level alarm or take action
0.001	Fails to isolate pipeline at planned shut-down
0.005	Fails to isolate pipeline at emergency shut-down (time available, 30 min)
0.0025	Misvalving in changeover of two-pump set (stand-by pump left valved open, working pump left valved in)
0.01	Pump stopped manually without isolating pipeline
0.003	General human error of commission, e.g. misreading label and therefore selecting wrong switch
0.01	General human error of omission, where there is no display in the control room of the status of the item omitted, e.g. failure to return manually operated test valve to proper configuration after maintenance
0.003	Errors of omission, where the items being omitted are embedded in a procedure rather than at the end as above
0.03	Simple arithmetic errors with self-checking but without repeating the calculation by re-doing it on another piece of paper
≅ 1.0	Operator fails to act correctly in the first 60 seconds after the onset of an extremely high stress situation
0.9	Operator fails to act correctly in the first 5 minutes after the onset of an extremely high stress situation
0.1	Operator fails to act correctly after the first 30 minutes in an extremely stress situation

estimated in this case as the product of the factors $K1$ to $K5$ in Table 6.4. The numerical values in the table are orientative, and on occasion must be modified to match specific circumstances. Thus, for example, stress does not always increase the probability of human error. Swain and Guttman [19] consider that an optimum level of stress exists that corresponds to values of moderate stress. Above this level, human reliability decreases because of the accumulation of tension, but this also happens if the level is too low, due to a monotonous task.

In any case, performing a human reliability analysis beyond the assigning of orientative values is normally the work of specialists. The CCPS [3] sketches a procedure of 12 stages for human reliability analysis:

1. Familiarisation with plant (operation, displays and controls, etc.).
2. Review information from FTA (check branches of fault trees for human failures affecting the top event.
3. Talk-through (familiarisation with relevant procedures).

Table 6.4 Parameters used in the TESEO method (18), to calculate error probabilities (as quoted in Kletz [17]

TYPE OF ACTIVITY	$K1$
Simple, routine	0.001
Requiring attention, routine	0.01
Not routine	0.1
TEMPORARY STRESS FACTOR FOR ROUTINE ACTIVITIES	$K2$
Time available: 2 seconds	10
Time available: 10 seconds	1
Time available: 20 seconds	0.5
TEMPORARY STRESS FACTOR FOR NON-ROUTINE ACTIVITIES	$K2$
Time available: 3 seconds	10
Time available: 30 seconds	1
Time available: 45 seconds	0.3
Time available: 60 seconds	0.1
OPERATOR QUALITIES	$K3$
Carefully selected, expert, well-trained	0.5
Average knowledge and training	1
Little knowledge, poor training	3
ACTIVITY ANXIETY FACTOR	$K4$
Situation of grave emergency	3
Situation of potential emergency	2
Normal situation	1
ACTIVITY ERGONOMIC FACTOR	$K5$
Excellent microclimate, excellent interface with plant	0.7
Good microclimate, good interface with plant	1
Discrete microclimate, discrete interface with plant	3
Discrete microclimate, poor interface with plant	7
Worst microclimate, poor interface with plant	10

4. Task analysis (tasks are analysed and broken down into smaller, discrete units of activity).
5. Develop human reliability event trees (each unit task is expressed sequentially as binary branches of an event tree).
6. Assign human error probabilities.
7. Estimate the relative effects of performance shaping factors.
8. Assess dependence between tasks.
9. Determine success/failure probabilities.
10. Determine the effects of recovery factors (operators may recover from errors before they have an effect).
11. Perform a sensitivity analysis, if warranted.
12. Supply information to fault tree analysis.

One must also bear in mind that, in modern industry, automation has eliminated a substantial part of the repetitive human tasks, so that the job of the operator is more and more directed to supervision, and decision-making as problems arise that are not automatically resolved. Analysis in these cases involves information processing at various levels: diagnosis of the situation, evaluation of objectives, establishment of priorities and planning of corrective actions.

When an operator is confronted by an abnormal situation, for which there is no routine treatment, he has to develop a procedure to deal with the situation, and usually has to do so in a very limited time. He will make decisions based on his knowledge of the system responses, and start actions that are in fact experiments in real time. On assessing the results obtained he will persist in his course of action or try alternative ones. In this way, the errors are, inevitably, part of the learning mechanism when confronting new situations, and certain actions which when judged in retrospect are often classified as errors, are in reality reasonable attempts to obtain information about the state of events and their possible evolution. Thus, according to Rasmussen [20], in complex situations it would be more appropriate to consider human errors as 'unsuccessful experiments in an unfriendly environment', and concentrate the design effort in the development of systems capable of tolerating human error. In this kind of system errors should produce observable results, and the response of the system should be such that errors can be corrected before they produce unacceptable consequences. In other words, it is not possible to change the nature of people to reduce their errors beyond a certain limit but we can conceive designs in which human errors arise with less frequency, or have less prejudicial results. It is also necessary to help the operator to make decisions in non-programmed situations. Thus for instance, expert systems have been developed capable of assisting the control room supervisor by estimating the increase in operating risk as new events occur, e.g. the sequential or simultaneous failure of equipment after the initial failure of a unit [21].

6.6   Consideration of external agents

In addition to equipment and human error, which have already been discussed, a risk assessment study should take into account the additional risks originating from external agents, whenever these are relevant. Understood as external agents are all those that do not have a direct relationship to the process carried out in the plant, but are, however, capable of significantly increasing the possibility of an accident in it. Included in this group are impact by aircraft, ship or ground vehicles, wars and terrorist attacks, industrial sabotage, severe meteorological disturbances (storms, floods, hurricanes, etc.), impact from meteorites, fires initiated outside the plant, seismic activity, volcanic activity, etc.

Obviously, although the probability of such events is in most cases very small, they have the potential to cause major accidents or at least to start them, often involving common cause failures. The risk analyst should, therefore, decide which external agents to include in the analysis and which to leave out, justifying his decision in both cases.

In general, industrial installations are designed to withstand a certain intensity of the above-mentioned phenomena. Thus, Federal Safety Standards in the US require that the design of LNG facilities should resist critical ground motions with an annual probability of 10^{-4} or less, the worst flood registered in a 100-year period or the most critical combination of wind velocity and duration having an annual probability of 10^{-4} or less [3].

Quantifying the probability of external events is often difficult, because of the wide variation that exists with time and location. Thus, the probability of an aircraft impact varies according to the distance from an airport, that of earthquakes upon the seismic activity in the area and that of war or terrorist attack can vary widely depending on the changing political situation in a given country. In spite of this, there are probability estimates available, often with sufficient precision for deciding whether or not to include a particular external agent in a risk assessment study: There are statistics giving the probability of aircraft impact in a particular area, meteorological registers with wind histories, temperatures, precipitation, etc., as well as geological studies and seismic activity records (and volcanic where necessary), for the majority of areas with industrial interest. The probability figures quoted for aircraft impact in a one-year period are in the range 10^{-6} to 10^{-7}, depending on the locality [2, 5], and for meteorite impact or lightning strike, in the order of 10^{-11} and 10^{-7}, respectively.

6.7 Uncertainty of data and parameter sensitivity

When looking for failure frequency data to carry out a reliability assessment on a specific piece of equipment, one tries, whenever possible, to select data obtained from similar components, working in similar conditions. Often the data found is scarce and widely dispersed, and as a consequence the estimates of probability or frequency of failure for a given confidence level vary over wide intervals.

This fact has sometimes been used to put forward the case against quantitative risk assessment, with the argument that it is not worth the very considerable effort that is needed for a rigorous analysis, when all that will be achieved is a not very accurate probability estimate. In this respect it should be stated that, in spite of uncertainty in the data which the risk analyst often has to confront, estimates of the reliability of systems have been shown to be fairly accurate. Thus, the majority of the reliability estimates for systems are considered to be within a factor of 2 [22] or within a factor of 4 [2] from the real values. This is due in part to the fact that, in most cases, the greater part of the frequency or probability of the top event in a fault tree is contributed by a reduced group of events. This implies that great accuracy is not required when estimating the probability of every event in a fault tree, but only for those events having the greatest influence on the final result.

A **parametric sensitivity analysis** provides information on the influence that the probability of each of the events has on the top event. This is interesting not only to decide when it is necessary to make a special effort to increase the accuracy of the data, but above all to be able to concentrate the risk reduction efforts in the areas that reduce global risk more efficiently. The parametric sensitivity relating to the probability of a particular event j is defined as

$$S_j = \Delta P_T / \Delta P_j \tag{6.27}$$

that is, as the ratio of the change of the probability of the top event to the change of the probability of event j, which is the subject of the parametric sensitivity study.

Example 6.8

Find the parametric sensitivity of the top event in Example 6.4 with respect to each of the basic events involved, for a variation level of 50%. Assume that the estimates of probabilities for events A, B, C and E are, respectively, 0.02, 0.005, 0.02 and 0.0001.

As shown in Example 6.4, the Boolean expression that represents the fault tree is

$$T = C\,[AE + B)$$

therefore, the probability of the top event is $T = 1.004 \times 10^{-5}$. To calculate the parametric sensitivity we assume that the probability increases by a specified percentage (in this case 50%) in turn for each of the above named events, while the rest maintain their original values and we calculate the corresponding probability for the top event. The sensitivities are then calculated using equation (6.27). The following table lists the calculations.

Parameter modified	Initial value	Modified value	New value of T	$\Delta P_T(\%)$	S_j
P_A	0.02	0.03	1.006×10^{-5}	0.2	0.004
P_B	0.0005	0.00075	1.504×10^{-5}	49.8	0.996
P_C	0.02	0.03	1.506×10^{-5}	50	1.0
P_E	0.0001	0.00015	1.006×10^{-5}	0.2	0.004

In this case it can be seen that the sensitivity of T with respect to the probability of events B and C is much greater (some 250 times) than to the probability of events A and E. This result indicates that the efforts to improve the accuracy of the probability estimates should be concentrated on events B and C, because A and E have hardly any influence on the probability of the top event (in spite of A having a probability 40 times greater than E). In the same way, it would be of little use to spend money to reduce the probability of A and E (e.g. installing redundant systems, etc.), while it would be beneficial to reduce B or C, which have a direct influence on T.

6.8   Acceptability of risk

The final result of an FTA is a value of probability or frequency that can be expected for the top event. This information is combined with the estimates of consequences discussed in Chapters 3 to 5, to give a value of the statistically expected loss. The result of the risk analysis could be something like 'Under scenario B considered, the explosion of reactor R3 takes place, with the instantaneous release of its contents. If the explosion occurs a 60% probability of human fatalities exists (the reactor operator), and economic losses of approximately \$20 million can be expected. The estimated frequency for scenario B is 2.3×10^{-3} years^{-1}, that is once every 435 years.'

Risk analysis calculations end here (after a confidence interval for the expected top event frequency has been estimated). The decision as to whether the level of risk is too high or on the contrary may be considered acceptable (or tolerable) requires considerations beyond the merely technical. Discussed in Chapter 1 were some of the difficulties that may be encountered when deciding which levels of risk are tolerable. Risk perception (and therefore risk tolerability) is strongly influenced by factors such as the degree to which the risk is assumed voluntarily by the passive subjects, the direct benefits that the activity implies for the subjects at risk, whether the possible harmful effects are immediate or delayed, evident or concealed, and whether the consequences are reversible and the risk is known with sufficient precision.

Because the perception of risk depends largely on subjective factors it appears impossible to establish levels of tolerability that have universal validity. In fact, social susceptibility to industrial hazards has increased considerably over time, and can be very different in different countries or regions. In spite of this, some general guidelines exist that are useful to establish risk acceptability criteria.

6.8.1   Risks involving only material losses

Because it is impossible to reduce risks to zero level, any industrial company must protect itself from economic loss deriving from accidents. The normal way to transfer risk is through insurance. This may be a traditional insurance with an established insurance company, or can be of the self-insurance type. The work of Natale [23] discusses this and other ways of risk financing. Taken into account must be not only damage caused directly to equipment and installations but also the costs due to the interruption of production and the costs of civil responsibility, which can be quite high, especially when the accident involves injuries to people. A well-known example of the latter situation is the accident at Bhopal, after which Union Carbide had to confront litigation amounting to some 3000 million dollars.

When only material losses are expected, the combination of the concept of accident insurance and risk analysis permits the establishment of criteria for deciding risk acceptability. A certain risk level is not acceptable if there exist means of reducing it, whose implementation is profitable, taking into account the insurance premiums required to cover the predicted losses. Thus, if in the current state of the plant the cost of insurance premiums is M \$/year and after the installation of certain safety measures they would be reduced to N \$/year, the present level of risk is not acceptable if the cost of the safety measures is equal to or less than $(M–N)$ \$/year.

A similar concept is based on the calculation of the average annual cost of an accident. Let us assume that in the case mentioned at the beginning of this section there was no possibility of human injury and the only danger was the explosion of the reactor causing a total damage of \$20 million, with an estimated frequency of once in 435 years. The average annual cost would then be nearly \$46 000. Suppose that if a redundant safety system is added the total losses are maintained, but the frequency is reduced to once every 980 years. The average annual cost after the installation of safety measures would be some \$20 400. The difference, \$25 600/ year, is the saving obtained after the installation of safety measures. Therefore, the current risk level is not acceptable if the cost of the measures is lower than this figure.

Obviously both criteria are simplified, and the amounts mentioned can be corrected by taking into account interest, depreciation and maintenance of equipment, discounts for future expenses, additional benefit for improved company image, etc. However, the main obstacle to the installation of supplementary safety measures is not based on economic calculations,which can be made as sophisticated as necessary, but in the inertia of those who do not perceive the economic cost of a potential accident as a genuine cost, because, 'after all we are dealing only with approximate risk estimates', 'perhaps an accident will never occur (during the useful life of the plant)', and 'we have been working for 5 years and nothing has happened'.

6.8.2   Accidents with a potential for loss of life

When a significant probability of human injury exists, a decision on the level of tolerable risk becomes considerably more complicated. When dealing with the **risk to employees**, Kletz [2] suggests using the FAR (Fatal Accident Rate) index, which is the number of fatal accidents in a group of 1000 workers throughout their collective working life (approximately 10^8 hours). If we accept a FAR of 4 as being representative of conditions in the chemical industry, then the average worker is exposed to a FAR of 2 as a consequence of specific risks, i.e. those derived directly from the process in which he works (the other half of the FAR comes from the contribution of ordinary industrial risks such as falling down the stairs or being run over by a fork lift). Kletz suggests removing as a matter of priority all those specific risks as a consequence of which the FAR number for a particular job is higher than 2. This implies, as a starting hypothesis, that we have identified and evaluated all the risks to which the workers in the plant are exposed. If this is not the case, a conservative response would be to reduce any individual specific risk that gives an FAR value greater than 0.4.

As to risks for the public in general, it seems reasonable that people who are not plant workers have the right to expect that the additional risk they confront because of the industrial installation is considerably less than for those who work in it. The reason for this is that the general public are not aware of the risks involved nor have voluntarily accepted them and obviously do not have direct control upon the way plant employees handle the risks in the plant.

Again following Kletz [2], the frequency of fatal accidents for the general public (i.e. outside the plant boundaries), due to the activities of a particular industrial installation should be less than 10^{-7} years^{-1}. This figure is arrived at via various considerations: for example, the total probability of death for a young person considering all possible causes, is approximately 1 in 1000 per year (10^{-3} years^{-1}). An additional increase in the probability of death from industrial risks of 10^{-7} years^{-1} means an increase of 0.01% in the risk to which the average young person is already exposed. This value (0.01%) is certainly orders of magnitude below the uncertainty with which the previous value of 10^{-3} years^{-1} was estimated. On the other hand, a risk level of about 10^{-7} years^{-1} is attributed to very unlikely events (such as someone being injured by lightning, death on the ground by the crash of an aircraft, etc.), and is of course much less than other risks voluntarily assumed (e.g. some 1000 times less than the probability of dying in a car accident or 40 000 times less than the risk assumed by climbers).

Example 6.9

Apply the criterion proposed by Kletz to determine if the risk of death for the employee in the case discussed at the beginning of this section can be considered acceptable.

As shown at the beginning of this section the scenario assumed (explosion of reactor R3), has an estimated frequency of 2.3×10^{-3} years^{-1} with a 60% possibility of death. This gives us an estimated frequency of human loss of 1.38×10^{-3} years^{-1}. The working year considered in the FAR calculation has 2000 working hours, therefore the estimated frequency expressed as occasion/hour would 6.9×10^{-7} hours^{-1}. Extrapolation to 10^8 hours gives us a FAR number of 69, i.e. some 172 times greater than the recommended value of 0.4.

Note that this criterion must be applied to the worker exposed to the greater risk, not to the average worker in the plant. As Kletz [2] remarks, it is of little consolation to say to a worker 'do not worry, although the risk for you is high, the average risk for you and your fellow workers is low'.

6.9   Questions and problems

6.1   (a) Discuss the advantages and disadvantages of using a Weibull distribution instead of an exponential one. (b) Assume that the failure rate of an equipment is given by the Weibull distribution with $v = 3 \times 10^{-5}\,h^{-1}$, and $\alpha = 1.30$. What is the reliability after 4000 working hours? (c) What would be the failure rate of an exponential distribution giving the same reliability at this time? (d) What would be the reliability in cases (b) and (c) for $t = 6000$ hours? (e) Discuss the relationship between the parameters of both distributions.

6.2   The exit signal of an on-line gas analyser is used to initiate the emergency stop of a reactor whenever concentrations of a specified level are detected. A large investment has been made to ensure that, once the signal is received, the reliability of the emergency shut-down system is very high. A study of the system shows that a failure in the emergency stop procedure would almost certainly be due to a failure in the gas analyser. The catastrophic failure rate of the analyser (obtained from generic data) is 20.8 failures for each million working hours (this value also includes the automatic sampling system within the equipment's boundary). However the conditions of service for the analyser are of a severity higher than the average, operating at high temperatures with corrosive fluid. (a) Estimate a new failure rate, in years^{-1}.(b) Calculate the hazard rate for monthly and weekly servicing, if hazardous concentrations are reached once a month on average.

6.3   Calculate the hazard rate for the previous problem if two analysers were installed in parallel.

6.4   A HAZOP analysis of the system in Example 6.4 revealed the basic event E (a catastrophic failure in a control valve) can also cause the intermediate event 3, so that the inputs to the corresponding OR gate are now A, B and E. Obtain the new initial Boolean expression, the reduced expression and the equivalent fault tree.

6.5   Discuss what type of failure probability distribution would be adequate to account for human errors in simple tasks (e.g. open and close valves in the discharge of a cistern truck). How could you characterize human error in more complex tasks? How would the learning curve affect the observed failure rate?

6.6   The reactor of a hydrocarbon oxidation plant has a pressure relief system that uses two piloted valves in parallel. It was found that the catastrophic failure rate of this type of valve was 4 (fails to open on demand) per 1000 demands. Calculate the reliability of the system for a demand rate of 0.2 per month, assuming a six-monthly maintenance of the valves.

6.7   An FTA is performed on the system described in the previous problem, assuming that the annual probability of the reactor collapsing due to the most probable external event (aircraft impact) is about 5×10^{-7}. Discuss how many relief valves in parallel are required so that the probability of the external event is comparable to the probability of the reactor exploding due to overpressure.

6.8   Kletz [2] asks the following question: Consider the case of an accident A, that can result in the death of a person, with a probability of once a year, and of accident B, in which 100 persons could lose their lives, that has a probability of occurring once every 100 years. In both cases the risk of loss of life is the same. Should preference be given to preventing B over A? Why?

6.9   In 1991, the journal *Chemical Engineering Progress* carried out a survey on ethical attitudes of chemical engineers (CEP *87*(4), 62, 1991), proposing two hypothetical situations to be considered. In one of the cases, Tom, a young engineer, is promoted and transferred to a plant where safety measures are not too strict. In particular, there is a danger of a runaway reaction if the reactor temperature reaches 180°C. An emergency stop system is provided but is not very reliable and in fact several dangerous incidents have already taken place, including spillage of the reactor contents as a consequence of internal pressure increase due to a runaway reaction. Real danger of an accident exists, which could result in injury or death to the reactor operators. Tom brings this to the notice of his boss, presenting abundant documentation, and proposing an investment to improve the instrumentation of the reactor. The reply is that the cost is not justifiable due to the economic situation of the company, although Tom does not agree. In particular, the situation is difficult because of the attitude of his immediate boss, a man close to retirement who has no intention of changing things before he leaves. Discuss the actions Tom may take which were proposed in the Chemical Engineering Progress survey: (a) Do nothing, wait until his boss retires in three or four years (it is probable that Tom will then be promoted) and then solve the problems. (b) Report the situation to the safety inspector (to whom a false account of the past incidents was given) although it seems clear that this action would finish Tom's career with the company. (c) Try to convince his boss with more technical data. (d) Go over his boss's head, ignoring the normal procedure, and speak directly with the group vice president. (e) Look for another job.

6.10   References

1. Frankel, E. G. (1988) *Systems Reliability and Risk Analysis*, 2nd edn, Kluwer Academic Publishers, Dordrecht.

2. Kletz, T. (1992) *Hazop and Hazan. Identifying and Assessing Process Industry Hazards*, 3rd edn, The Institution of Chemical Engineers, Rugby.
3. CCPS (Center for Chemical Process Safety) (1989) *Guidelines for Chemical Process Quantitative Risk Analysis*, American Institute of Chemical Engineers, New York.
4. CCPS (Center for Chemical Process Safety) (1989) *Guidelines for Process Equipment Reliability Data*, American Institute of Chemical Engineers, New York.
5. Lees, F. P. (1980) *Loss Prevention in the Process Industries*, Butterworth-Heinemann, London.
6. Hauptmanns, U. (1986) *Análisis de Arboles de Fallos*, Ediciones Bellaterra, Barcelona.
7. Walpole, R. E. and Myers, R. H. (1991) *Probabilidad y Estadística*, 4th edn, McGraw-Hill/Interamericana de México, Mexico.
8. O'Mara, R. L., Greenberg, H. R. and Hessian, R. T. (1991) Quantified risk assessment, in *Risk Assessment and Risk Management for the Chemical Process Industry* (eds H. R. Greenberg and J. J. Cramer), Van Nostrand Reinhold, New York.
9. Medhekar, S. R., Bley, D. C. and Gekler, W. C. (1993) *Process Safety Progress*, **12**(2), 123.
10. Billington, R. and Allan, R. N. (1983) *Reliability Evaluation of Engineering Systems: Concepts and Techniques*, Plenum Press, New York.
11. Henley, E. J. and Kumamoto, H. (1981) *Reliability Engineering and Risk Assessment*, Prentice-Hall, Englewood Cliffs.
12. McCormic, N. J. (1981) *Reliability and Risk Analysis*, Academic Press, New York.
13. Fussell, J. B. (1976) Fault tree analysis. Concepts and techniques, in *Generic Techniques in Systems Reliability Assessment* (eds E. J. Henley and J. W. Lynn), Noordhoff International Publishing.
14. Battelle Columbus Division-AIChE/CCPS (1985) *Guidelines for Hazard Evaluation Procedures*, American Institute of Chemical Engineers, New York.
15. Schreiber, A. M. (1982) Using event trees and fault trees. *Chem. Eng.*, **89**(20), 115.
16. O'Mara, R. L. (1991) Calculation of human reliability, in *Risk Assessment and Risk Management for the Chemical Process Industry* (eds H. R. Greenberg and J. J. Cramer), Van Nostrand Reinhold, New York.
17. Kletz, T. (1991) *An Engineer's View of Human Error*, 2nd edn, The Institution of Chemical Engineers, Rugby.
18. Bello, G. C. and Columbiori, V. (1980) *Reliability Engineering*, **1**(1), 3.
19. Swain, A. D. and Guttmann, H. E. (1983) *Handbook of Human Reliability Analysis, with Emphasis on Nuclear Power Plant Applications*. US Nuclear Regulatory Commission, NUREG/CR-1278, Washington, DC.
20. Rasmussen, J. (1987) *Approaches to the Control of the Effects of Human Error on Chemical Plant Safety*. Proceedings of the International Symposium on Preventing Major Chemical Accidents (ed. J. L. Woodward), American Institute of Chemical Engineers, New York.
21. Arendt, J. S., Lorenzo, D. K., Montague, D. F. and Dycus, F. M. (1987) *Ensuring Operator Reliability during Off-normal Conditions Using an Expert System*. Proceedings of the International Symposium on Preventing Major Chemical Accidents (ed. J. L. Woodward), American Institute of Chemical Engineers, New York.
22. European Federation of Chemical Engineering (1985) *Report of the International Study Group on Risk Analysis. Risk Analysis in the Process Industries*, The Institution of Chemical Engineers. Rugby.
23. Natale, M. J. (1991) Risk financing, in *Risk Assessment and Risk Management for the Chemical Process Industry* (eds H. R. Greenberg and J. J. Cramer), Van Nostrand Reinhold, New York.

7 Risk reduction in the design of chemical plants

But he suddenly remembered something that one of the ship's designers had once said to him, when discussing 'fail safe' systems: 'We can design a system that's proof against accident and stupidity; but we can't design one that's proof against deliberate malice...

Arthur C. Clarke, 2001 A Space Odyssey, Chapter XXVIII

7.1 Introduction

There is a well-known proverb that says: 'Prevention is better than cure', and as already indicated in this book, the best way of preventing accidents of any type is by eliminating the possibility of them taking place. The reduction of risk should begin with the conception of a new process, designing intrinsically safe and easy to control plants.

Kletz [1] proposes six sequential steps to control the risks arising from the handling of hazardous materials:

- Do not use them (substitution)
- Use less quantity (intensification)
- Use them in conditions which make them less dangerous (attenuation)
- Confine them (to prevent leaks)
- Control the leaks (emergency blocking, facilitate dispersion, etc.)
- Defend against the consequences of leaks (fire protection, fire brigades, explosion-resistant buildings, etc.).

The application of these rules throughout all stages of a project, from research and development of the product and the process, to detailed engineering, passing through process engineering, can help to minimize the risks that necessarily, to a greater or lesser degree, are inherent in a chemical plant.

7.2 Research and development

This is the first stage in the development of any process on an industrial scale. If, from the start, the objectives of the investigation are established, and the intrinsic safety of the process is included in these, a considerable advance will have been made.

Frequently, processes which at laboratory level are not dangerous, due either to the small quantities of product used, to the type of material used (glass, special steels, copper, etc.) or due to the separation processes used, may be much less safe on a larger scale (large quantities, catalytic effect of some materials, impurities in raw materials, leaks, etc.). All of these aspects should be carefully considered in the change of scale, ensuring that the extrapolation implied is the minimum possible.

A number of cases exist of processes with improved safety through the investigation of alternatives, by working in conditions of moderate reaction, lower pressures or temperatures, elimination of dangerous intermediate products, a higher catalyst selectivity, with lower reaction times (and, therefore, lower quantities in the reactors), etc.

To be considered generally in this stage, within the degree of conciseness which can be attained, are aspects such as the chemical products to be handled, reactions which take place, working conditions, the basic operations necessary and the quantities of the products involved (inventory).

7.3   Project development

For the successful development of any project it is useful to establish clearly defined phases, which ensure that all of the information and necessary documents for the design are generated in time, and that they undergo the established revisions to verify that they fulfil the objectives of the process.

There are opportunities to reduce the risk in all of the stages, although the changes to be introduced are more effective and less expensive in the early stages than later on, when they require the repetition of work already carried out, the generation of new P&ID revisions or, even worse, when the equipment has already been bought.

The correct techniques, which have been described in the preceding chapters, should be applied at each stage in order to identify and evaluate the risks and determine the consequences. According to the results obtained, the appropriate actions can be taken in order to reduce the risk to an admissible level by changes in the process, design, control, equipment distribution or safety and protection measures.

At the end of each one of the stages which are dealt with from now on, an exhaustive check should be carried out on all safety aspects. In this check or audit people who are not directly involved with the project team should take part, and bring a fresh perspective to it. Experts in process technology and safety specialists should also participate. An appropriate team may include the project leader, the process and the project engineer, a manufacturing representative, the safety officer and a process or safety expert from a similar plant.

7.3.1   Scope definition

Study of alternatives

Before carrying out any project it is necessary to clearly establish both the desired product and the technology that will to be used to obtain it. Normally a project begins with an idea launched by a director such as, 'How much would it cost to make product X here?' or 'When could we have a Y plant running here?'. Then the person delegated sets the process engineering department to work to obtain an estimate of investment and a preliminary plan for the project in the shortest possible period of time.

Often there are significant obstacles to the complete development of this phase. Some of the most important are:

- The desire for a proven and 'safe' process, doing things 'as they have always been done', using well known, already developed process engineering.
- The necessity of offering the cheapest project in order to bring it to one's own plant (in large companies or multinational groups).
- The urgency to finish this first stage, which is often only considered as a bureaucratic step in order to begin the following stages.

At this stage important information for the plant design is still unknown or undefined. This phase is therefore often disregarded because of this uncertainty, when really it is the most critical for the final optimization of the plant. The use, traditionally extended, of the criterion of minimum cost of investment is the greatest enemy at this stage, and generally speaking in all of the stages. Although energy costs are more frequently included, maintenance expenses, the environmental impact and process safety should all be considered in these first stages. The evaluation of the different alternatives available is essential, but can be complicated; therefore, the critical aspects to be considered and their priorities should be previously established. This point is important due to the fact that usually a compromise between two or more of them will have to be reached.

Some aspects to be especially borne in mind are:

- Process safety: high pressures, exothermic reactions, high speed rotating machines, inventory of dangerous products, etc.
- Raw materials: toxicity, flammability, required storage capacity, etc.
- Storage of intermediate and final products
- Process controllability
- Liquid effluent production, gaseous emissions and solid wastes
- Community relations
- Space.

7.3.2 Process engineering

With the process now chosen, the range of possible improvements is greatly reduced, with numerous restrictions appearing. This greater certainty facilitates, on the other hand, the study of tangible alternatives to the weakest points.

In general at this stage one tries to design the process using safer equipment or basic operations, easier working conditions, more stable compounds, smaller and more compact equipment and any other measure which helps in our final objective [2].

There are difficulties to be overcome in all of the fields, for example: the belief in the necessity of the existence of large inventories to guarantee a good process control. Certainly large surge tanks mitigate fluctuations in the properties of streams, but often one can operate the processes without these inventories and the added risk they entail.

Some of the tools cited in previous chapters are of great help at this stage. HAZOP analysis, applied to the different stages of a project, is especially helpful in detecting the weak points of a design in time.

We shall concentrate on the peculiarities of some basic equipment and operations, beginning with a topic which affects all of them, the control of the plant.

Control strategy.   From the beginning of the design of the plant, the way in which it is to be controlled should be taken into account. Difficult start-up conditions or low process stability could cause a plant, which initially could appear to be intrinsically safe, to be more difficult to operate and, therefore more likely to suffer an accident. The risks can in general be reduced through an appropriate control system, which takes into account the predictable circumstances.

Increasingly plants are regulated by complex programmable control systems using computers. These systems help enormously in the control of plants but introduce a certain amount of complexity which make them rather inaccessible to a layman. Mistakes in the operation of the plant are often attributed to 'human' errors, when they are really due to the design of the control system [3], as discussed in Chapter 8. The incidents are often produced by small errors in the program, normally in areas which are not used very frequently (as they are usually the least tested). It is necessary that the person who carries out the programming has sufficient knowledge of the process, whether on his own or with the help of another project team member.

This is why the process engineer should bear in mind the existing control possibilities at all times. At the beginning of the project the basic control strategy should be established, including the critical parameters and the variables on which action will be taken. All conditions which could reasonably be expected in the plant operation, including start-up, normal operation and emergency shut-downs have to be considered.

The control systems and elements should always follow a **fail-safe** criterion, i.e. in the case of any instrument failure or lack of any supply they should remain in the safest situation. The typical case is that of control valves which in the case of power failure or lack of compressed air should remain open or closed, whichever is the most convenient.

Interlocks can be used to avoid carrying out incorrect operations. They prevent an operation taking place if a specified condition is not present, or activate a specific action if another condition is present. They are used frequently during start-up or shut-down procedures, loading and unloading operations with many valves involved and, in general, whenever there is an established operational sequence.

Also to be considered are the possible failures of instruments, whether owing to normal use or to installation errors, maintenance errors, bad atmospheric conditions, etc. Equally for the normal or emergency functioning of equipment, a study should be made to determine from the beginning which instruments should be considered as critical and which should be duplicated or triplicated for total reliability.

Alarms can often become a double edged sword. It is necessary for the control system to advise the worker when there is a fault or a decision is required. But, if there are too many at one time, or if they are repetitive, they could be ignored by the worker, with the consequent risks.

Risks with potentially more serious consequences can be prevented by using automatic action systems that actuate in an emergency **(trips)**. These systems should be especially reliable, and therefore they should undergo a special maintenance programme. Their function is to stop the process before it loses control. They can affect individual units or whole plants.

Example 7.1

A distillation column has a steam heated boiler, while the condenser is cooled by water coming from a cooling tower. In the case of a failure in the water supply, the pressure would increase until the opening of the safety valve, as the distillate cannot be condensed, producing a vapour emission. If a control system is installed which, before reaching the set pressure, closes the steam valve to the boiler, it will prevent or at least reduce the problem.

Example 7.2

In a storage tank an exothermic reaction may occur, which could cause the opening of the safety valve and the consequent venting. The temperature of the tank is controlled by a water cooled jacket. A system could also be installed which, when the temperature or pressure goes above a certain value, would open a valve at the bottom of the tank directing the contents to a distant controlled area. Another possibility is the addition of an inert compound which stops the reaction or cools the medium (quenching).

Design codes do not allow the substitution of regulatory emergency relief devices by control systems. In any case its size can be reduced through elimination of a grave scenario by being able to consider it improbable. It is the process engineer's responsibility to determine when a risk is sufficiently probable, in accordance with the rules described further on.

Pumps

In any chemical plant, however small, there are tens if not hundreds of pumps. The incorrect selection of the kind of pump to be used for a flammable or toxic fluid is a certain cause of leaks, due to the presence of rotating parts. The most critical point of a pump is the seal, the part that prevents leakage of the pumped fluid around the rotating shaft that connects the drive with the impeller. There are several types of pumps according to the kind of seal they use [4]:

- Pumps with single mechanical seal.
- Pumps with double mechanical seal: in tandem or series.
- Seamless or 'canned' pumps: both the motor and the pump are enclosed in the same casing, in such a way that sealing around the shaft is not necessary. The motor is immersed in the fluid being pumped.
- Pumps with magnetic coupling. As the name suggests, the shaft is substituted by a magnetic coupling between the drive and the impeller, eliminating the necessity of a seal. The motor is not in contact with the pumped fluid.

The use of seamless and magnetically driven pumps is becoming more and more frequent for hazardous fluids, as their higher initial cost is greatly compensated for by their extended life, low maintenance and safety. Single seals do not allow the detection of a leak, while with double seals it is detected because of the increase in the level of sealing fluid.

Under certain conditions the heat generated by the pump can be sufficient to start an undesirable reaction, therefore it is important to pay attention to situations in which the pump may be working for a long time with its discharge line blocked or at a very low flow.

Example 7.3

A centrifugal pump was handling a fluid which at 100°C decomposes exothermically. To prevent the pump being damaged if a valve downstream was inadvertently closed, a safety valve was installed in the discharge so that, in the case of overpressure, the pressurized fluid was returned to the pump input. When this situation happened, the pump recycled and heated the same fluid until an explosion occurred.

A safety valve should never be allowed to be used as a control element: it is emergency equipment. If it is desired to maintain the pump running without net forward fluid flow, the necessity of refrigeration should have been considered and provided through a set of valves and a cooler.

Motors and electrical equipment

Electrical equipment is a potential ignition source for explosive mixtures. In accordance with the existing fire and explosion risk in the place where it is to be installed an adequate protection level should be used. The classification of electrical motors and equipment according to their level of safety is defined in European Standards (EN) based on the recommendations of the International Electrotechnical Committee (IEC), which regulates the characteristics that each class should fulfil. In Europe the equipment installed in certain risk areas should be backed by a certificate of conformity with the corresponding European standard. The cost of electrical equipment increases rapidly as its level of safety increases.

Three classes of electrical areas are described according to the products which could be present in them:

- Class I: Flammable liquids (flash point below 37.8°C)
              Combustible liquids above their flash point
              Flammable gases
- Class II: Combustible or explosive dusts
- Class III: Fibres.

In Class I areas, two or three types of electrical zones are distinguished, depending on the standard being considered. According to the IEC there are three areas, and this is how they are defined in some European standards (UNE 20.322-86, VDE 0165), while others including the United States (API RP 500A and NFPA 70/497 A), do not define zone 0. These zones are:

- Zone 0: Areas in which there is an explosive atmosphere always or for long periods of time. The safety requirements for these areas are maximum. The use of intrinsically safe equipment (Ex)i is required. This kind of equipment is unable to cause ignition because the energy from any spark produced would be lower than that required to ignite whatever flammable mixture is present in the zone.
- Zone 1: Areas where there is frequently or occasionally an explosive atmosphere. Equipment with an anti-explosion enclosure (provided with a casing which prevents the spreading of an explosion), (Ex)d, is normally used. Another alternative is the (Ex)e type, of increased safety (with systems which reduce the probability of arcing or sparks forming in abnormal conditions).
- Zone 2: Zones where there is rarely an explosive atmosphere, and if there is, it is only for a short period of time. The same types as before are allowed.

Other types of electrical equipment exist: pressurized internally with gas, (Ex)f; those immersed in oil, (Ex)o, etc. Generally, the requirements for equipment of the same category to be installed in different areas are different and also depend on the products that are present. For a correct selection it is necessary to specify, apart from the type of electrical equipment, the area where it is to be installed and the type of gases or vapours to which it will be exposed.

Heat exchangers

Heat exchangers (HEs) are among the most frequently found equipment in a chemical plant. Use of the most suitable type and the correct allocation of fluids are of vital importance, searching for optimum compromise between cost, head loss and inventory.

Among the most common types are the shell and tube units and the plate heat exchangers (PHEs), the latter usually having a lower volume for the same heat exchange area, allowing the use of materials which are more resistant to corrosion (due to the lower quantity of material required for their construction) and are easier to clean. Their disadvantage is the need to find a suitable material for the joints and the low working pressures permitted (up to about 30 bar).

HEs are often designed for different duties, the final choice being made in the light of the capacity to handle the most unfavourable conditions. These cases are especially frequent in batch plants, where heating and cooling are done with the same equipment, or the process is carried out at different temperatures according to the type of product to be produced. In these cases it is necessary to carry out a simulation of the behaviour of the final design under all possible conditions. Also the effects of process or fluid changes have to be considered in advance to avoid dangerous situations.

More and more complex heat exchange networks are being designed to improve the energy efficiency of the plants. The methods used for its design, 'pinch' or similar, always assume steady-state situations. In all cases, and even more so in complex systems, it is essential to bear in mind the start-ups, partial shut-downs and particular situations of the different process units during its design.

Example 7.4

An exothermic reaction is carried out in a fixed bed reactor. To eliminate generated heat, the products are cooled in a HE with the feed, preheating it in this way to the necessary temperature to start the reaction.

If the reaction rate is not high enough at the temperature of the cold feed, it will be necessary to provide alternative heating media during the start-up, and design the system for both situations, not just for the worst case thermally speaking. On the other hand, if the reaction rate is significant at low temperatures, it will be necessary to bear in mind the reaction that happens inside the HE.

The use of fouling or safety factors is widespread. It is very common that plants are able to increase their capacity up to 50% or more by simply changing critical equipment, such as distillation towers, large compressors, reactors, etc. This is due to the fact that the less expensive elements are usually designed much larger than necessary. HEs are usually among these elements. Abuse of these factors can be very negative for the operation of the HEs, leading to:

• Inadequate speeds of the fluids, which, if too low, might lead to more fouling than normal.
• Accelerated fouling, caused by low speeds or high temperatures of the tube wall.

- Tube temperatures different to those expected, which favours fouling, formation of hot points or creation of cold areas which the high viscosity of the fluid will turn into dead pockets.

One of the weakest points of shell and tube heat exchangers are the expansion joints. When there is an important difference in temperature between the shell and the tubes, the difference in thermal expansion creates tensions which, unless flexible elements are introduced in the shell, can seriously damage the HE, producing deformation of the tubes, leaks in the joints between tube and tubesheet or even broken tubes. At the design stage the use of expansion joints can be avoided in most cases by:

- Changing the allocation of the fluids, in such a way that the shell and the tubes are at a similar temperature. It is very important to make a rigorous calculation of the temperatures.
- Using floating heads or 'U' tubes instead of fixed tubesheets. In this way the problem is eliminated, by leaving space for thermal expansion at one end of the tube bundle.
- Considering changing the materials, in cases where the tubes expand much more than the shell, for others with thermal expansion coefficients that compensate for this fact.

When one of the fluids is especially hazardous or toxic or they are incompatible, there exists the possibility of using a double tubesheet which detects immediately any leak at the tube to tubesheet joints and in which the two fluids cannot be mixed (Figure 7.1).

One of the most frequent causes of tube breakage in HEs is vibration. In most cases suitable design can eliminate this kind of problem. The TEMA [5] standards provide the rules to follow to detect possible vibration problems at the design stage.

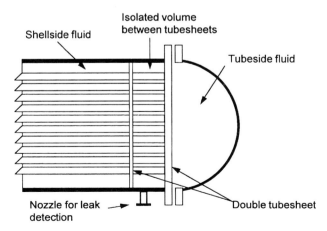

Figure 7.1   Detail of the head of a shell and tube heat exchanger with double tubesheet.

Normally the highest pressure fluid is placed on the tube side. When a tube breaks, pressurization of the shell takes place. Sometimes, when depressurizing the fluid on the tube side, a sudden boiling will occur. This possibility should be borne in mind by increasing the design pressure of the shell or in the design of the required safety valve.

In the case of incompatible fluids (for contamination, chemical reaction or toxicity reasons) the possibility of using an intermediate closed circuit with a heat exchange fluid compatible with both process streams should be considered. Periodic control of the quality of the heat exchange fluid allows leak detection before process contamination occurs. A higher pressure in the process side will prevent contamination of the process stream with the heat exchange fluid.

Shell and tube HEs are considered to be pressure vessels, for which it is legally required [6] to have a safety valve on each side (the design of these valves will be discussed in detail later in the book). In the case of plate HEs the installation of safety valves is not a legal requirement. This does not mean that they should not exist, in order to protect their integrity and for safety reasons, an aspect covered by existing regulations. They are not regulated because the most important design codes do not consider them as pressure vessels, because they are a group of plates and joints held together in a frame by bolts.

Thermal and refrigerant fluids.   Although usually separated physically from the process, in certain circumstances they can be in contact with it. The correct selection of the heat exchange fluids and their temperatures can completely eliminate some dangerous situations.

Example 7.5

A polymerization reactor works at 120°C, heated by saturated steam at 13 bar. At 150°C an undesirable reaction commences, and therefore it is especially important to stay below this temperature.

The use of saturated steam at 13 bar (190°C) to heat the reactor can overheat the reactant mixture to above the starting temperature of the undesired reaction, so that it would be safer to reduce the steam pressure, in such a way that it could never exceed the temperature considered as safe.

It is thought [1] that something similar could have occurred in the Seveso accident in 1976 (see Appendix). The starting temperature for the undesired exothermic reaction of the reactant mixture was about 190°C. The reaction was done in a stirred jacketed reactor, heated with steam at 190°C and at atmospheric pressure so that the boiling point of the liquid was 160°C and therefore the starting point of the undesired reaction could not be reached. When the agitator was stopped and the reactor was left without cooling over a complete weekend there were probably points (especially those closer to the liquid surface where the metal wall was hotter, as it was in contact with the vapour phase which has a lower thermal conductivity than the liquid) where the temperature exceeded the allowable limits and a runaway reaction was started. The safety measures installed worked, discharging a cloud of vapours of some two tonnes, containing approximately 1 kg of dioxin, which gave rise to the consequences, which are well-known.

Whenever possible the use of flammable oils or hydrocarbons as thermal fluids should be avoided. They can often be satisfactorily replaced by other types of non-flammable compounds or even by steam or hot water. The latter is usually the easiest and the cheapest solution. One must bear in mind that the thermal oils used as a heating media usually need a special boiler, apart from the steam boiler normally present in all plants, with the additional risk of a duplicity which, sometimes, may be unnecessary. In many cases the possibility of increasing the overall steam production efficiency with the installation of a plant for the co-generation of heat and power (CHP) can be economically beneficial.

The ethylene or propylene used in some cooling systems can be replaced by ammonia and other compounds from the freon family. In this way the inventory of flammable products can be considerably reduced, although preventive measures to protect the environment are still necessary.

Distillation

Distillation columns contain large quantities of boiling liquid, which can be a real hazard if it is flammable or unstable. In a column, the liquid accumulates in three zones: the condenser and reflux tank, the column bottom and the plates or packing.

Depending on the kind of packing or plates in the column, there can be significant differences in liquid retention. Kletz [1] gives values of 40–100 mm per stage for plates, 30–60 mm for normal packing and less than 20 mm for film plates. This factor must be taken into account when selecting the kind of packing for a distillation column, in addition to the efficiency and pressure drop. In general, the lower the retention of liquid per stage, the lower the inventory, although the column diameter also has an influence.

The reflux tank and the condenser can sometimes be integrated in one piece of equipment, with the reflux pump connected directly to the condenser level. This alternative leads to the elimination of the 'traditional' reflux tank. Although this has a function, acting as a feed tank for the reflux pump and reducing flow fluctuations, the correct design of the condenser, providing a sufficient level of liquid on the shell side and passing the coolant inside the tubes, can also ensure a steady flow. Its size can also be reduced without eliminating it. Another alternative is to use dephlegmators (HEs inside the distillation tower which are also additional equilibrium stages).

The integration of the reboiler at the base of the column or the reduction in diameter (and, therefore, the capacity) of the tower bottom will help to reduce the product inventory. Bayonet-type HEs, inserted laterally at the base of the column, can be used, or even thermo-siphons, although these still maintain an important inventory. Reducing the diameter at the base of the column hardly affects its mechanical design, and maintains the same pump suction head with a much lower amount of product.

Alternatives to distillation which operate in less hazardous conditions, such as absorption and membrane processes, are being developed. However, their application is still infrequent, and the day of their use by large ethylene plants to substitute in the separation section the distillation process by other unit operations is still not close. Meanwhile it is necessary to increase the safety levels in the design of distillation columns or even use non-conventional distillation (for example, distillation in a centrifugal field [1]).

Storage

Special attention should always be paid to the possibility of eliminating or reducing storage of dangerous products. The reasons usually given for the necessity to store large quantities of products (whether final or intermediate or raw materials) in process or in reserve are flexibility of production, ease of control or strategic. Apart from the reduced risk, financial reasons and production management (Just in time systems) justify, including economically, the reduction of stocks. It has been shown that the loss of flexibility is minimal and what is really necessary is a change in the way of working.

A certain storage capacity is usually required. From this point the method of minimizing risk consists of choosing the safest storage conditions, including the materials to be stored and the installations:

- Storage of another less reactive chemical compound that may be easily converted to the active agent required, in order to avoid risks of uncontrolled chemical reaction. Storage of pastes or slurries instead of dusts to avoid the risk of dust explosions. Storage in a different concentration or with another solvent.
- The pressure, temperature, inerting, isolation from the environment, type of tank, and other design elements can be selected in such a manner that the final risk is minimized.
- Containment systems can impede the extension of the consequences to other areas of the plant.

There are local regulations on the safety measures to be taken for different kinds of products (flammable, corrosive, etc.). The Seveso 2 Directive [7] establishes different requirements depending on the amount of product stored or in process.

Reaction

This constitutes the centre of the majority of chemical processes. All of the aspects which have been covered up to now are applicable to the design of reactors. In addition there are aspects exclusive to these, such as stability, kinetics and heat generation. Due to the extreme complexity involved in the selection of the most suitable reactor and the great variety of factors to be considered in each case, we will not give a description of the characteristics of each reactor type. There are several chemical reaction engineering texts that discuss this subject in depth [8–10].

In some cases the high reactivity of a compound is reduced thanks to the use of large quantities of solvents. When these solvents are flammable, the risk of fire is greater with this option, so that it is convenient to determine which risk is preferred.

Gas-phase reactors usually contain a lower inventory than liquid-phase reactors, the same applies to continuous versus batch reactors .

Very often different criteria oppose their respective solutions. This is the typical case of an exothermic reaction showing a runaway decomposition above 180°C. Operation well below this temperature (120°C) requires that the continuous stirred tank reactor (CSTR) used has a greater volume in order to maintain the desired conversion, with the resulting increase in risk, the inventory could be reduced working at 160°C, although the risk of losing control increases. In almost all cases concerning safety of reactors, the control strategy plays an important role, allowing the control of the relevant reaction parameters within the safe zone.

Many serious accidents have occurred due to loss of control of reactors, usually with exothermic reactions. In these processes the energy liberated by the reaction increases the temperature of the reacting mixture, increasing even more the reaction rate and thus the temperature, reaching a point at which an explosion or a loss of vessel containment can occur if adequate emergency systems do not exist.

The characteristics of reacting systems and the design of pressure relief systems for runaway chemical reactions will be discussed in detail later on.

7.3.3 Detailed engineering

This is the final stage in the design of a plant. Here the calculations, specifications, diagrams and drawings pass from paper to reality. Knowing how to make this transformation in the best and safest manner is always the goal for the project engineer.

This job is done by the specialists in each section: pressure vessels, piping, instrumentation, rotating equipment, electricity, civil, etc. Many of the procedures to be followed are covered by codes or design regulations.

At this point co-ordination reaches its maximum importance. The quality and punctuality of the information needed by each team for their work are critical. At any given moment, various specialists could be working at the same time on the same piece of equipment. If there is no fluid exchange of information, errors can occur, apparently small, but that could affect the safety of the plant in the future.

Minor details.   Many minor details of design or installation can pass unnoticed and can cause problems later in the operation, or cause later modifications, always more costly. They are normally obvious observations, but which, however, can pass unnoticed due to the great volume of information managed in a project and the ease with which things are done on paper. A typical case is the accessibility to equipment requiring maintenance, such as instrumentation, safety valves, manholes, etc.

A few examples are given below. This is not a comprehensive list, which is outside the scope of a book like this.

1. A typical case is the tank in Figure 7.2. If a small hole is not made in the highest point of the filling pipe of the large tank, its contents can be siphoned to the small one when the pump stops.
2. The thermocouples should never be introduced directly into the process if it is a flammable fluid. Many accidents have occurred when unscrewing a thermocouple for maintenance a leak has occurred with a consequent fire. The use of a welded or flanged thermal jacket in which the thermocouple is inserted, without direct contact with the process fluid, reduces the probability of these incidents.
3. The use of threaded connections should be reduced to the minimum, because of the greater risk of leaks and breakage. Equally, the use of small diameter pipes at ground level (25 mm or less) for hazardous fluids should be reduced, because of their fragility.
4. Check valves (one-way) are often used to prevent flow in a given direction. If it is imperative to prevent this situation, these valves should not be relied on, because of the high risk of leaks, and should be substituted by an alternative system.
5. Much equipment can be mounted in two positions: control valves, check valves, flow meters, etc. In the case of wrong mounting, the majority of them lose precision or are useless.
6. Isolation of control instrumentation should be foreseen for safe and accurate calibration and adjustment. One way is the block and bleed system shown in Figure 7.2.
7. Always when practicable, items made of glass such as sight glasses, tube levels, rotameters, etc. should be avoided because of their fragility.

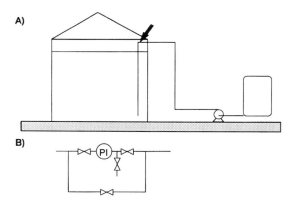

Figure 7.2   (A) Siphon breaking hole; (B) double block and bleed.

8. Static electricity is a source of ignition for flammable mixtures. The spillage of liquids should be avoided, using filling tubes that enter below the minimum level of liquid in tanks and process vessels. When necessary the system should be grounded, or it may even be necessary to install humidifying equipment to increase the conductivity of the air.

Very often, when designing modifications for existing installations all their details are not known and errors are committed. Haste, or the absence of good documentation, are frequently the cause of these minor errors.

A typical case is the spillage at Sellafield [3]. Probably due to a modification, on occasion a 50-mm line was used to return to the plant waste that could not be sent to the sea, sharing part of another 250-mm line used to dump acceptable waste into the sea. In the 250-mm section, part of the solids decanted, due to the low fluid velocity in this area, and were later entrained into the sea. The official report stated 'human error'.

Layout

The layout, i.e. the distribution of equipment in the space available, is the result of the correct interpretation of a large number of variables. It is therefore difficult to define rules that systematically lead to the optimum distribution of a plant. From the wind, passing through maintenance, access, neighbourhood, plants or equipment already existing, storage, loading and unloading, offices, to the interrelation of the different units, all aspects are united under the global safety of the plant or complex to determine the best way of distributing the process units of the plant. Mecklenburg [11] and Kern [12] have covered this theme thoroughly.

The first enemy of a safe layout is the lack of space. Adequate separation is necessary, not only between units and offices but also between the plant and its neighbours. Moreover easy access to all points of the plant where maintenance, fire prevention or transport is required must be guaranteed. The process units must be accessible by at least two routes. It is recommended that all zones requiring access by moving vehicles are located on the periphery of the complex.

Safety concerns in the layout development, in a first instance, are commonly expressed in the form of minimum distances between process units. In the first stage it is necessary to group the equipment in process groups that should be together, e.g. a distillation unit, a hydrogenation unit, etc. An order of priority between the process groups should be established according to their safety requirements. Within these groups a distribution according to recommended distances is elaborated (Table 7.1).

In Table 7.2 the recommended distances when making a preliminary layout are given. These are estimated distances, and should always be revised when there are reasons which indicate it is convenient to do so. Especially in the case of storage tanks, reactors or special fire or explosion hazards, the distances should be carefully calculated, according to the scenario foreseen. In this way a more rational basis for the layout is given if the distance from a unit is calculated bearing in mind, for example, the shock wave or flying debris formed if it should explode.

Table 7.1   Minimum recommended distances between units (from Wells and Rose [13])

	From similar unit (m)	From the closest units (m)	From a possible ignition source (m)
Pressurized storage			
Ethylene	15*	60	60
C3	15*	45	45
C4	15*	30	30
Storage of flammable liquids			
<400 m³, flash point <66°C	5+	15	30
>400 m³, flash point <66°C	8+	45	60
flash point >66°C	5+	45	30
Oil pipe	7.5	8	30
Distillation of light hydrocarbons	7.5	8	60
Catalytic reforming	15	30	30
Alkylation unit	30	45	60
Autoclaves of >20 bar and >0.3 m³	7.5	15	30
Hydrogenation >68 bar	15	15	20
Catalytic polymerization	15	15	45
Thermal polymerization	15	15	30
Naphtha caustic washing	7.5	7.5	30
SO₂ extraction	15	15	30
Wax centrifugation	15	45	60
Boiler	NL	30	NL
Fire-fighting pumps, hoses	NL	45	NL
Cooling tower	NL	30	NL
Waste water decanter	NL	20	20
Venting chimneys	NL	15–30	60
Buildings in general	NL	10–60	NL
Flares	NL	60	NL
Loading terminals	NL	15–60	60
Offices and canteens	NL	30	NL

NL - No limits
* Use 1/4 of the sum of the diameters of the adjacent tanks if it is greater.

Table 7.2   Recommended distances between equipment for preliminary layouts. (From Wells and Rose [13])

	From similar units (m)	Free horizontal space (m)	Free vertical space (m)
Pumps	0.8–1.5	0.8–1.5	4
Compressors	average width	3	4
Distillation column	5	1.5	–
Vertical vessels	1/2 diam.	1.5	–
Horizontal vessels	1/2 diam.	2	1.5
Heat exchangers	1–1.5	1.5	1
Furnaces	3	3	–
Reactors, stirred vessels	7.5	1	–
Centrifuges, mills	5	3	–
Main streets to battery limits	–	9	5.5
Secondary streets to battery limits	–	7.5	5
Rail roads to battery limits	–	4.5	7
Main pipe rack	–	4.6	4.9
Secondary pipe rack	–	3	3.7
Other elevated pipe racks	–	3	2.1

More and more, a detailed hazard evaluation is required instead of the application of these tables. Once one or more alternative preliminary layouts have been developed, an exhaustive risk analysis should be carried out. According to the conclusions of the risk analysis the best layout is selected, or the original proposal is improved. The process is stopped when it is considered that the risk is sufficiently acceptable, or it cannot be reduced any more.

To calculate the risks or the safe distances in a thorough way the consequence analysis techniques described in previous chapters can be used. For instance, the Dow Fire and Explosion Index can be used to determine the risk existing in different distributions of the plant, and discriminate between the different possibilities.

Some practices widely accepted in the past are being substituted by safer ones. Normally, storage tanks used to be surrounded by a dike of adequate height to contain any spillage that could be produced, plus an estimated volume of fire-fighting water. The problem with this design is that it confines the fire close to the tank or tanks within the dike. Now the best practice is the substitution of dikes by adequate slopes and trenches to conduct the spill to an area sufficiently away from other equipment, where it can be better controlled.

Design codes. Throughout the chapter numerous design codes and standards applicable to the design of a chemical plant have been cited. In each country there are some which are obligatory, like the Pressure Vessels Code [6, 14], regulations for the storage of chemical products, electrical equipment regulations, etc. These should be clearly distinguished from those that are merely recommendations or good practice codes. In Chapter 10 of this book, together with some examples of current legal requirements, some standards and recommended practices that could be of interest are included. Addresses of institutions and professional organizations that can give information and Internet sites are also included.

7.4   Emergency relief systems

When it has not been totally possible to eliminate a hazardous situation during design, it is necessary to adopt safety measures that prevent major accidents. A typical case is the increase in pressure above that which a vessel can withstand.

The pressure vessels used by the chemical industry (tanks, heat exchangers, distillation columns, gas–liquid separators, etc.) are designed to withstand a certain pressure (called design pressure) at a certain temperature (also called design temperature). Although the codes and standards used for the calculation of their resistance use safety factors of four or more, the majority of them require the installation of pressure relief devices to prevent dangerous situations.

Essentially, a pressure relief device consists of a closing mechanical element, elastic or rigid, whose resistance is surpassed at a certain fixed pressure, allowing the opening of a path sufficient to provide passage of a certain flow of fluid, which has to be enough to prevent the pressure from continuing to rise above the acceptable limit.

Spanish [6] and most European law demands that all pressure vessels within its scope are equipped with emergency relief elements designed according to the design code chosen (ASME, DIN, AD-Merkblatter, BS, etc.). An element is normally defined as the equipment or group of equipment not separable by devices that permit the closure of its connection. This would be the case of a distillation column and its condenser if no valve existed between them.

Generally an emergency relief element is required:

- In all vessels or piping elements that may be exposed to a pressure or vacuum exceeding their design limits.
- In storage tanks (either for pressure or for vacuum situations).
- For the discharge of positive displacement pumps, compressors or turbines.

7.4.1   Terminology

There exists a series of specific terms, some of them defined in the standards and regulations applicable to design, that it is necessary to define. Shown in Figure 7.3 are the equivalencies and interrelationships of the different parameters.

The pressure at which an emergency device is designed to open is called the **set pressure** (P_{set}).

The maximum permissible pressure in the vessel at the design temperature is the **design pressure** (P_d) or **maximum allowable working pressure** (MAWP). The maximum pressure increase above the design pressure, normally expressed as a percentage over the manometric design pressure, that is permitted during an emergency venting process is called **accumulation**. The ASME code allows 33% for pipes, 21% in the case of fire and 10% for other cases. The equivalent of the accumulation, but referred to the set pressure, is called **overpressure.** The pressure existing in the discharge side of a safety valve is the **backpressure** (P_b). It can be superimposed or due to the flow. The first is constant, for example, the atmospheric pressure, or that existing in a flare header. The second is that due to the head loss in the discharge line of the valve, and only exists during a venting situation. The difference between the set pressure and the design backpressure is the **differential set pressure**. The difference between the set pressure and the pressure at which the valve closes again (i.e. the pressure decrease necessary for the valve to re-close), expressed as a percentage of the set pressure (manometric), is called **blow-down**. A typical value is 6% of the differential set pressure (or the manometric set pressure, if the only backpressure is the atmosphere).

7.4.2   Determination of the design scenario

The first step to be considered in the design of an emergency relief system is the scenario: for what situation are we designing the system? The design scenario

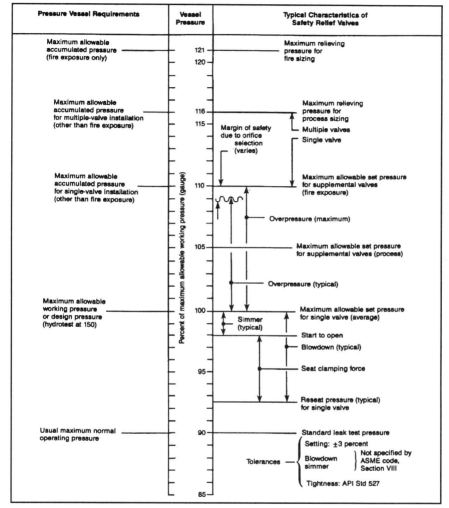

Figure 7.3 Terms used and relationships between the different pressures used in the design of safety relief systems. According to API RP 521, 1990 [15] (courtesy of the American Petroleum Institute, Washington, DC). The operating pressure may be any lower pressure required. The set pressure and all other values related to it may be moved downward if the operating pressure permits.

establishes the conditions under which we design the emergency relief devices. For the final system to be safe, the final scenario should be the event or chain of events that, being credible, require the biggest emergency relief element.

There are usually several possible scenarios for each equipment. But with which criteria do we determine whether a scenario is 'reasonable' or not? Complex and time-consuming methods exist for calculating the probability of an event occurring, such as FMEA, FTA and ETA (Chapter 2). It is important to make clear at this time that we are not looking for accuracy, because we will always select the worst case, although it may be less likely than others less serious, thus we simply wish to eliminate really remote possibilities.

In the majority of cases it is more practical to establish general criteria to be followed, which permits a quicker discrimination of the credibility of a scenario. A detailed probability analysis is left exclusively for the specially grave cases, whose importance requires more time dedicated to analysis.

Presented below are some general rules that may be of use, bearing in mind the greater care necessary in specific cases that should be treated according to the methods described in Chapter 6:

- All single events that may occur are credible scenarios.
- Those requiring the simultaneous coincidence of two or more independent events are not credible scenarios.
- Scenarios requiring a sequence of more than two independent events are not credible.

For example, credible scenarios are:

- A fire that affects the exterior of the vessel under study, provoking the boiling of its contents.
- The water inlet valve of a reactor's cooling system becomes blocked in the closed position.
- There is a fire while a vessel is full and blocked, with consequent thermal expansion.
- That, as the consequence of a fire, a control valve and a pump fail (simultaneous but not independent).
- The breaking of a control valve, causing it to open completely allowing ingress of a fluid at high pressure.
- The breakage of a tube of a heat exchanger.
- An operator makes an error.

Highly improbable scenarios would be:

- A fire is declared and at the same moment a control valve fails for a different reason.
- Two valves fail simultaneously for different reasons.
- Two alternative cooling systems fail simultaneously.

Shown in Table 7.3 are some causes that may be used as a scenario development aid. The development of a good group of credible scenarios demands intuition and knowledge of genuine problems in plant operation. A series of 'standard scenarios' exists that can always be applied to each kind of equipment, but normally the most important are specific to a process or a chemical reaction. The HAZOP methodology applied during the design stage can help in the detection of hazards not previously considered and in the development of new scenarios.

Once the apparent worst scenarios have been identified, the relief flow required for each of them must be evaluated.

Table 7.3   Some typical design scenarios for pressure relief systems.

Concept	Causes	Required capacity
Chemical reaction	• Runaway reaction • Stirrer failure • Lack of inhibitor • Mixing of incompatible compounds • Air intake • Generation of non-condensable gases	Quantity of gas or vapour generated by the reaction or associated with the heat released in the reaction
External heating	• External fire • Fully open steam valve • Electrical heating	Quantity of vapour or steam generated
Loss of cooling	• Loss of coolant • Loss of reflux • Lack of power supply to air-cooler	Quantity of vapour or steam generated and not condensed
Line blocking	• Blocked line • Outlet pump failure • Outlet valve closed	Outlet flow
Excessive flow	• Broken control valve • Controller failure • Broken tube	Difference between operating and emergency conditions
Thermal expansion or contraction	• Process blocked and cooling (or heating) working • Temperature changes	The required according to the variation of volume per unit of time
Entrance of volatile products	• Entrance of water or light hydrocarbons in tanks or units filled with heavy oils	Quantity of vapour generated
Utility failure	• Power failure • Cooling water failure • Steam failure	Difference between the vapour generated in operating and emergency conditions

The basic principle to apply is that the system pressure does not exceed the allowed accumulation. The practical application of this principle is usually simple, provided that the system is simple. The API Recommended Practice 521 (RP521) [15] gives a guide for the selection of scenarios and calculation of required relief flows. In many cases guaranteeing the elimination of the vapour or gas flow generated at accumulation pressure is enough. In cases of fire, the NFPA 30 standard [16] includes the method of calculating the heat input to the affected equipment.

Where complex systems are involved, with significant variations of temperature or composition, a specific approach to the problem is necessary, including material and energy balances, and from this calculating the relief flow. When calculated in this way, one finds that the required flow is not always the same, but varies as the venting takes place. It is important to remember that greater flow does not always bring the greater relief element (due to the influence of pressure, temperature, composition, vapour quality, etc.), so that the criteria to bear in mind to determine the design conditions should be the largest orifice area required.

Example 7.6

Calculate the relief flow required in the case of a fire for a horizontal cylindrical tank 5 m long and 3 m in diameter.

$\Delta H_v = 300\,kJ/kg$. The tank is surrounded by a dike, it is not insulated and fire protection systems have not been installed.

Following the NFPA 30 standard [16], we first calculate the external surface (assuming the heads are approximately hemispherical):

$$A = \pi DL + \pi D^2 = 75.4\,m^2$$

According to the type of vessel, the surface exposed to fire (A_{exp}) which should be used in the following calculations is a percentage of the external surface, as shown below

Sphere	55%
Horizontal tank	75%
Vertical tank	100% up to 9.1 m height, excluding top head.

In our case $A_{exp} = 0.75 \times 75.4 = 56.6\,m^2$

Now we calculate the heat absorbed from the fire. There are five different formulas, depending on the area exposed, which give Q (W):

$A_{exp} < 18.6\,m^2$	$Q = 63\,092\,A_{exp}$
$18.6\,m^2 < A_{exp} < 92.9\,m^2$	$Q = 224\,168\,A_{exp}^{0.566}$
$92.9\,m^2 < A_{exp} < 260\,m^2$	$Q = 630\,353\,A_{exp}^{0.338}$
$A_{exp} > 260\,m^2$ and $P_d > 0.07$ bar rel.	$Q = 44\,192\,A_{exp}^{0.82}$
$A_{exp} > 260\,m^2$ and $P_d < 0.07$ bar rel.	$Q = 4\,103\,000\,W$

Applying the second of these we obtain

$$Q = 2.20\,MW$$

Now, the mass flow to be relieved (W) is calculated as

$$W = FQ/\Delta H_v$$

where ΔH_v is the enthalpy of vaporization of the product and F is a 'credit factor', depending on the safety measures adopted:

Adequate drainage, according to NFPA 30 (includes a
  minimum slope of 1% directed to a remote area) and
  A_{exp} more than 19.6 m² $F = 0.5$

Drainage plus fire-resistant water spray (NFPA 15) $F = 0.3$
Fire-resistant insulation (NFPA 30) $F = 0.3$
Drainage plus spray plus insulation $F = 0.15$
None of the above $F = 1$

Taking $F = 1$ (the dike is not valid as protection) we obtain the flow that would be necessary to vent in the case of fire as

$$W = 7.33 \text{ kg/s} = 26\,400 \text{ kg/h}$$

Example 7.7

For cooling the contents of a tank, a coil of 20 mm internal diameter is used. Calculate the relief flow necessary in case the coil should break if the set pressure of the tank safety valve is 4 bar g and the water pressure is 7 bar g (density $= 1000 \text{ kg/m}^3$).

We will assume a total rupture of the coil which will produce a flow from both of the open parts. We will calculate for one of the openings and at the end will double the result.

For a sudden expansion the head loss is given by the equation of Borda–Carnot, therefore for expansion to zero velocity ($A_1/A_2 = 0$)

$$-\Delta P = \rho \, (u_1^2/2) \, (1 - A_1/A_2)^2 = \rho \, u_1^2/2$$

where $(-\Delta P)$ is the pressure drop available, ρ is the fluid density and u_1 is the velocity in the pipe.

Taking an overpressure of 10%, the relief pressure will be 4.4 bar, therefore in our case $-\Delta P$ will be $(7 - 4.4)$ bar, equivalent to 260 kPa.

Substituting and resolving, u_1 is 22.8 m/s, equivalent to, multiplying by the section, 0.0072 m³/s, or 25 800 kg/h.

Although the velocity is very high, this is because it corresponds to the instant immediately after the rupture.

Bearing in mind the two broken ends, the total flow to be relieved will be 51 600 kg/h.

Example 7.8

Calculate the relief flow required should a water control valve break and become completely open, supplying water at 10 bar absolute to a vessel with a safety valve set to relieve at 6 bar absolute. C_v of the valve = 0.000015 m³/s Pa$^{1/2}$. Density 985 kg/m³.

The flow through a control valve can be calculated using a similar equation to that of flow through an orifice, i.e.

$$Q = C_v F(x) \sqrt{\frac{\Delta P}{\rho / \rho_{ref}}}$$

where $F(x)$ is a function of the valve opening that varies from 0 to 1. ρ_{ref} is, if not specified otherwise, the density of water, $1000 \, kg/m^3$. C_v is the valve capacity coefficient (dimensional), a value normally supplied by the manufacturer. If unknown, it can be estimated from the seat diameter (D) as

$$C_v \; (m^3/s \; Pa^{1/2}) = 0.022 \, D^2 \quad (D \text{ in m})$$

The relief pressure, using an overpressure of 10% will be

$$P = (6-1.013) \, 0.1 + 6 = 6.5 \, bar$$

because the overpressure is always in gauge pressures. Applying the formula for $F(x) = 1$, $\Delta P = 350\,000$ Pa, $\rho/\rho_{ref} = 0.985$ and a value of $0.0090 \, m^3/s$ or $31\,900 \, kg/h$ is obtained.

7.4.3 Selecting the type of element

There are many kinds of pressure relief elements. All of them have advantages and disadvantages. Selecting one over another depends on the characteristics of the system in which it will be installed. Making a mistake in this selection can seriously affect the safety of the plant. Presented below are the characteristics of the most important types.

Safety valves

These consist basically of a spring that maintains a disc pressed against a seat, closing the exit to the fluid until the system pressure is capable of overcoming the pressure exercised by the spring. In Figure 7.4 a conventional safety valve is shown. The main problem with this type of valve is that, because the disc area is greater on the outlet side than on the inlet side, any backpressure has a dual effect: delaying the opening of the valve (superimposed backpressure) and causing it to close too soon (flow backpressure). A typical situation when the backpressure is too high is the rapid opening and closing of the valve, very quickly causing significant damage. This is because when the valve is closed and there is no flow backpressure the valve opens, but when flow is established, the flow backpressure increases enough to reclose the valve.

To reduce the influence of backpressure on the operation of safety valves, balanced bellows valves were developed (see Figure 7.5), which make use of a bellows connected to the atmosphere that leaves part of the disc area free from backpressure.

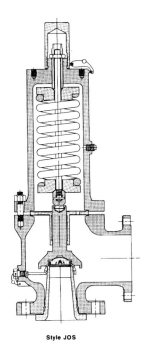

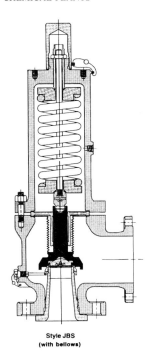

Figure 7.4   A conventional safety valve (courtesy of Walthon Weir Pacific SA).

Figure 7.5   A balanced bellows safety valve (courtesy of Walthon Weir Pacific SA).

When precise opening of a valve independently of the backpressure is desired, pilot-operated valves have to be used (Figure 7.6). These consist of an auxiliary valve (or pilot) that, while the system pressure is below set pressure, lets the spring be assisted by the system pressure to close. At the moment when the pressure rises above the set pressure, the auxiliary valve opens, leaving the main valve without pressure to assist the spring, thus opening. When the pressure drops and the auxiliary valve recloses, the pressure on the spring side is re-established and the principal valve closes again.

Whilst for conventional valves the relief capacity drops drastically with a backpressure above 10% (calculated on the manometric set pressure), for balanced valves it drops more slowly, the loss being significant above 20–30%. For pilot valves the limit is about 50%, which is the approximate value at which the flow starts being lower than the choking flow.

There is another classification for safety valves, according to the fluid handled:

- Pressure safety valve (PSV). Designed for gas or vapour flow. They open completely at 10% above set pressure.
- Pressure relief valve (PRV). Designed for liquid flow. Usually open completely at 25% overpressure.

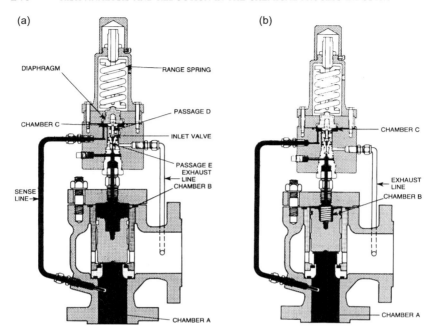

Figure 7.6   Operation of a pilot-operated safety valve (courtesy of Walthon Weir Pacific SA). (a) While the pressure is less than the set pressure, chambers A, C and B remain in communication through passages D and E. Under these circumstances the pilot valve is closed and the system pressure helps the pressure of the spring to maintain the main valve closed. (b) When the system pressure reaches the set pressure, the pilot valve opens, closing passages D and E and opening the exhaust line to chamber B. When chamber B depressurizes, the piston loses the pressure that maintains it closed, so that it opens.

- Safety relief valve (SRV). Designed to work with any flow: liquid, gas, vapour or two-phase.

Rupture discs

A rupture disc is a sheet of material designed to withstand a certain pressure. Above this it breaks, opening completely the section of the pipe where it is installed. The main difference between a rupture disc and a safety valve is that, while the latter close when the pressure returns to normal, the former remain open, allowing the total contents of the vessel to escape if a safety valve is not installed in series with it.

The principle advantages of rupture discs are that they are not subject to leaks, having no moving parts, their maintenance is cheap, they are made in a great variety of corrosion resistant materials and they have sizes much greater than safety valves at a lower cost. The main disadvantage is that they have to be replaced after each incident, although even with this disadvantage a cost analysis normally favours them.

There are six main types of rupture discs according to their design. Some of them are shown in Figure 7.7.

1. *Conventional:* The cheapest, they consist of a convex metal plate with the concavity on the process side and adequate thickness to withstand an established pressure. The operating pressure can be up to 70% of the rupture pressure. They suffer from fatigue if the operating pressure undergoes frequent cyclic variations. When it opens, it breaks into fragments which can damage equipment placed downstream, such as safety valves. If incorrectly installed (e.g. upside down) the opening pressure is lower than desired.

2. *Conventional, pre-scored*: Similar to the conventional, but weakened by several scores to control opening and avoid fragmentation. As for the conventional, if installed upside down the opening pressure is lower.

3. *Conventional, composite:* As above, but made of several layers, normally three, of different materials, thus adapting better to the conditions of corrosive fluids. The two outer layers are usually pre-cut, and the inner layer seals the openings. If these discs and the previous ones are going to be exposed to vacuum, they need support to prevent their collapse.

4. *Reverse buckling, scored*: Conversely to those above, these are installed with the convex face on the pressure side. They are the most expensive. They do not fragment and can work up to 90% of the desired opening pressure. Generally they are more reliable than conventional discs because they do not break due to material resistance, but through collapse of convexity, although if installed upside down the rupture pressure increases. Vacuum support is not required.

5. *Reverse buckling with cutting blades.* They have on the outlet side blades that cut the disc on opening, avoiding fragmentation. They can work at up to 90% of the rupture pressure. One important problem is that if the blades suffer corrosion the rupture pressure will be higher than desired.

6. *Flat, graphite*: The principal advantage is their resistance to corrosion at a reduced cost and their low rupture pressure. They disintegrate on rupturing and so may cause damage to downstream equipment if precautions are not taken, and therefore should not be used upstream from safety valves.

The scored reverse buckling discs are the most reliable, although their price is an inconvenience. To avoid mounting problems (reverse installation) they are usually designed so as to make this impossible.

Most commonly used are the pre-scored conventional discs. The use of reverse buckling discs with cutting blades is usually avoided because they provide more opportunities for incorrect installation and failures on opening. In all cases it is better to ask the manufacturer, who can help to choose from the wide range of rupture discs available.

Pressure-vacuum relief valves

This type of valve (Figure 7.8) is normally used for atmospheric storage tanks. They have two orifices, one for overpressure cases and the other for vacuum. They are normally used when filling or emptying the tank, and not in cases of fire or chemical reaction.

(a)

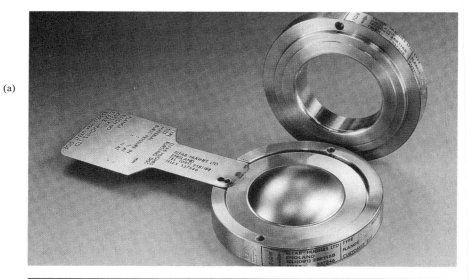

(b)

(c)

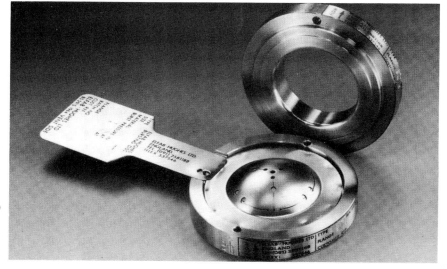

(d)

Figure 7.7   Different types of rupture discs (courtesy of Elfab-Hughes Ltd). (a) Conventional; (b) conventional, pre-scored; (c) composite, scored; (d) reverse buckling with cutting blades.

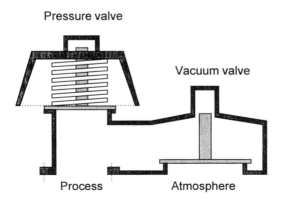

Figure 7.8   A pressure-vacuum relief valve.

Example 7.9 (Kletz [17])

We are very often not aware of the real mechanical resistance of storage tanks, as we will now show.

An atmospheric storage tank is normally designed for an internal pressure (on top of the liquid column) of 203 mm water column (mm w.c.) (the pressure in the bottom of a pint of beer) and 63.5 mm w.c. of vacuum, which are equivalent to 0.02 and 0.006 bar gauge, low pressures when compared with what a conventional pump can do.

A normal person can blow and suck air with the lungs at a pressure of up to 0.06 bar, which means that we could create vacuum or pressure sufficient to break or collapse a storage tank using only with lungs (if we were able to blow for a long enough time). Of course any pump taking fluid in or out of the tank would do it more easily than us.

If we compare the mechanical resistance of a small can of baked beans with a 1000 m³ tank, the walls are eight times weaker, and the top 57 times. It is easy to bend the can just with two fingers. For a 100 m³ tank, the factors are respectively 4 and 11.

The mechanical resistance of storage tanks is enough for their normal design conditions, but the use of pressure and vacuum relief devices is required to avoid working outside these conditions, as we have seen that even small deviations can very easily cause damage to the tank.

Emergency vents

When a pressure-vacuum relief valve is not sufficient to evacuate the flow necessary in a specific case, for example fire, an emergency vent can be used. This is just a manhole with the weight of its cover calculated so that it opens at a certain pressure, usually a little higher than the set pressure of the pressure-vacuum relief valve.

Pin valves

These are safety valves in which the spring has been substituted by a pin with material, length and diameter calculated to collapse at the set pressure. Once open it will not close again, like a rupture disc. Such valves allow very low pressure ratings (down to 0.01 bar) and suffer less failures due to fatigue than rupture discs.

Other

Explosion panels are zones in the walls of tanks, buildings or conductors which are substituted by a weaker material so that overpressure is liberated through them and does not damage the rest of the equipment and installations. They are frequently used for dust explosions in silos and in transport of solids.

In some storage tanks weaker welds are used between the roof and the walls so that in the case of overpressure the tank breaks at this point first.

Combinations

Combinations of rupture discs and safety valves are often used, especially when the fluid in the vessel or the environment has corrosive, toxic or incompatible characteristics:

1. Rupture disc upstream from a safety valve. The rupture disc is usually used to protect the valve from the process fluid and avoid leaks or corrosion: the valve will also avoid losing all of the vessel's contents in each relief. It is necessary to ensure that no fragment from the rupture of the disc can block the valve and that there are no leaks through the disc that can damage the valve and prevent the opening of the disc (a manometer is usually installed between both to detect any increase in pressure). When carrying out the design, the ASME code [14] establishes that the capacity of the system is 0.8 times that of the valve if there are no certified trials of the disc–valve combination.
2. Rupture disc and valve in parallel. Normally the valve is designed for minor incidents. The set pressure of the disc is somewhat higher and is designed for a greater flow. This is the typical case for the storage of a reactive product in an inert atmosphere. The disc is designed for a runaway reaction, while the valve is of a much smaller size and is designed for the case of a failure in the pressure control of the nitrogen used for inertization. In other cases the disc is a backup for the valve, should it fail.
3. Rupture disc downstream from the safety valve. Similar to case (1), but when the compound that can attack the valve is on its discharge side. It is important to pay attention to the backpressure created by the discharge line, which is increased in the rupture disc.

Selection

Figure 7.9 shows a flow diagram for the selection of the type of relief element or the combination of them best suited to each situation, proposed by Parry [18]. It is based on the criteria described above.

7.4.4 Pressure relief element design

Design data

Once the relief flows required for the different scenarios are known, it is necessary to calculate the size of the relief element. The main information needed is:

1. *Required relief flow.* Derived from the worst-case relief scenarios. How to calculate them has been described in previous sections.
2. *Composition, physical state, physical and thermodynamic properties of the fluid.* The composition and the physical state are usually known with sufficient precision through the material and energy balances of the plant or process data. The data on physical and thermodynamic properties may be what most limits the reliability of the results. Reliable values and correlations exist in numerous

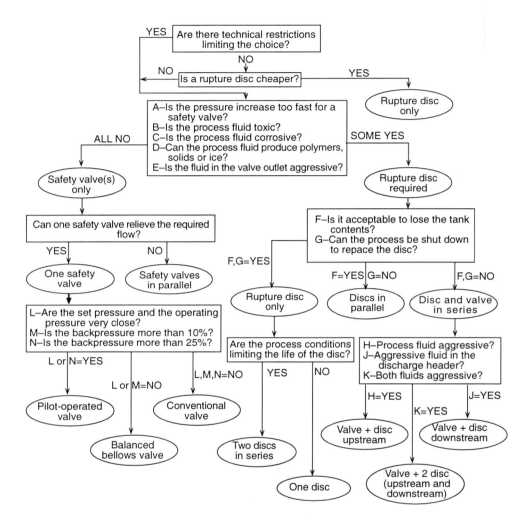

Figure 7.9   Flow diagram for the selection of pressure relief elements (adapted from Parry [18]).

databases for most of the compounds involved in the production of the most important chemical products. If there is no available data, which often happens with less common products, it is necessary to estimate in order to complete the design [19].

3. *Type of relief element and configuration of inlet and outlet pipes.* Often during the design process, the configuration of the inlet and outlet lines is unknown. Due

to its great importance for a correct design (principally because of its influence on the superimposed backpressure, and of this on the valve capacity) it is necessary to have a reliable estimate so as not to have to modify designs. Good co-ordination between the process engineer and the piping design team is especially important so that the influence of any change that could affect the design is evaluated.

The head loss in the valve inlet line should not, as a general rule, exceed 50% of the blow-down, which is normally equivalent to 3% of the differential set pressure. For pilot-operated valves following the manufacturer's recommendations is advised.

As for the discharge line, a maximum backpressure of 10% of the set pressure is recommended in the case of conventional valves for gases, in order to ensure that the valve works in choked flow and at full capacity. However, in the calculation formulae for valves this correction factor is usually already included. If the backpressure is high or the set pressure low, the system may enter the non-choked compressible flow regime, which would cause any small increase in backpressure to decrease the capacity of the valve.

4. *Set pressure, design pressure, superimposed backpressure.* These are items that, although often given as fixed, can be varied throughout the design of the relief system, because they have much influence on the size of the relief element to be installed. Especially when the possibility of a runaway reaction with vapour generation exists, consideration is recommended of the installation of at least one unit that opens well below the design pressure, to allow the start of evaporation to eliminate heat from the system as soon as possible.

In the simplest cases (liquid or gas flow), the manufacturers give tables and formulae in their technical sheets to calculate the capacity of their products. One must be very careful when using this data because, although usually reliable, they do not always reproduce the conditions in which the equipment is to be installed, or include data for reference fluids in the constants used.

Numerous references exist where relatively sophisticated methods are described, with more or less complicated final formulae. Discussed here are the methods considered as industry standards that are accepted by regulatory bodies such as ASME or API. In general, for liquids or vapours, there are no discrepancies between the different methods and basically all are derived from the flow through an orifice. Two-phase flow is described in a later section because of its special importance and complexity.

In order to define the selected element the main parameter is the orifice size. For safety valves a series of standard orifice sizes exists, which are identified by means of a letter according to the API 526 standard [20] (Table 7.4). This standard also regulates the main characteristics of the valves for each orifice: inlet and outlet port sizes, type of connections and design pressures.

Selection of safety valves for gas flow

The first thing to check is if the system is within the range of choked or critical flow. This situation arises if the discharge pressure P_2 is less than the pressure

Table 7.4   Nominal orifice sizes for safety valves [20]

Identification	Nominal area (cm²)	Nominal area (in²)
D	0.710	0.110
E	1.26	0.196
F	1.98	0.307
G	3.25	0.503
H	5.06	0.785
J	8.30	1.287
K	11.9	1.838
L	18.4	2.853
M	23.2	3.60
N	28.0	4.34
P	41.2	6.38
Q	71.3	11.05
R	103	16
T	168	26

required to establish critical flow, P_{cf}. The overpressure produced by the flow through the discharge line must be included in P_2:

$$P_2 = P_{\text{Flow backpressure}} + P_{\text{Superimposed backpressure}}$$

The passage time through the valve is usually sufficiently short that the process can be considered adiabatic. In these conditions,

$$P_{cf} = P_1 \left[\frac{2}{k+1} \right]^{\frac{k}{k-1}} \qquad (7.1)$$

where P_1 is the system relief pressure (set pressure plus overpressure). The value of k is calculated for ideal gases as

$$k = C_p/C_v \qquad (7.2)$$

The heat capacity ratio should never be used for a real gas at high pressures, because large errors can be introduced. In this case the adiabatic or polytropic expansion coefficient must be used.

When the set pressure is very low, it is impossible to work within the region of critical flow. In these cases, just as for large backpressures, the size of the valve is much greater than that necessary for critical flow.

The expression to be used can be deduced from the flow through an orifice. If the system is in choked flow, the relief area required (A) is

$$A = \frac{W}{C_D \chi K_b P_1} \sqrt{\frac{Tz}{M}} \qquad (7.3)$$

where W is the mass flow to be relieved, M is the molecular weight of the vented stream, z is its compressibility factor and T the relief temperature. χ is a coefficient calculated by the expression

$$\chi = \sqrt{\frac{k}{R}\left(\frac{2}{k+1}\right)^{(k+1)/(k-1)}} \qquad (7.4)$$

where R is the ideal gas constant.

C_D is the discharge coefficient for the flow through the valve orifice. One usually takes the value of 0.95 for a gas flow, if there is no data on the specific valve, although whenever possible manufacturer's data should be used.

K_b is a backpressure correction factor for conventional and balanced valves, which is displayed in the form of graphs (API RP 520 [21], see Figures 7.10 and 7.11).

Although the previous expression includes the effect of backpressure through K_b, when the flow is not critical the graphs in Figures 7.10 and 7.11 cannot be used, and calculations should be carried out using the expression for compressible flow, which is somewhat more complex and depends on the pressure in the discharge of the valve P_2:

$$A = \frac{W}{C_D P_1}\sqrt{\frac{Tz}{M}}\left[\frac{R}{2}\frac{(k-1)}{k}\frac{1}{\left(\dfrac{P_2}{P_1}\right)^{2/k}-\left(\dfrac{P_2}{P_1}\right)^{(k+1)/k}}\right]^{1/2} \qquad (7.5)$$

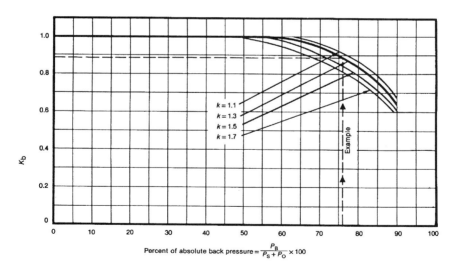

Figure 7.10 Backpressure correction system (K_b) for the selection of conventional safety valves for gas or vapour flow according to API RP 520 [21] (courtesy of the American Petroleum Institute, Washington, DC). This graph should only be used when the make of the valve or the actual critical flow pressure point for the vapour or gas is unknown. Whenever possible, the manufacturer should be consulted for specific data. P_B = backpressure; P_s = set pressure; P_o = overpressure.

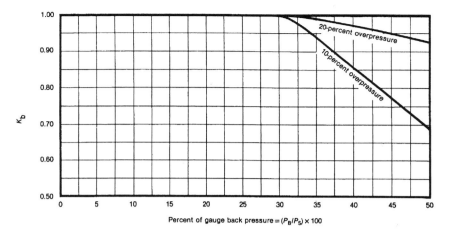

Percent of gauge back pressure $= (P_B/P_S) \times 100$

Figure 7.11 Backpressure correction system (K_b) for the selection of balanced bellows safety valves for gas or vapour flow according to API RP 520 [21] (courtesy of the American Petroleum Institute, Washington, DC). These curves are for set pressures of 3.5 bar g and above. They are limited to backpressure below critical flow pressure for a given set pressure. For subcritical flow backpressures below 3.5 bar g the manufacturer should be consulted for values of K_b. P_B = backpressure; P_s = set pressure.

Example 7.10

(a) Determine the orifice size of the safety valve necessary to vent 8000 kg/h of an organic vapour with a molecular weight of 153 and z = 1. The relief pressure is 6 bar g, the temperature 160°C and the backpressure 0.1 bar g ($k = C_p/C_v = 1.3$). (b) What if the backpressure was 4 bar g?

(a) It is clear that in the first case the backpressure is much less than 50% of the set pressure and the expressions for critical flow must be used. A conventional valve would be adequate.

Applying equation (7.4) with the data we have and $R = 8314$ J/kmol K we obtain $\chi = 0.007318$. Now using equation (7.3), where $K_b = 1$ according to Figure 7.10, we have that

$$A = \frac{8000/3600}{0.95 \times 0.007318 \times 1.0 \times 701300} \sqrt{\frac{433 \times 1.0}{153}}$$

$$A = 0.000767 \, \text{m}^2 = 7.67 \, \text{cm}^2$$

Looking at Table 7.4, an orifice type J, with 8.30 cm³ is sufficient.

(b) In this case we will calculate the maximum backpressure which gives us choked flow, which, from equation (7.1), is

$$P_{cf} = 7.013 \ (2/2.3)^{(1.3/0.3)} = 3.83 \ \text{bar}$$

The backpressure in our case is 5.013 bar, higher than critical, so that the valve must be designed for sub-critical flow, using equation (7.5):

$$A = \frac{8000/3600}{0.95 \times 701300} \sqrt{\frac{433 \times 1.0}{153}} \sqrt{\frac{8314}{2} \frac{(1.3-1)}{1.3}} \frac{1}{\left(\dfrac{501300}{701300}\right)^{2/1.3} - \left(\dfrac{501300}{701300}\right)^{(1.3+1)/1.3}}$$

$$A = 0.000824 \ \text{m}^2 = 8.24 \ \text{cm}^2$$

Although the area is almost 8% larger, the same orifice, size J, is valid. However, in this case a balanced or pilot-operated valve must be used.

Selection of safety valves for liquid flow

In this case the expression to use according to the API RP 520 [21] is that for incompressible flow through a valve, with some modifications:

$$A = \frac{Q_v}{C_D K_v K_p K_b \sqrt{2}} \sqrt{\frac{\rho}{1.25 P_{set} - P_2}} \tag{7.6}$$

where Q_v is the design volumetric flow and the pressures are manometric. Normally relief valves are designed for an overpressure of 25%, so that P_1 is 1.25 P_{set}. The correction factors K_v, K_p, K_b correspond respectively to viscosity, overpressure and backpressure and are calculated using Figures 7.12, 7.13 and 7.14. In the case of the viscosity correction factor, its value depends on the Reynolds number, which in turn depends on the orifice diameter. An iteration is therefore required. For the first trial the smallest orifice available is assumed (this type of valve is normally used in cases of fluid thermal expansion resulting in very small sizes, C or D), the calculation is repeated until the assumed value coincides with the obtained result. K_b is always 1 for conventional valves. For balanced bellows valves the graph is used.

The manufacturer's discharge coefficient should be used. If unknown, a conservative value of 0.61 may be used.

Example 7.11

Select a relief valve for liquid flow (C_D = 0.73 from the catalogue) for 100 m³/h of water (20 °C, 998 kg/m³, viscosity 1 mNs/m²). The valve set pressure is 7 bar and the vessel design pressure is 8 bar. An accumulation of 10% is allowed and the backpressure in the discharge vessel is 2 bar.

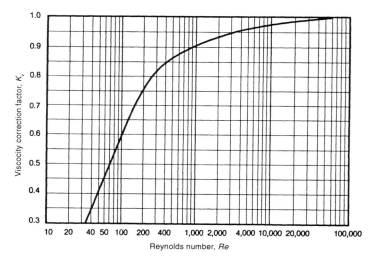

Figure 7.12   Viscosity correction factor (K_v) for the selection of liquid relief valves according to API RP 520 [21] (courtesy of the American Petroleum Institute, Washington DC).

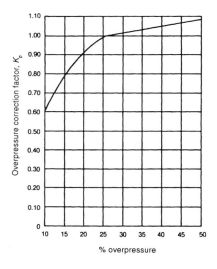

Figure 7.13   Overpressure correction factor (K_p) for the selection of liquid relief valves according to API RP 520 [21] (courtesy of the American Petroleum Institute, Washington DC).

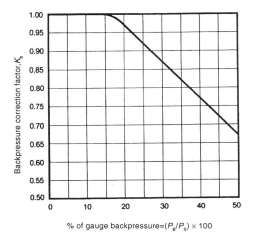

Figure 7.14 Backpressure correction factor (K_b) for the selection of balanced liquid relief valves according to API RP 520 [21] (courtesy of the American Petroleum Institute, Washington DC). The curve represents values recommended by various manufacturers and may be used when the manufacturer is not known. Otherwise, the manufacturer should be consulted for the applicable correction factor. P_B = backpressure; P_s = set pressure.

For a design pressure of 8 bar, the relief pressure with an accumulation of 10% (always with reference to the relative pressure) will be

$$P = (8 - 1.013) \times 0.1 + 8 = 8.7 \text{ bar}$$

and the overpressure

$$\% \text{ overpressure} = (8.7 - 7)/(7 - 1.013) = 28\%$$

This result is higher than 25%, and therefore the valve will be completely open. To apply equation (7.6), we first look for the coefficients in the graphs.

K_b: A backpressure of $2 - 1.013 = 0.987$ bar g is over a set pressure of $(7 - 1.013) = 5.987$ bar g, 16.5%, so it is convenient to use a balanced bellows valve. Looking on the graph we find a value of 0.99.

K_v: The graph is a function of the Reynolds number, therefore it is necessary to proceed iteratively. As water has a low viscosity, we assume as a first approximation that $K_v = 1$ and later will test the assumption.

K_p is 1.01, corresponding to 28% overpressure. Substituting in SI units in equation (7.6)

$$A = \frac{100/3600}{0.73 \times 1 \times 1.01 \times 0.99\sqrt{2}} \sqrt{\frac{998}{1.25 \times 598700 - 98700}}$$

$$A = 0.00106 \text{ m}^2 = 10.6 \text{ cm}^2$$

We select therefore, an orifice K, which has an area of 11.9 cm². We then
check the assumption of $K_v = 1$:

$$D = \sqrt{4A/\pi} = 3.89 \text{ cm} = 0.0389 \text{ m}$$

The Reynolds number will be

$$Re = Du\rho/\mu = D(Q/A)\rho/\mu = 0.0389 \times ((100/3600)/0.00106) \times$$
$$998/0.001$$

$$= 1.02 \times 10^6$$

For this value of the Reynolds number, K_v is in the unity range, and it is
not necessary to repeat the calculation, because the initial assumption was
correct.

Rupture disc calculation.   Rupture discs, because of their special characteristics,
deserve a separate section. It is always better to use capacity data from the
manufacturer.

Suppliers often offer design and selection software for personal computers,
which can be very useful, although one should check that they include all the
design factors. When determining the size two methods can be considered:

1. As pipe sections, with a certain equivalent length, usually between 8 and 16
   diameters. This length is added to the entrance and exit sections of piping and the
   relief flow for the total length and the relationship P_1/P_2 are calculated. When
   dealing with gas flow, this calculation can be made using the Lapple–Levenspiel
   charts [4]. One or two iterations are normally required, because the final diameter
   should be the same as that used in the calculation of the equivalent length and the
   friction factor f.
2. As an ideal orifice. The same formula as for safety valves is used, but with a
   lower discharge coefficient (if the certified value is not known, use 0.62). For the
   case in which critical flow exists:

$$A = \frac{W}{C_D \chi P_1} \sqrt{\frac{Tz}{M}} \tag{7.7}$$

Example 7.12

*Select a rupture disc to vent at 60°C 50 000 kg/h of a gas with a molecular
weight of 70 and k = 1.4. The rupture pressure is 9 bar and the backpressure
1.4 bar.*

*Consider a venting pipe of the same diameter as the rupture disc, 50 m
long, and with four 90° elbows (K = 0.8) and a sudden entrance. ρ = 22.75
kg/m³ and μ = 0.012 mN/sm². The manufacturer's catalogue gives L/D = 16
for the rupture disc.*

Given that the discharge line is quite long, we will design the system using the Lapple–Levenspiel charts, as shown in the *Chemical Engineer's Handbook* by Perry [4].

To calculate all the parameters we need an initial estimated diameter, which we will take as 100 mm = 0.10 m. First we will calculate the area mass flow G:

$$G = W/A = (50000/3600)/(\pi \times 0.1^2/4) = 1768 \text{ kg/sm}^2$$

$$G^* = \sqrt{kP\rho \left(\frac{2}{k+1} \right)^{\frac{(k+1)}{(k-1)}}}$$

$$G^* = \sqrt{1.4 \times 900000 \times 22.75 \times (2/2.4)^6} = 3098 \text{ kg/sm}^2$$

Therefore $G/G^* = 0.57$ and $P_2/P_0 = 1.4/9 = 0.16$

Looking in the Lapple chart for G/G^* and P_2/P_0 we find a value of N available of approximately 3. We now determine the value we have in reality. First we calculate Re:

$$\text{Re} = DG/\mu = 0.1 \times 1768/0.000012 = 1.5 \times 10^7$$

As we are at a highly turbulent flow, which is normal in these cases, we can apply the limit friction factor (f) which for carbon steel is 0.016. Next we calculate the factor N, which includes the velocity head loss in the pipe (50 m) and accessories (4 elbows of $K = 0.8$ and a sharp contraction $K = 0.5$) and the rupture disc ($L/D = 16$).

$$N = (fL/D)_{\text{total}} = 0.016 \times 50/0.1 + 4 \times 0.8 + 0.5 + 0.016 \times 16 = 12$$

which is greater than the maximum allowable, so we must install a larger rupture disc. To make N close to 12 in the graph we must go to a G/G^* of approximately 0.34, so that we require a diameter of 0.13 m.

So we will install a disc of 150 mm diameter, to use a standard size pipe.

Pressure-vacuum relief valves and emergency vents.   The equations for the design of conventional safety valves with non-critical compressible flow can be applied (7.5), because in these cases the differences between the exterior pressure and the set pressure are normally very small. A discharge coefficient of 0.5 can be used, if no manufacturer's data is available.

7.4.5   Two-phase flow

One of the most frequent errors that occurs in the design of emergency relief systems is not considering the possibility that two-phase flow may exist. Two-phase flow can require an area up to 10 times greater than that necessary for gas-phase flow. If the relief system has not been designed for these circumstances, a high overpressure can be generated because of the lack of the necessary relief capacity, with the resultant risk.

Both the prediction of possible situations of two-phase flow and then the design of the relief system are technically complex. Important developments have been achieved in recent years, especially since the creation of the DIERS (Design Institute for Emergency Relief Systems) programme of the AIChE (American Institute of Chemical Engineers). The methods developed by DIERS have replaced the most rudimentary and inaccurate that were applied in the past, introducing a significant improvement in the tools available to the process engineer for identifying and reducing risks.

Two-phase flow originates from a phenomenon very similar to what happens when opening a bottle of champagne. The sudden vaporization within the liquid mass provokes an increase in the total volume, producing the emission of a mixture of liquid and gas or liquid and vapour.

Once the two-phase flow is established, the vent stream is normally considered to correspond with the average contents of the tank with respect to its composition and vapour fraction. This is known as the homogeneous model (in the simpler models liquid flow is assumed, as a worst-case situation). The more complex models consider a gradual increase in the vapour fraction of the vent stream due to the return of a part of the entrained liquid to the vessel (usually assigning lower velocity for the liquid than for the vapour through a **slip velocity**) or even a step change to vapour flow as the level in the vessel decreases. The homogeneous model is usually the most conservative.

DIERS two-phase flow test

The DIERS programme developed a method [22, 23] that can be used to determine if two-phase flow will occur or not, and in what conditions it takes place.

The basis for this test is the following: when the volume fraction of vapour or gas bubbles held up in the system (α) is greater than the volume fraction of the vessel which is available for expansion (α_0, free vessel volume), there will be two-phase flow (or boilover). The vapour hold-up depends on the volumetric flow of gas generated, the flow area of the vessel, the density and the surface tension of the liquid, the density of the gas and the flow regime.

The calculation of α is based on a flow model for the bubbles that rise through the liquid. This permits calculation of the ascensional speed and, therefore, the quantity of bubbles in transit through the liquid at a given moment, which gives the volume hold-up. Thus the first selection is the flow model:

- For high or medium viscosities (>100 cP), and when the liquid foams on boiling, the **bubbly** model is used.
- For low viscosities, and when it is known that there is no foaming tendency, the **churn-turbulent** model is used.

In the calculation sequence, first the factor $j_{g\infty}$ which is the surface velocity of the vapour or gas generated, is determined, calculated on the horizontal section of the vessel (A_{horiz}).

$$j_{g\infty} = \frac{W}{\rho_g A_{horiz}} \qquad (7.8)$$

where ρ_g is the density of the gas or vapour produced and W its mass flow, calculated as a function of all the heat supplied (Q_e), either from external sources or chemical reaction, as

$$W = \frac{Q_e}{\Delta H_v} \qquad (7.9)$$

Then u_∞, the bubble rise velocity, is calculated, according to the formula for the flow model chosen. For bubbly type flow

$$u_\infty = 1.18\left[\sigma g \frac{1}{\rho_1}\left(1 - \frac{\rho_g}{\rho_1}\right)\right]^{1/4} \qquad (7.10)$$

and for churn-turbulent

$$u_\infty = 1.53\left[\sigma g \frac{1}{\rho_1}\left(1 - \frac{\rho_g}{\rho_1}\right)\right]^{1/4} \qquad (7.11)$$

where σ is the surface tension of the fluid, g is the acceleration due to gravity and ρ_1 is the liquid phase density .

The quotient of $j_{g\infty}$ and u_∞ has been correlated with α for cases of uniform vapour generation in the whole volume of liquid, obtaining the two equations shown below. If the quantity of vapour or gas generation were constant, as in the case of gas injection from the bottom of the vessel, one would have to substitute the value of $j_{g\infty}$ by $2 j_{g\infty}$, because as the quantity of vapour is the same at all levels of the vessel, there would be twice the volume than if it were generated by homogeneous boiling (almost zero at the base and W on the upper surface, i.e. $W/2$ on average). For churn-turbulent flow

$$\frac{j_{g\infty}}{u_\infty} = 2\frac{\alpha}{1 - C_o\alpha} \qquad (7.12)$$

where C_o takes the value of 1.5, although 1 can be taken if a conservative calculation is desired. For bubbly flow

$$\frac{j_{g\infty}}{u_\infty} = \frac{\alpha(1-\alpha)^2}{(1-1.2\alpha)(1-\alpha^3)} \qquad (7.13)$$

In this last case it is necessary to calculate the value of α iteratively.

Once α is known, conclusions can be drawn. If the vessel has a free space α_o lower than α, there will two-phase flow. In this case it is necessary to design the relief system for these circumstances or to reduce, if possible, the level of liquid through reliable control systems.

This test is not adequate in the case of large storage tanks, where convection currents and local effects in the areas closer to the walls are extraordinarily important. In these cases boiling is normally produced on the walls and the surface. Models exist in the bibliography [24, 25] that account for these peculiarities.

Example 7.13

Determine if there will be two-phase flow if a fire affects a tank 80% full of a flammable hydrocarbon. The area exposed to the fire is 80 m² and the horizontal section of the tank is constant and equal to 10 m². There is no fire protection of any kind.

$$\rho_l = 740 \ kg/m^3 \quad \rho_v = 3 \ kg/m^3 \quad \sigma = 0.019 \ N/m$$
$$\Delta H_v = 400 \ kJ/kg \quad C_p = 2 \ kJ/kg \ K$$

(The fluid has a low viscosity and does not foam on boiling)

We will carry out the fire heat input calculations according to the NFPA 30 standard (see Example 7.6).

$$Q = 224168 \ A^{0.566} = 2677 \ kW$$

The resulting vapour flow is, with $F = 1$

$$W = QF/\Delta H_v = 2677/400 = 6.7 \ kg/s$$

Now we will calculate $j_{g\infty}$ and u_∞ with equations (7.8) and (7.11), choosing churn-turbulent flow in view of the fluid characteristics

$$j_{g\infty} = W/(\rho_v A_{horiz}) = 6.7/(3 \times 10) = 0.223 \ m/s$$

$$u_\infty = 1.18 \left[0.019 \times 9.81 \frac{1}{740} \left(1 - \frac{3}{740} \right) \right]^{1/4}$$

$$u_\infty = 0.149 \ m/s$$

Therefore $j_{g\infty}/u_\infty = 1.50$

Solving equation (7.12) a value for α of 0.37 is obtained for $C_0 = 1$. As this is much greater than α_0 (0.2) the design will be for two-phase flow.

Capacity calculations for two-phase flow

Once the flow to be relieved and its quality are known, the characteristics of the required relief system must be determined.

Relief system capacity calculations for two-phase flow are much more complicated than for gas or liquid flow. In most cases there is a significant flash in the valve, so that it is necessary to take into account the effects of phase change, both thermal and volumetric.

Numerous equations exist based on flow models which are simplifications of the general case. The most important models are the ERM (Equilibrium Rate Model) and HEM (Homogeneous Equilibrium Model). The main difference between them is that the HEM assumes instantaneous thermodynamic equilibrium at all times, while the ERM is a non-equilibrium model, that assumes certain kinetics to reach it. Figure 7.15 shows the comparison of different flow models for the case of styrene. The results of the application of the HEM model are more conservative. Leung [26] has shown that the HEM choked flow is equivalent to 90% of that calculated according to the ERM.

Fauske [27] has developed the following expression to calculate the critical mass flow (G_c, kg/s m^2) according to the ERM for an ideal orifice. It is generally conservative, but more suited to vapour fractions close to zero.

$$G_c = \frac{\Delta H_v}{\upsilon_{fg}} \sqrt{\frac{1}{C_p T_s}} \tag{7.14}$$

where υ_{fg} is the difference of the specific volumes of the liquid and vapour phases, T_s is the fluid saturation temperature in the conditions at the orifice inlet and C_p the liquid heat capacity.

To convert to pipe flow a correction coefficient Ψ is used, that is a function of its length (L) and diameter (D), through the non-dimensional factor L/D. The function Ψ is shown in Figure 7.16 for a friction factor of $f = 0.016$.

The resulting expression is, for the HEM model:

$$G_c = 0.9\Psi \frac{\Delta H_v}{\upsilon_{fg}} \sqrt{\frac{1}{C_p T_s}} \tag{7.15}$$

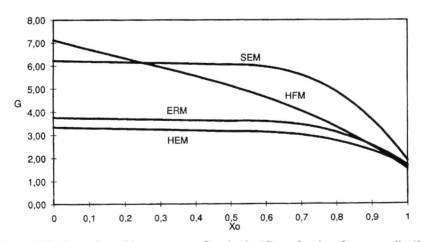

Figure 7.15 Comparison of the styrene mass flow density (G) as a function of vapour quality (X_0) calculated according to different models (adapted from Leung [26]). ERM = Equilibrium Rate Model; HEM = Homogeneous Equilibrium Model; SEM = Slip Equilibrium Model; HFM = Homogeneous Non-equilibrium Model.

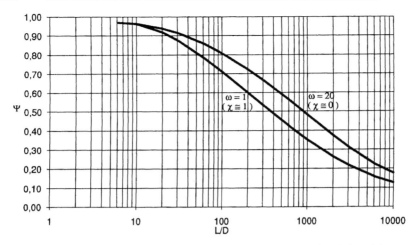

Figure 7.16   Pipe length correction factor for two-phase critical flow in pipes (adapted from Leung and Grolmes [28] for $f = 0.016$).

An equivalent expression that may be useful on occasions can be derived by substituting the Clapeyron–Clausius equation:

$$\frac{\Delta H_v}{v_{fg}} = T_s \frac{dP}{dT} \tag{7.16}$$

substituting also the derivatives by finite increments

$$G_c = 0.9 \Psi \frac{\Delta P}{\Delta T} \sqrt{\frac{T_s}{C_p}} \tag{7.17}$$

In this case ΔP is the overpressure during venting and ΔT is the boiling temperature increase corresponding to this overpressure (Figure 7.19).

Less simple expressions exist that take into account the influence of more parameters and make fewer assumptions. For example, Leung [26] has developed a correlation for the choked flow in HEM, based on a non-dimensional ω

$$\omega = \frac{x v_{fg}}{v} + \frac{C_p T P}{v} \left(\frac{v_{fg}}{\Delta H_v} \right)^2$$

where x is the vapour fraction at the valve inlet and v is the specific volume of the liquid–vapour mixture. The pressure (P) and temperature (T) are taken at relief conditions.

For $\omega \geq 4.0$, corresponding to low vapour fractions, the expression below is used:

$$\frac{G_c}{\sqrt{\dfrac{P}{v}}} = \frac{0.6005 + 0.1356 \ln \omega - 0.0131 \, (\ln \omega)^2}{\sqrt{\omega}} \tag{7.19}$$

For $\omega < 4.0$ the expression is simpler:

$$\frac{G_c}{\sqrt{\dfrac{P}{v}}} = \frac{0.66}{\omega^{0.39}} \qquad (7.20)$$

To determine if we are in choked flow, the critical pressure ratio can be calculated [26] according to

$$\eta = P_{fc}/P = 0.6055 + 0.1356\ \ln\omega - 0.0131\ (\ln\omega)^2 \qquad (7.21)$$

If the pressure at the outlet is greater than the critical pressure P_{cf}, the capacity can be corrected to obtain the sub-critical flow value, using the correlation developed by First and Huff [29], shown in Figure 7.17, which gives the ratio G/G_c as a function of the vapour quality at the entrance and the parameter $(1 - \eta)/(1 - \eta_c)$.

When working with three-phase flow (vapour, liquid and non-condensable gas) more complex expressions must be used. Leung and Epstein [30] have described a method appropriate for such cases.

Normally from these methods a sufficiently accurate result can be obtained. However, advanced mathematical models exist that describe two-phase flow phenomena with much more precision.

Example 7.14

Determine what will be the flow of a stream with $x = 0.2$ through a pipe of 50 mm diameter and 15 m of equivalent length. The temperature is 90°C, the discharge pressure is 1.5 bar and in the vessel 5 bar.

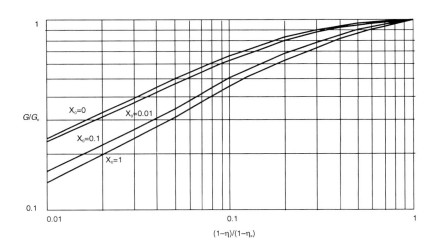

Figure 7.17   Non-critical flow correction factor for two-phase flow in pipes (adapted from First and Huff [29]).

$$\rho_l = 820 \ kg/m^3 \quad \rho_v = 3 \ kg/m^3 \quad \mu = 0.019 \ mN/s \ m^2 \quad \Delta H_v = 400 \ kJ/kg$$

$$C_p = 2 \ kJ/kg \ K$$

(a) First we will calculate for HEM flow using equation (7.15) (Fauske model). To obtain Ψ, we will first calculate $L/D = 15/0.05 = 300$ and from Figure 7.16, $\Psi = 0.62$ (interpolating for $x = 0.2$)

$$\upsilon_{fg} = (1\rho_v) - (1/\rho_l) = (1/3) - (1/820) = 0.332 \ m^3/kg$$

Substituting in equation (7.15)

$$G_c = 0.9 \times 0.62 \frac{400000}{0.332} \sqrt{\frac{1}{2000 \ (273 + 90)}}$$

$$G_c = 789 \ kg/sm^2$$

Multiplying by the area

$$W = 789 \ kg/sm^2 . \ (\pi \times 0.05^2/4) = 1.55 \ kg/s = 5580 \ kg/h$$

b) Now we will do the calculation according to Leung's method. To obtain ω, we must first calculate the specific volume of the mixture

$$\upsilon = (1 - x) \ \upsilon_l + x\upsilon_v = 0.8 \ (1/820) + 0.2 \ (1/3) = 0.0676 \ m^3/kg$$

Substituting in equation (7.18), we have that

$$\omega = \frac{0.2 \times 0.332}{0.0676} + \frac{2000 \times 363 \times 50000}{0.0676} \left(\frac{0.332}{400000} \right)^2$$

$$\omega = 4.68$$

Substituting in equation (7.19), we obtain

$$G_c/\sqrt{(P/\upsilon)} = 0.3621$$

$$G_c = 0.3621 \ \sqrt{(P/\upsilon)} = 0.3621 \ \sqrt{(500000/0.0676)} = 985 \ kg/s \ m^2$$

This value is somewhat higher than before. Because the vapour quality is relatively high, the value given by the first equation (Fauske) is probably conservative.

7.4.6 Runaway reactions

The term **runaway** is applied to chemical reactions, whether desired or not (normally decompositions or polymerizations), that self-accelerate in a spiral process with a heat generation greater than the system's cooling capacity (because the former grows in an exponential manner with temperature and the latter only linearly). In principle any exothermic reaction can give rise to a runaway if not properly controlled. The temperature at which the heat generation equals the cooling capacity is called temperature of non-return (TNR), and above it control of the system is lost. A system should never be allowed to reach its point of no return. In practice, reactor stability is a complex field, and its study requires a careful simulation of the system dynamics, taking into account the control algorithms implemented. A classic case that represents the simplest situation is that of a continuously stirred tank without automatic temperature control. For an exothermic reaction there are three possible operating points that correspond to the equilibrium solutions between heat generation by the reaction and removal by cooling. Figure 7.18 shows the curves of generation and elimination of heat for a typical system.

Example 7.15

Calculate the operation points of a continuous stirred tank reactor (CSTR) of 25 m³, cooled by a heat exchanger with water at a constant temperature of 10°C . The space time (τ) is 800 s, the concentration of reactant is 7 kmol/m³ and the kinetics are of the first order with k_o = 200 000 s⁻¹ and E_a = 51 000 kJ/kmol. The exchanger has an area of 10 m² and U can be estimated as 2500 W/m²K. ΔH_r = –10 000 kJ/kmol.

The heat generated by the reaction (Q_{gen}) can be expressed as (Levenspiel [8])

$$Q_{gen} = V_o\, C_o\, (X_{As})\, (-\Delta H_r) = (V/\tau)\, C_o\, (X_{As})\, (-\Delta H_r)$$

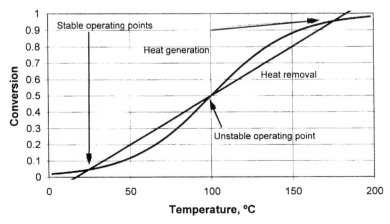

Figure 7.18 Heat generation by a chemical reaction and removal by cooling for a continuous stirred tank reactor in which an exothermic reaction is taking place.

where V is the reactor volume, X_{As} is the conversion reached, C_o the concentration of the reactant, $-\Delta H_r$ the reaction heat and V_o the flow fed to the reactor. The conversion for a CSTR can be calculated as

$$X_{As} = \tau\, k_o \exp(-E_a/RT_r)/(1 + \tau\, k_o \exp(-E_a/RT_r))$$

T_r being the reation temperature (K). Substituting the values in SI units in the previous expressions, we obtain

$$X_{As} = 1.6 \times 10^8 \exp(-6134/T_r)/(1 + 1.6 \times 10^8 \exp(-6134/T_r))$$

$$Q_{gen} = (25\ \mathrm{m^3/800\ s}) \times 7\ \mathrm{kmol/m^3} \times X_{As} \times 10000\ \mathrm{kJ/kmol} = 2.1875 \times 10^6\, X_{As}\ (W)$$

The cooling (Q_{ref}) can be calculated as

$$Q_{ref} = U A\,(T_r - T_{ref}) = 2500\ \mathrm{W/m^2K} \times 10\ \mathrm{m^2}\,(T_r - 283)\ \mathrm{K} = 25000\,(T_r - 283)\,W$$

The possible operation points are given at the moment that the cooling capacity is equal to the heat generated by the reaction. Setting one equation equal to the other $(Q_{ref} = Q_{gen})$ and solving graphically or iteratively three values are found

$$T_{r1} = 20.1°C$$

$$T_{r2} = 44.5°C$$

$$T_{r3} = 84.2°C$$

The two extremes correspond to stable operation temperatures. The middle value is an unstable operation point. If this temperature is exceeded, the system evolves spontaneously towards the higher operation point, producing a runaway reaction.

When evaluating the probability of running into a runaway four factors must be taken into account:

1. The initial temperature, which can be the normal operation temperature, the temperature after another type of incident or the heating media temperature.
2. The energetic potential of the system, i.e. the quantity of heat that could be released by the chemical reactions that are considered possible. Other possible sources of heat.
3. The reaction kinetics, which determine the release velocity of the system's energy potential. Knowledge of the heat release as a function of temperature and conversion is necessary.
4. The system's heat removal capacity, either by heat exchange with the environment or by boiling of the reactant mixture.

It is generally difficult to predict accurately the behaviour of a system. The consequences of a calculation error can be, moreover, very serious from the point of view of safety. For this reason, experimental tests are normally carried out that enable obtaining data on the system reactivity when the kinetics of the reaction that takes place are not well-known. There are a number of ways to determine experimentally the reactivity of a system. Most of them use calorimeters that show the variation of exothermal activity with temperature in conditions close to adiabatic (DSC, differential scanning calorimetry and ARC, adiabatic reaction calorimetry). Figure 7.19 shows the typical curves for pressure and temperature of a runaway reaction, with and without a venting system.

The most advanced system developed for diagnosis of runaway situations, although at a relatively high cost, is the VSP (vent sizing package). VSP is an adiabatic calorimeter with a very low thermal inertia due to its special design. From this apparatus the curves T–t and P–t can be obtained, and from them dT/dt at the set pressure and the temperature increase of the reactant mass corresponding to the overpressure during venting [31, 32]. With these parameters, as we shall see later, the pressure relief system can be designed.

According to its behaviour during venting, three types of system are identified:

1. *Vapour systems.* Also called boiling or vapour pressure systems. Named as such because the system pressure is the vapour pressure of the vessel contents at the system temperature. During venting, considerable heat elimination due to the latent heat necessary to produce a phase change of the reacting mixture takes place. This is the case in the polymerization of styrene in an organic solvent.
2. *Gassy systems.* In this system gases are generated as a consequence of the reaction, therefore no heat is removed during venting because of boiling. This is the case of the decomposition of an organic peroxide producing gases well below the boiling point of the decomposition products.

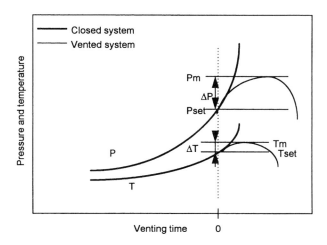

Figure 7.19   Temperature and pressure evolution in runaway reactions with and without venting (adapted from Huff [33]).

3. *Hybrid systems.* They have components of both the previous systems. For example, the decomposition of a peroxide dissolved in a low boiling point organic solvent.

The basic method for solving the three cases is the same. It involves the solution of the dynamic mass and energy balances of the system. These balances must include the description of:

- The chemical reactions that are taking place, in the form of kinetic equations.
- The flow leaving the vessel.
- The required physical and thermodynamic properties.

The global resolution of the system is usually very complex, so that the problem is normally approached with two groups of assumptions:

- The separation of the fluid dynamics problem and the kinetic problem.
- The assumption of constant values for some variables such as the relief mass flow (W), the total heat input per unit of mass in the reactor (q) (either from exothermal reaction, from external fire or other sources) and the physical and thermodynamic properties.

Applying these assumptions some authors have developed expressions for the required relief flow in several situations. They are approximations and should be treated as such. Whenever possible, the results obtained with these methods should be subjected to verification, either experimentally or against more sophisticated models.

For vapour systems, Leung [34] assuming W and q to be constant, as well as the physical properties and also assuming that the vent stream has the same composition and vapour quality as the contents of the vessel, arrives at the expression:

$$W = G_c A = \frac{m_0 q}{\left[\left(\frac{V}{m_0} \frac{\Delta H_v}{v_{fg}} \right)^{1/2} + (C_v \Delta T)^{1/2} \right]^2} \qquad (7.22)$$

where m_0 is the initial mass in the vessel, C_v is the constant volume liquid heat capacity and ΔT the temperature increase corresponding to the overpressure (ΔP), which in vapour systems, for which this formula has been developed, can be obtained from the system's vapour pressure curve .

It is recommended to calculate q as the arithmetical average of the heat generation during venting, if no experimental data are available. If experimental data are available on the temperature increase over reaction time (t) (for example from the VSP), the use of the expression below is recommended:

$$q = \frac{1}{2} C_v \left[\left(\frac{dT}{dt} \right)_s + \left(\frac{dT}{dt} \right)_m \right] \qquad (7.23)$$

where m corresponds to the saturation temperature of the vessel contents under the maximum allowed overpressure and s to the set pressure (Figure 7.19).

Also for vapour systems, Fauske [27, 35] has developed a very simplified nomograph which is based on the following expression for the calculation of the area:

$$A = \frac{V\rho}{G_c \Delta t_v} \tag{7.24}$$

where the density, ρ, corresponds to the contents of the reactor, G_c is the mass flow density through the vent (according to the simplified expressions seen in a previous section for the HEM model) and Δt_v is the venting time, calculated as:

$$\Delta t_v \cong \frac{\Delta T C_p}{q_{set}} \tag{7.25}$$

where q_{set} is the heat generation per unit of mass at the set point, and C_p is the heat capacity of the reactor contents. Combining equations (7.14), (7.24) and (7.25), the expression below is obtained:

$$A = V\rho\ (T_{set}\ C_p)^{-1/2}\ (q_{set}/\Delta P) \tag{7.26}$$

This expression usually gives conservative results. However, it should not be used as more than a first approximation to obtain an order of magnitude of the required area. Any final calculation must be based on a more strict calculation.

For hybrid and gassy systems, Leung [36] describes an approximate method, although somewhat more complex than the previous ones.

The most complete methods currently are based on the joint resolution of the differential equations that constitute the system material and energy balances. The best known software is the SAFIRE code, developed by DIERS [37]. Equally, Huff [22] describes a numeric method for the simultaneous solution of the balances.

Example 7.16

Calculate for two-phase flow, according to the Leung and the Fauste methods, the relief area required to protect a 15 m³ reactor that contains 11000 kg of an unstable product A that can decompose exothermally. The set pressure of the relief system must be 15 bar. The parameters obtained from a VSP experiment are given.

Data: *Specific heat = 3.2 kJ/kgK $T_{saturation}$ at 15 bar = 167°C*
VSP: *$\Delta P/\Delta T$ = 16 000 Pa/K (average between 15 and 17 bar)*
       dT/dt (15 bar) = 0.14 K/s   dT/dt (16.5 bar) = 0.16 K/s

We will take an overpressure of 10%, so that the pressure during venting will be

$$P = (15-1.013) \times 10\% + 15 = 16.4\ bar$$

Therefore $\Delta P = 1.4$ bar

As $\Delta P/\Delta T = 16\,000$ Pa/K, for $\Delta P = 140\,000$ Pa

$$\Delta T = 140\,000 / 16\,000 = 8.75 \text{ K}$$

The heat generated per unit mass during venting will be, on average, from equation (7.23)

$$q = 0.5 \times 3200 \text{ J/kgK} \times (0.14 + 0.16) \text{ K/s} = 480 \text{ W/kg}$$

On the other hand, if the system has no heat losses, for a temperature increase velocity at the set point of 0.14 K/s, we can write

$$q_{set} = 3200 \times 0.14 = 448 \text{ W/kg}$$

(a) First we will apply the Fauske method, using equation (7.26) $(V \rho)$ is the reactant mass, i.e. 11 000 kg

$$A = 11\,000 \text{ kg} \times [(167 + 273) \text{ K} \times 3200 \text{ J/kgK}]^{-1/2} \times (448 \text{ W/kg} / 140\,000 \text{ Pa})$$

$$A = 0.0297 \text{ m}^2$$

which is equivalent to a diameter of 194 mm, so that a 200-mm diameter rupture disc will be necessary (the calculations only assume the flow through an orifice. If there were some additional piping, the area would have to be divided by the factor Ψ from Figure 7.16 to consider its effect on the flow).

(b) Applying now the Leung equation (7.22), with the data we already have, and taking into account equation (7.16), so that :

$$\Delta H_v / \upsilon_{fg} = T_s (dP/dT) = (273 + 167) \text{ K} \times 16\,000 \text{ Pa/K} = 7.04 \times 10^6 \text{ Pa}$$

we have that

$$W = (11000 \text{ kg} \times 480 \text{ W/K}) / \{(15 \text{ m}^3 \times 7.04 \times 10^6 \text{ Pa}/11000 \text{ kg})^{1/2} + (3200 \text{ J/kgK} \times 8.75 \text{ K})^{1/2}\}^2$$

$$W = 75 \text{ kg/s}$$

We now calculate the required area using equation (7.17) to calculate G_c taking $\Psi = 1$ for comparison with the previous case:

$$A = W/G_c = 75/\{0.9 \,(16000 \text{ Pa/K}) \,[(273 + 167)/3200]^{1/2}\} = 0.0140 \text{ m}^2$$

which would require a diameter of 134 mm.

As can be seen, the two methods differ by a factor of 2 in the area necessary. The most conservative value of 194 mm should be used.

7.4.7 Dust explosions

When handling combustible solids the venting systems usually consist of explosion panels. For their design semi-empirical equations are usually employed, that normally include constants that should be determined experimentally. The NFPA

68 standard [38] gives the recommended methods of calculation. Schofield [39] discusses them extensively.

Low pressure structures

The Swift and Epstein equation [40] is intended for low pressure structures (designed to withstand less than 0.1 bar g) and calculates the venting area required (in m²) as

$$A = C_{vent} A_s / P^{1/2} \tag{7.27}$$

A_s is the internal system area (m²) and P is the maximum pressure (kPa) that the weakest element of the installation can resist. C_{vent} is a dimensional constant whose value for different products or types of solids is given in Table 7.5. The values of this constant must be determined experimentally. In the bibliography values for other gases and dusts can be found.

In Table 7.5 the kinds of dust are classified into three types, depending on the value of the constant K_{st}. This constant comes from the 'cubic law', which determines that for certain conditions the maximum pressure increase rate is related to the system volume (V), so that:

$$(dP/dt)_{max} V^{1/3} = K_{st} \tag{7.28}$$

The constant K_{st}, when measured under standard conditions, is a good yardstick of the violence of the explosion that a dust can cause. Given in Table 7.6 are details of the principal characteristics of each dust class. To find the value of k_{st} one may look in tables, such as those in the previously quoted references, or send samples to a qualified laboratory to have it determined.

High pressure structures

High pressure structures are defined as those that can withstand more than 0.1 bar g. The calculation procedure is based on the cubic law and is described in detail in NFPA 68. It has been collected in a group of nomographs shown in Figures 7.20, 7.21 and 7.22.

Table 7.5 Characteristics of some combustible dusts according to NFPA 68 (1988)

Combustible	C (kPa$^{1/2}$)
Dust St-1	0.26
Dust St-2	0.30
Dust St-3	0.51

Table 7.6 Definitions of the dust explosion classes based on a 10 kJ ignition source and tests in a volume of 1 m³

Dust explosion class	K_{st} (bar m/s)	Characteristics
St-0	0	No explosion
St-1	0–200	Weak explosion
St-2	200–300	Strong explosion
St-3	>300	Very strong explosion

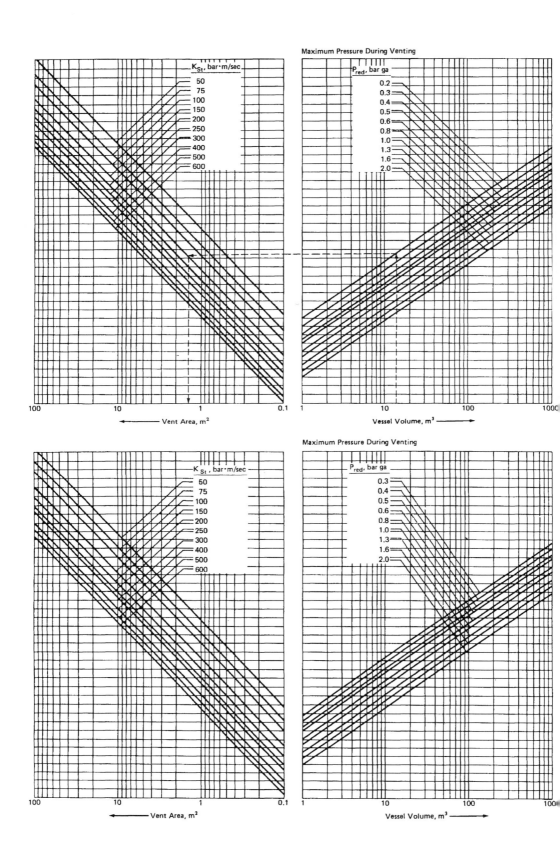

Maximum Pressure During Venting

K_{St}, bar·m/sec
- 50
- 75
- 100
- 150
- 200
- 250
- 300
- 400
- 500
- 600

P_{red}, bar ga
- 0.2
- 0.3
- 0.4
- 0.5
- 0.6
- 0.8
- 1.0
- 1.3
- 1.6
- 2.0

Vent Area, m²

Vessel Volume, m³

Maximum Pressure During Venting

K_{St}, bar·m/sec
- 50
- 75
- 100
- 150
- 200
- 250
- 300
- 400
- 500
- 600

P_{red}, bar ga
- 0.3
- 0.4
- 0.5
- 0.6
- 0.8
- 1.0
- 1.3
- 1.6
- 2.0

Vent Area, m²

Vessel Volume, m³

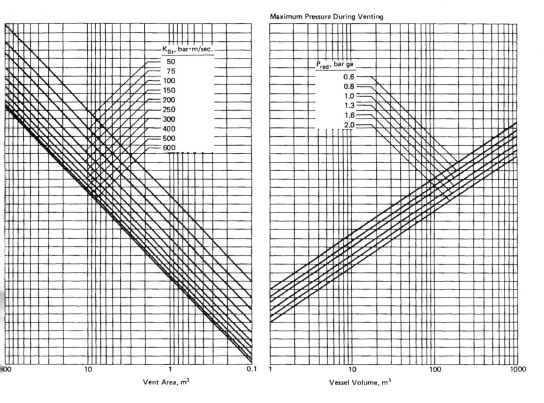

Maximum Pressure During Venting

K_{St}, bar·m/sec.
50
75
100
150
200
250
300
400
500
600

P_{red}, bar ga
0.6
0.8
1.0
1.3
1.6
2.0

Vent Area, m² 00 10 1 0.1

Vessel Volume, m³ 1 10 100 1000

Figure 7.22 Nomograph for the design of relief openings for combustible dusts according to the standard NFPA 68 [38], P_{stat} = 0.5 bar g. Reproduced with permission from NFPA 68: 'Deflagration venting' Copyright © 1988 National Fire Protection Association, Quincy, MA 02269. This reprinted material is not the complete and official position of the National Fire Protection Association on the referenced subject, which is represented only by the standard in its entirety.

Figure 7.20 (page 270, top) Nomograph for the design of relief openings for combustible dusts according to the standard NFPA 68 [38], P_{stat} = 0.1 bar g. Reproduced with permission from NFPA 68: 'Deflagration venting' Copyright © 1988 National Fire Protection Association, Quincy, MA 02269. This reprinted material is not the complete and official position of the National Fire Protection Association on the referenced subject, which is represented only by the standard in its entirety.

Figure 7.21 (page 270, lower) Nomograph for the design of relief openings for combustible dusts according to the standard NFPA 68 [38], P_{stat} = 0.2 bar g. Reproduced with permission from NFPA 68: 'Deflagration venting' Copyright © 1988 National Fire Protection Association, Quincy, MA 02269. This reprinted material is not the complete and official position of the National Fire Protection Association on the referenced subject, which is represented only by the standard in its entirety.

Each graph is constructed for a different value of the vent opening pressure (P_{stat}). The graph is entered by the horizontal axis of the graph on the right with the system volume. One moves upwards until finding the maximum pressure that the system is allowed to reach (P_{red}). Now horizontally one passes to the graph on the left, until the K_{st} value of the dust (there are also graphs with the type of dust only). From this point one drops vertically until reading the value of the venting area necessary in the horizontal axis.

7.4.8   Design documentation

Pressure relief systems are critical equipment, with a very important role in plant safety. Therefore it is essential that a complete record exists of all the reference information for each one of them. All the records should be filed so they are easily accessible in case of modifications or design reviews. The minimum to be included:

1. Design basis:
    *   Schematic diagram of the protected unit.
    *   Process conditions (normal operation, start-up, etc.).
    *   Physical and thermodynamic properties, as well as the source from which they came.
    *   Reactivity (reactions, kinetic constants, VSP, ARC results, etc.), if applicable.
    *   Scenarios considered.
    *   Scenario selected and justification.
2. Calculations:
    *   Complete calculation of the required flow, with reference to the codes and standards applicable.
    *   Selection of the type of element and justification of the decision.
2. Conclusions:
    *   Specifications sheet of the equipment to be installed. (Shown in Figure 7.23 is an example for safety valves adapted from the requirements of the API 526 standards. The sections that should be filled in by the Process Engineer for requesting bids from manufacturers are marked with an X.)

If any kind of additional protection is necessary to reduce the size of the relief element, this should be clearly specified in the documentation.

Emergency relief systems should be designed by qualified and trained personnel, who are aware of the applicable legislation and who have been supervised for a time by an experienced designer. Especially in the cases of two-phase flow or chemical reaction, experienced personnel should be selected.

7.4.9   Vent treatment

During venting, toxic fluids or combustibles must not be allowed to escape into the atmosphere or the drainage systems, so it is necessary to provide conduction

PRESSURE RELIEF VALVE
SPECIFICATION SHEET

Sheet No. _____
Requisition No. _____
Job No. _____
Date _____
Revised _____
By _____

General

1	Item No.							
2	Tag No.	×						
3	Service, line, or equipment no.	×						
4	Number required	×						
5	Full nozzle, semi-nozzle, or other							
6	Design type							
	A. Safety, relief, or safety relief	×						
	B. Conventional, bellows, or pilot operated	×						
7	Bonnet type							

Connections

8	Size (inlet/outlet)	×//×						
9	Flange class, ANSI or screwed	×//×						
10	Type facing							

Material

11	Body/bonnet	×//×						
12A	Seat/disk	×//×						
12B	Resilient seat seal							
13	Guide/rings							
14	Spring							
15	Bellows							

Accessories

16	Cap, screwed or bolted							
17	Lever, plain or packed							
18	Gag							
19								
20								

Basis of Selections

21	Code	×						
22	Fire	×						
23	Other	×						

Service Conditions

24	Fluid and state	×						
25	Required capacity per valve and units	×						
26	Molecular weight or specific gravity at flowing temperature	×						
27	Viscosity at flowing temperature and units	×						
28	Operating pressure, psig/set pressure, psig	×/×						
29	Operating temperature, F/flowing temperature, F	×/×						
30	Constant back pressure, psig	×						
31	Variable back pressure, psig	×						
32	Differential set pressure	×						
33	Allowable overpressure, percent	×						
34	Compressibility factor	×						
35								

Orifice area

36	Calculated, square inches	×						
37	Selected, square inches	×						
38	Orifice designation	×						
39	Manufacturer's model no.							
40	Manufacturer							

Notes:

Figure 7.23 Pressure relief valve specification sheet (according to API Standard 526 [20]). (Courtesy of the American Petroleum Institute, Washington, DC.)

and treatment systems for these effluents. When dealing with air, steam or water, it is normally sufficient to protect the safety valve exit to avoid direct injury to personnel. For the rest of the cases different alternatives exist, depending on the process and the equipment. These are generally limited because of the unpredictability of venting and the abnormal conditions in which it normally occurs. The most frequent way is to conduct the fluid to an adequate treatment (flare, treatment pool, absorber, etc.). In the API Recommended Practice 521 [15] criteria are given for the treatment of these effluents and rules for the design of the equipment.

Headers

When designing a vent header one must take into account:

1. *The composition and physical state of the fluid.* If there are liquids or very cold currents it is necessary to vaporize them (to protect the flare) or heat them (to avoid material brittleness, as normally carbon steel is used). In these cases a special collector for the cold streams and the liquids is usually designed, because special materials are needed.

   All these streams are fed to a heating or vaporization system, and from there to the common header.

   It is necessary for the vaporization system to react quickly at the start of venting, before it is completely full of liquid. One possibility is that shown in Figure 7.24. The system uses a thermal fluid, that can be methanol, ammonia or another compound with the correct boiling point. When, due to venting, cold liquid reaches the vaporizer, the pressure of the thermal fluid drops because of condensation, causing the steam valve to open. The thermal fluid vaporizes in the boiler going from there to the heater and to the process vaporizer, where it condenses, returning to the boiler to repeat the cycle. The liquid coming from the

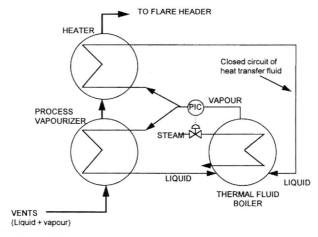

Figure 7.24   Vaporizing and heating system for liquid vent systems.

venting boils in the process vaporizer and passes to the heater to reach the required temperature before going to the common header.

The vapours can be reused after condensation many times, avoiding their loss in the flare. The design of condensers and vaporizers is described, among others, by Coulson [41, 42] and Kern [43].

There is a certain risk that some liquid can reach the flare due to entrainment or condensation, so that drop separators or water traps are installed as close as possible to the flare to remove them. The design, including typical drawings, is described in API RP 521, and also in the bibliography, e.g., Coulson [41] and Crowl and Louvar [44].

2. *The reactivity and compatibility between the different streams that can end up in the header.* It can be the case that the combination of two of them can create a reaction risk. Whenever this possibility exists it has to be avoided by the design of the system. The normal way is to take the most reactive components to a different system, with equipment for their treatment or neutralization. Frequent equipment are washers or gas absorbers that, with a suitable fluid, separate the hazardous components, sending the rest to the flare or to the atmosphere. The design of this equipment has been described by Coulson [41, 42], Treybal [45] and others.

The presence of reactive or corrosive compounds in the header during venting is to be avoided. A valve constructed in a material suitable for the process can be attacked and severely damaged in these circumstances by products from other valves.

3. *The flow of each valve and the possibility that several open simultaneously.* Just as a design scenario has been elaborated for each relief element, a global scenario has to be developed for the design of the general plant header. A basic criterion is that the design backpressure value for each safety relief element is not exceeded.

In addition to being able to handle the flow of each relief element individually, the system must be optimized to cope with all of the reasonable combinations of simultaneous opening of various elements. In this case the scenario is much more general, and affects almost all the equipment in the plant. Header design emergencies normally considered are failures in the power supply, fire affecting a determined area, lack of instrument air, cooling water failure, etc.

When considering this type of situation time has to be considered in the selection of the design scenario. For example, in the case of a failure in the cooling water system in an ethylene cracker, the opening of the compressor safety valves will be almost instantaneous, while the large distillation columns have a greater inertia due to their higher inventory. Normally when the column valves start relieving, the compressors are already shut down. If this consideration is omitted the flow can be overestimated by more than 100%, leading to headers much bigger than required.

The calculation of backpressures is usually made from the end of the header backwards, solving each branch until reaching the valves. If the head loss in a pipe section is not greater than 10% of the initial pressure, the equations for incompressible flow can be used. Equations for the calculation of head losses can be taken from standard references [42, 46]. The physical properties and the temperature should be taken for the mixture present at each moment. It is important to consider all the head losses caused by size increases and other accessories, not forgetting 'T' connections [47, 48]. They are very often as high as or higher than straight pipe sections.

For ratings of existing designs, the resolution is direct. With one calculation the backpressures for each valve are obtained. Repeating the calculation for all the possible scenarios one can verify if the design backpressure of each valve is not exceeded. If this is not the case for each valve it is necessary to consider replacing the affected valves by balanced or pilot operated ones (where the backpressure has less influence) or even modifying the header (this option is usually more expensive, and normally requires a general plant shut-down).

When designing a new system, the determination of the diameter of each section is not direct. It is necessary to follow an iterative process until finding possible combinations of diameters that do not cause the design backpressures to be exceeded. In deciding the final solution the minimum total cost is considered.

Flares

Flares are the most frequently used system to eliminate vent streams. Flaring is always wasteful, so only streams that cannot be recycled or for which there is no other alternative should be sent to the flare. The most common type is the elevated flare, although ground flares, and even enclosed flares exist, but these need much more free space.

An elevated flare is a burner supported by metal guys (up to 150 m high) or by structure in cases of greater heights. The burner is located at the top with a seal to prevent the entrance of air into the flare head. The main characteristic of a flare is its capacity to burn flows variable in ratios of 1 to 30 000, while a normal burner is usually designed for a relation of 1 to 5–10. They normally have a steam injection system to reduce smoke formation.

The main flare design parameters are the height and the diameter. Flare design is a highly specialized discipline, and specially when determining the diameter, it is convenient to consult with the manufacturers and take a joint decision.

The diameter not only influences the head loss in the flare. If the exit velocity of the gases is too low, a flame backflow (flashback) might occur, while if it is too high the flame can be blow out. The diameter is usually fixed so that the velocity during normal venting is around 0.2 Mach. In shorter, less frequent situations it can be allowed to reach up to 0.5 Mach.

The radiation level on the ground during venting depends on the height. The allowable values (excluding solar radiation) are 1.58 kW/m^2 for areas where people are continually exposed and 9.46 kW/m^2 for areas accessible to personnel. Various methods exist for calculating the flame length and tilting as a function of the heat generated and the velocities of the flame and the wind [15, 44]. Normally a punctual emission model is considered, with the centre at the middle point of the flame. From here the distance for the previously mentioned radiation levels can be calculated. The available methods have to be used with precaution, as they have limited intervals of application, because numerous factors are included, such as wind, fraction of heat radiated, length and inclination of the flame, etc.

7.5 Questions and problems

7.1   A chemical plant receives liquefied chlorine that must be vaporized before use. Discuss the different possibilities existing for the design of the vaporizer, including:

(a) Type of heat exchanger.
(b) Heating media.
(c) Control.
(d) Emergency relief and protection systems.

Select the best system and produce a diagram of the equipment, piping and instrumentation showing the details of the installation.

7.2   Study the different alternatives for the storage of 1000 MT of propylene oxide (PO).
Consider, among others, the following aspects:

(a) The physical state and type of tank.
(b) Design pressure and temperature.
(c) Auxiliary systems (condensers, compressors, etc.).
(d) Pressure relief systems.

7.3   Consider the possible causes of vacuum in a volatile liquid storage tank and determine the required air inlet flow to prevent the collapse of the tank. Discuss the intrinsic safety of systems with vacuum safety valves.

7.4   Shown in Figure 7.25 is a distillation system for a flammable fluid in order to separate the high boiling point compounds (relief elements not shown).

(a) Discuss generally the control strategy and the risks it presents. Make the necessary modifications to the diagram.
(b) Study the relief elements necessary, their type and location.

7.5.   In a stirred batch reactor a highly exothermic reaction is carried out. The process stages are:

(a) Loading of reactants and solvents (hexane).
(b) Initiator addition.
(c) Heating to the reaction start temperature.
(d) Cooling to maintain the temperature stable during the reaction.
    Condensation of the evaporated solvent and return of the condensate to the reactor.
(e) Cooling.
(f) Discharge of the reactor to the separation unit and purification of the product.

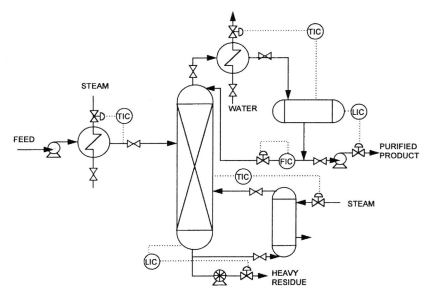

Figure 7.25   Scheme of a distillation unit (Problem 7.4).

Discuss the different ways to carry out the operations: equipment, controls, piping and instrumentation. Draw a P&ID with the final design.

7.6   A tank 2 m in diameter and 4 m high is used for in-process storage of benzene. Normally the level is at 70% and the working pressure is 3 bar. Design the safety valve necessary for the case of an external fire. The design pressure of the tank is 5 bar.

Because of the actual layout of the equipment, it is necessary that the valve is located at a distance of 1.2 m from the tank. From this point up to the branch of the nearest flare header 6 m of pipe will be necessary, including three 90° elbows.

7.7   Select the safety valve necessary to protect a stage of a centrifugal compressor with the capacity–pressure curve given. The fluid is propylene at 10°C (at relief conditions), the set pressure is 1.3 bar g and the backpressure existing in the header is 0.7 bar g .

Capacity (m³/h)	0		30 000	50 000	70 000
Pressure (bar)		1.6	1.35	1.2	1.1

7.8   A tank is cooled with water at 25°C through a 10 m² coil. The temperature in the tank is 140°C and the global heat exhange coefficient is 2000 W/m²K. Calculate the relief valve size required to protect the coil in the case of blocking and thermal expansion. The design pressure of the coil is 4 bar.

7.9   What if in the above case the temperature of the water should be 125°C and the tank temperature 165°C ?

7.10   In a 10-m³ tank a highly reactive compound dissolved in isooctane is stored at 20°C. The total quantity of mixture is 5000 kg and the set pressure of the relief element is 5 bar.

Determine the size and type of the pressure relief system required. To reach the nearest collector 6 m of piping and two 90° elbows are needed.

For calculating the physical and thermodynamic properties the system can be considered as pure isooctane.

The system characteristics with regard to reactivity are:

$$dT/dt = 0.75 \text{ K/s} \qquad \text{at 5 bar abs.}$$
$$dT/dt = 1 \text{ K/s} \qquad \text{at 5.4 bar abs.}$$

7.11   Figure 7.26 shows the evolution of temperature and pressure with time for an exothermal polymerization in a closed system. Select the design pressure and temperature for a storage tank containing 100 MT of this product and design a suitable pressure relief system.

Data: $C_p = 2400 \text{ J/kgK}$ $\rho_1 = 710 \text{ kg/m}^3$

Assume that the system pressure is determined exclusively by the vapour pressure.

7.12   Determine the pipe diameter required to vent 14 000 kg/h of hexane at its saturation temperature and with a vapour fraction (quality) of 0.05, along a line 10 m long with four 90° elbows. The system pressure is 5 bar and the backpressure 3.5 bar.

Study the influence of the inlet vapour quality on the required diameter.

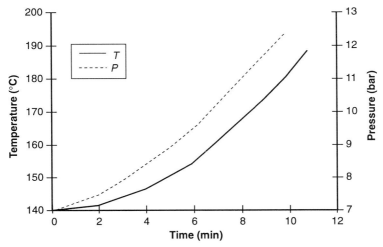

Figure 7.26   Evolution of temperature and pressure with time during the runaway reaction of Problem 7.11

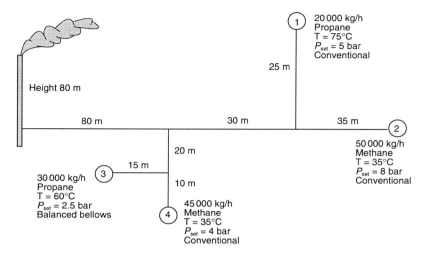

Figure 7.27   Group of safety valves for Problem 7.13.

7.13   Determine the diameter of different sections of the flare header in Figure 7.27. Take as the maximum backpressure for a conventional valve 10% of the gauge set pressure, and for balanced bellows 40%. Use $0.015 \, mNs/m^2$ as an average viscosity.

7.14   A flash dryer for maize starch ($K_{st} = 202 \, bar \, m/s$) is 20 m high and has a diameter of 0.8 m. The volume of the cyclones is $3 \, m^3$. Calculate the venting area for a venting opening pressure of 0.2 bar g and a maximum pressure of 0.4 bar g.

7.6   References

1. Kletz, T. (1985) *Cheaper, Safer Plants or Wealth and Safety at Work. Notes on Inherently Safer and Simpler Plants*, Institute of Chemical Engineers, Rugby.
2. CCPS (Center for Chemical Process Safety) (1993) *Guidelines for Engineering Design for Process Safety*, American Institute of Chemical Engineers, New York.
3. Kletz, T. (1991) *An Engineer's View of Human Error*, 2nd edn, The Institution of Chemical Engineers, Rugby.
4. Perry, R. H. and Green, D. (eds) (1984) *Perry's Chemical Engineer's Handbook*, 6th edn, McGraw-Hill, New York.
5. Tubular Exchangers Manufacturers Association (TEMA) (1988) *Standards*, 7th edn, Tarrytown, NJ.
6. Spanish Department of Industry and Energy, Publications Centre (1989) *Reglamento de aparatos a presión e instrucciones técnicas complementarias* (Pressure vessels code and related technical instructions), Madrid.
7. Directive on the Control of Major-accident Hazards Involving Dangerous Substances. 96/82/EC of 9 December 1996. Modified later by 87/216/EC and 88/610/EC.
8. Levenspiel, O. (1972) *Chemical Reaction Engineering*, 2nd edn, Wiley, New York.
9. Levenspiel, O. (1979) *Chemical Reaction Omnibook*, Corvallis, OSU book centre.
10. Aris, R. (1969) *Elementary Chemical Reactor Analysis*, Prentice-Hall, Englewood Cliffs.
11. Mecklenburg, J. (1983) *Plant Layout*, Longman.

12. Kern, R. (1978) Series of articles on layout. *Chem. Eng.*, August (first chapter).
13. Wells, G. L. and Rose, L. M. (1986) *The Art of Chemical Process Design.* Elsevier.
14. The American Society of Mechanical Engineers (1986) *ASME Boiler and Pressure Vessel Code, Section VIII, Rules for construction of pressure vessels, Division I and Division II - Alternative rules.* ASME, New York.
15. American Petroleum Institute (1990) Guide for pressure relieving and depressuring systems. *API Recommended Practice 521*, 3rd edn, API, Washington, DC.
16. NFPA (National Fire Protection Association) (1988) NFPA 30: *Flammable and combustible liquids code*, NFPA, Quincy, Massachussets.
17. Kletz, T. A. (1987) *An Engineer's View of Human Error.* Proceedings of the International Symposium on Preventing Major Chemical Accidents (ed. J. L. Woodward), CCPS/AIChE, New York.
18. Parry, C. F. (1992) *Relief Systems Handbook*, Institute of Chemical Engineers, Rugby.
19. Reid, R. C., Prausnitz, J. M. and Poling, B. E. (1987) *The Properties of Gases and Liquids*, McGraw-Hill, New York.
20. American Petroleum Institute (1984) Flanged steel safety-relief valves. *API Standard 526*, 3rd edn, API, Washington, DC.
21. American Petroleum Institute (1990) Sizing, selection, and installation of pressure-relieving devices in refineries. *API Recommended Practice 520. Part 1 - Sizing and selection,* 5th edn, API, Washington, DC; (1988) *Part 2 - Installation*, 3rd edn, API, Washington, DC.
22. Huff , J. E. (1977) *A General Approach to the Sizing of Emergency Pressure Relief Systems.* Proceedings of Second International Symposium on Loss Prevention and Safety Promotion in the Process Industries, September 1977, Heidelberg, Germany. DECHEMA, Frankfurt, pp. IV 233–40.
23. Singh, J. (1990) Sizing relief vents for runaway reactions. *Chem Eng.*, August.
24. Fauske, H. K., Epstein, M., Grolmes, M. A. and Leung J. C. (1986) Emergency relief vent sizing for fire emergencies involving liquid-filled atmospheric storage vessels. *Plant/Operations Prog.*, 5(4).
25. Grolmes M. A. and Epstein, M. (1985) Vapor-liquid disengagement in atmospheric liquid storage vessels subjected to external heat source. *Plant/Operations Prog.*, 4(4).
26. Leung, J. C. (1986) A generalised correlation for one-component homogeneous equilibrium flashing choked flow. *AIChE J.*, 32(10), 1743.
27. Fauske, H. K. (1984) Generalised vent sizing nomogram for runaway chemical reactions. *Plant/Operations Prog.*, 3(4), 213.
28. Leung, J. C. and Grolmes, M. A. (1987) The discharge of two-phase flashing flow in a horizontal duct. *AIChE J.*, 33(3), 524.
29. First, K. E. and Huff, J. E. (1989) *Design Charts for Two Phase Flashing Flow in Emergency Pressure Relief Systems.* Proceedings of International Symposium on Runaway Chemical Reactions, CCPS/AIChE. New York.
30. Leung, J. C. and Epstein, M. (1991) Flashing two-phase flow including the effects of non condensable gases. *ASME Trans. J. of Heat Transf.*, 113(1), 269.
31. Fauske, H. K. (1985) Emergency Relief System (ERS) design. *Chem. Eng. Prog.*, August, 53–6.
32. Fauske, H. K. and Leung, J. C. (1985) New experimental technique for characterising runaway chemical reactions. *Chem. Eng. Prog.*, August, 39.
33. Huff, J. E. (1982) Emergency venting requirements. *Plant/Operations Prog.*, 1(4), 211.
34. Leung, J. C. (1986) Simplified vent sizing equations for emergency relief requirements in reactors and storage vessels. *AIChE J.*, 32(10), 1622.
35. Fauske, H. K. (1984) A quick approach to reactor vent sizing. *Plant/Operations Prog.*, 3(3).
36. Leung J. C. (1992) Venting of runaway reactions with gas generation. *AIChE J.*, 38(5), 723.
37. Grolmes, M. A. and Leung, J. C. (1985) Code method for evaluating integrated relief phenomena. *Chem. Eng. Prog.*, August, 47–52.
38. NFPA (National Fire Protection Association) (1988) *NFPA 68: Guide for venting of deflagrations,* NFPA, Quincy, Massachussets.
39. Schofield, C. (1985) *Guide to Dust Explosion Prevention and Protection. Part 1 - Venting,* Institute of Chemical Engineers, Rugby.
40. Swift, I. and Epstein, M. (1987) Performance of low pressure explosion vents. *Plant/Operations Prog.*, 6(2).

41. Sinnot, R. K. (1996) *Coulson and Richardson's Chemical Engineering, Vol. 6, Chemical Engineering Design*, 2nd edn, Pergamon, Oxford.
42. Coulson, J. M. and Richardson, J. F. *Chemical Engineering*, Vols 1 and 3, 5th and 4th edn, Pergamon, Oxford.
43. Kern, D. Q. (1950) *Process Heat Transfer*, McGraw-Hill, New York.
44. Crowl, D. A. and Louvar, J. F. (1990) *Chemical Process Safety, Fundamentals with Applications*, Prentice Hall, Englewood Cliffs.
45. Treybal, R. (1980) *Mass Transfer Operations*, 3rd edn, McGraw-Hill, New York.
46. Hall, S. M. (1993) Size and design of relief headers. *Chem. Eng. Prog.*, **89**(3).
47. Miller, D. S. (1986) *Internal Flow Systems*, BHRA, The Fluid Engineering Centre.
48. Hooper, W. B. (1981) The two-K method predicts head losses in pipe fittings. *Chem. Eng.*, 24.
49. Wells, G. L. (1980) *Safety in Process Plant Design*, Godwin/IChemE, Rugby.
50. King, R. (1990) *Safety in the Process Industries*. Butterworth-Heinemann, London.
51. Lees, F. P. (1980) *Loss Prevention in the Process Industries*, Butterworth-Heinemann, London.
52. Englund, S. M. (1991) Design and operate plants for inherent safety (2 parts). *Chem. Eng. Prog.*, March–May.
53. Kletz, T. (1994) *What Went Wrong? Case Histories of Process Plant Disasters*, 3rd edn, Gulf Publishing Company, Houston.

8 Risk reduction in operation and maintenance

...neither should he lend too ready an ear to the terrifying tales which may be told him, but should temper his mercy with prudence, in such a manner, that too much confidence may not put him off his guard...

Nicolás Maquiavelo, The Prince, Chapter XVII

8.1 Risk reduction in operation

The design of a chemical factory can never be totally safe, completely 'human error proof'. The complexity of the work to be carried out, the wide variety of working conditions, the adaptation to the conditions of raw materials and the ever possible chance of an unforeseen failure are factors which make the correct management of a plant a factor of such importance as that of the initial design.

A well-designed, completely automated factory with the best existing technology cannot guarantee that no serious accidents will occur because of an error in communication, a start-up operation done in an incorrect sequence, insufficient control of the modifications or inadequate maintenance procedures, etc.

Equally, an originally safe factory could cease to be so if it were not managed and maintained with maximum care. The accidents which occurred at Flixborough and Bhopal are two typical cases caused by serious negligence, as described in the appendix.

To control the process we resort to ever more complex automated systems. The handling of these requires qualified and trained operators. The training and education programmes, together with dynamic simulations of the plant operation, are ever more vital elements of the management systems necessary to attain a high degree of safety in a chemical plant.

8.1.1 Safety and environmental management systems

Safety has traditionally been managed as a programme, similar to those existing in other areas of a company, based on slogans and objectives. The main problem with this is the amount of fashionable dogma that can be involved. The effectiveness of the programme could suffer, with the consequent risks, when other objectives distract attention from it.

The concept of a management system improves effectiveness through a different approach. Such a system is a form of working that secures, in a continuous and systematic way, the fulfilment of the established rules and procedures. The management system should be established by the head officer of a company who should assure that periodic revisions of its efficiency are carried out. In this way the to-and-fro movements of classical programmes can be successfully avoided.

More and more legislation is being introduced at national and international levels on safety and prevention of accidents in the chemical industry. The SEVESO EC Directive has recorded an historical milestone, demanding that those industries which handle dangerous compounds, above certain fixed quantities, must give notification to the corresponding government entity and set up internal and external emergency plans. Documents such as these have followed:

- The Royal Decree RD 886/1990, of 29.6.1990 adapted from the SEVESO Directive (82/501/CEE, 87/216/CEE and 88/610/CEE) to Spanish legislation.
- The American Law on 'Process safety management of hazardous chemicals' (OSHA 29CFR 1910.119 February 1992) [1, 2].
- The British Standard: BS7750 (Environmental Management Systems) [3] on 'Environmental management systems', based on the scheme for quality management in the series of standards ISO 9000 (UNE-EN-ISO 9000 [4]).
- The recommended practice API RP 750 'Management of process hazards' [5].

All of these standards, to a greater or lesser degree, require or define safety management systems. Some of them are limited to activities related to certain types of product (toxic, flammable, etc.) but their philosophies are very similar.

The Centre for Chemical Process Safety (CCPS) of the AIChE proposes [6, 7] twelve basic elements which should be specifically defined by a safety management system.

1 Responsibility, policy and objectives.
2 Process knowledge and documentation.
3 Project reviews and design procedures.
4 Management of process hazards.
5 Management of changes (modifications in the installation, in operating procedures, etc.)
6 Process and equipment adequacy.
7 Incident investigation.
8 Personnel training and performance evaluation.
9 Human factors.
10 Regulations, codes and legislation.
11 Audits and corrective action.
12 Improvement of knowledge on process safety.

These factors are not independent but reach maximum efficiency when carried out simultaneously. It is their inculcation to the company's philosophy that allows the system to attain its objectives. The suitability of the management system directly influences the safety results of the company through the creation of a positive, or negative, socio-technical environment. The quantification of the suitability of management systems has been much discussed by Jenssen [8]. Although techniques are still limited, some indices allow the correction of the frequency of normal failures in some equipment, depending on the management system of the company, revealing the influence a good management of risks has on the reduction in the number of accidents, including aspects attributable to equipment, due to good design and maintenance practice. Figure 8.1 shows how the different aspects of a management system affect achievement of a safe operation. The possible causes of accidents, from those more direct and immediate to those that are found at the root of a problem, are strongly related to the management system.

Another recently introduced concept is that of 'responsible care' [9] which goes one step beyond the safety regulations already mentioned, approaching the models of Total Quality. It is a voluntary programme promoted by the chemical industry associations in which are included all the activities of a company, and it guarantees that the environmental impact of the company, in the broadest sense, is maintained at a minimum level. It includes six aspects, each subject to a set of requirements that must be observed.

- Process safety (including hazard analysis).
- Pollution prevention (environmental management).
- Product stewardship (product life monitoring, recycling and environmental impact)
- Safety and hygiene at work
- Community right-to-know and emergency plans.
- Product distribution.

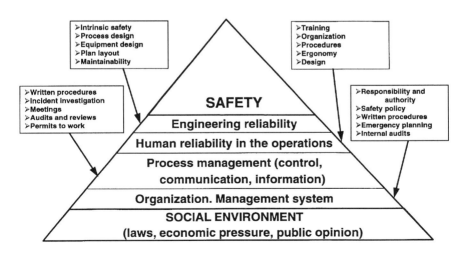

Figure 8.1 Pyramid showing how different aspects of a management system affect the achievement of a safe operation (based on [8]).

Principles of a safety control system

All safety management systems should start from a solid foundation, established in a written policy and signed by the company management. This policy is the cornerstone on which the system is based. In this case the strategic vision of management regarding the safety of employees, the installations and the community should be clearly defined.

The key to success when introducing a safety management system is the support of the management. Senior management should make known to all levels in the company organization their concern for security, and provide the human, financial and technical resources necessary. Periodical revisions, based principally on the audit reports, investigations of accidents and other statistics, should establish and review medium and short-term objectives.

In order to achieve these objectives it is not only important to fix a goal (if possible specified by a number) to be attained and a time frame, but also to establish control procedures for effectiveness. A system of this type should contain at least the following elements:

1. System procedures: These procedures define the safety management activities and the monitoring and correction of problems. They establish the mechanisms which let the system work, its control and day to day running. It includes audits, incident and accident investigations, training, emergency plans, documentation control, records, communications, etc.
2. Operating procedures or work instructions: They describe the manner in which a task or operation should be completed. They are directed at the operator.
3. Records: Documents containing information confirming that activities have been completed according to procedures or instructions.

The direct responsibility for safety is not of one person or department alone. It is of each employee and of all levels of supervision. Furthermore a safety co-ordinator normally exists whose function is to give technical support and training. He receives authority from the management to make the system effective, co-ordinating its operation and acting as an adviser. If this person's authority or prestige is not sufficient or he does not have direct communication with the senior executive of the company, the system will probably not work properly.

Figure 8.2 shows a scheme for the implementation and running of an environmental management system proposed by the British Standard BS 7750: 1992 [3]. Its basic principles are also applicable to safety management. It starts with management conviction that it is necessary to establish a management system. After evaluation of current conditions the company's policy is defined and the circle is entered. The first steps are the assignment of personnel and the necessary resources, and a compilation of the applicable legislation together with all available data on the actual conditions within the company. With this information to hand the objectives may now be determined. To achieve them a management programme must be established, detailed in a procedural handbook which will generate records showing that the procedures have been observed and allow an evaluation of the

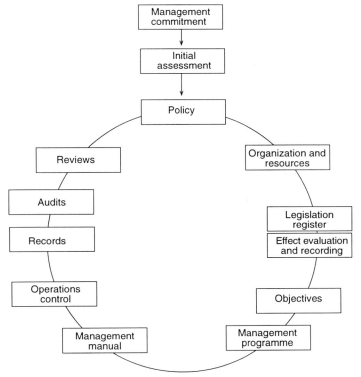

Figure 8.2   Scheme for the implementation and running of an environmental management system, proposed by the British Standard BS 7750: 1992 [3].

fulfilment of the objectives. Verification that the established procedures are being carried out is obtained through internal audits whose results are periodically checked. When the objectives are achieved, new ones are established and the circle is closed.

8.1.2   Control of the safety management system

The level of fulfilment of the procedures should be assessed with a frequency that has been previously established according to the importance of each activity. Through these checks weak points can be detected before problems occur. Immediately after this the necessary corrective actions should be taken to rectify the detected deficiencies. These controls also serve to [10]:

- Maintain awareness of the importance of following the established procedures.
- Identify hazards not considered or introduced by process or plant modifications.
- Verify observation of the law or external regulations to which the plant is subject.
- Improve the system: revising the suitability of operational procedures and proposing alternatives.

Whenever possible it is preferable to use a system which allows presentation of audit results by a number [8]: 'What gets measured gets done'. If the quality of a system can be measured by a numerical index it is much easier to assess the effort for improvement and its efficiency. It is not easy moreover, in the widest ranging revisions, because different elements that are checked have a different relative importance and, therefore, it is necessary to introduce balancing factors. In addition the numerical evaluation of the degree of fulfilment requires that the audit criteria have been very clearly defined, reducing subjective differences to a minimum.

Depending upon the extent, complexity and transcendence of the activities involved different types of evaluation may be performed [11, 12].

Safety audits

These consist of detailed revision and evaluation of the observance of regulations established for a determined activity. All system activities and procedures, operational procedures, training and education, emergency plans, etc. should be audited on a regular basis. Depending on the kind of audit, the frequency could vary considerably. Those which are confined to checking the observance of system procedures usually occur twice a year while those that affect procedural contents (process technology, design philosophy, etc.) normally take place every two or three years. The audit team should be made up of trained people independent from the area that is to be audited. The number could vary depending on the extent of the audit. Normally, for reasons of co-ordination, it is not practical for more than three people to participate, except in large audits which affect a whole plant and need two or three weeks to complete. A balance between safety experts and non-specialists with experience in similar processes, with good judgement skills should be sought. The presence of outside people favours the interchange of ideas and experiences.

It is advisable to prepare an annual audit plan so that dates and procedures to be audited are known in advance. In this way, people who are to be interviewed will be more readily available. Furthermore, knowledge of the date in advance will favour the fact that personnel involved will review the procedures to be audited prior to the event, detecting and correcting problems. It should not be forgotten that these audits are an internal exercise carried out in good spirit. The objective is not the discovery of non observance but improving the system.

The preliminary work of the audit team is its key to success. It is necessary to request in advance the main procedures, regulations and documents that are used in the area to be audited. Each auditor should request the information he needs, depending on his knowledge of the plant and its process. All relevant documents, without exception, should be made available. Based on this information the audit programme, and checklists which help to back it up, will be developed. The checklist is a useful tool but in no way is it a substitute for the investigative capacity and problem follow through possessed by a skilled auditor. It is usually better to choose a few subjects to be investigated thoroughly than to try and take on everything. An audit is an exercise with limited time and adapting to this circumstance is the mark of a good auditor.

The preparation phase ends here. In a strict sense the audit begins at a meeting with the person in charge of the activities to be audited, in which the purpose of the audit is reviewed and a programme with the maximum possible detail is established. After this meeting, the auditors revise in situ the application of the procedures, check the records and interview all necessary staff. A person from the department visited will accompany the team always to resolve doubts or give any explanations. All information about employees is confidential, with emphasis on the correction of the system's defects and not the search for negligent staff.

Once gathered, the data is discussed with the escort and is recorded for later discussion by the auditors. At the end of the audit, a meeting is held with the department involved and preliminary conclusions are presented prior to the final report which is presented at a later date. Any doubts can be clarified at this meeting. Equally, solutions to any problems encountered during the audit can be presented. In answer to the final report the audited department puts forward a plan of corrective action, with people responsible and a time frame. The efficacy of the action taken and the resolution of problems can be verified in the next audit, or in a safety revision directed only at the problem areas.

Figure 8.3 shows the results of an internal safety audit, evaluated as numbers.

Safety revisions

These are limited to more restricted areas than audits, normally of only one procedure. They allow more concrete revisions with more detail not attainable in audits which are too extensive. Some of those carried out are:

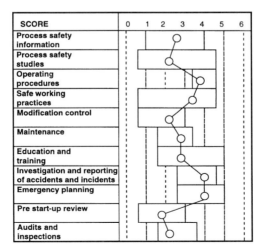

Figure 8.3   Numerical results of five internal safety audits for 11 areas, with the average values for the five audits and the range of variation in each area (adapted from [8]).

- Permit-to-work system
- Control of plant and process modifications
- Sampling procedures
- Procedures for specific operations: loading, unloading, installation start-up, etc.

Safety inspections

Different to those above in that they are carried out by plant personnel at a pre-established frequency. Items that should be subjected to periodic inspection are:

- Emergency relief systems
- Critical instrumentation
- Alarms and automatic shut-off systems (trips)
- Blocking and drainage systems
- Fire prevention and alarm systems
- Gas detectors

8.1.3   Accident and incident investigation

Another way of correcting problems and of preventing them in the future is to learn from the errors of the past. The investigation of incidents is based on this principle. Incidents do not occur spontaneously rather they are caused by a situation or an action sometimes very different from the apparent cause. Behind each human error or technical fault there is usually a problem of training, organization, maintenance, etc. or in general, a defective system. In all accidents there is some human error, whether it be at an operational or management level. Kletz [13] writes that 'to say accidents are due to human error is like saying falls are due to gravity'. However, it is normally easier to attribute an accident to human error for different reasons:

- At times the person responsible is more sought after than the cause of the accident. For the person in charge of the investigation, normally a supervisor or manager, it is easier to look for someone beneath him to blame.
- Asking personnel to be more cautious is a cheap and easy solution.

The limit that separates the concepts of accident and incident is not clearly defined. In general terms, an incident implies whatever event with or without damage to persons, the community or to property in which the consequences could have been much worse in other circumstances. It is the case of escapes, errors in manipulation of valves or equipment, overpressure due to safety valve failure, etc. The dividing line at which incidents should be subjected to investigation or not should be fixed by each organization.

Incidents generate information of inestimable value before a serious accident occurs. It is very rare that a serious accident has not been preceded by one or more minor incidents as a warning.

Following an incident it is advisable to begin an investigation as soon as possible, whilst everything is fresh, thus avoiding loss of potentially valuable information. It is important that the people involved have trust in the investigator because frequently they are reticent to give information that may damage their prestige. The investigator needs tact and discretion whilst being meticulous and able to discern the contradictions that are wont to appear in the testimony of those involved.

The conclusions should not give rise to speculation. If there are obscure points or various alternatives he should estimate which is the more likely using the methods presented in previous chapters. In general, the theory that contains least suppositions is more likely to be the correct one.

Once his work is finished, the investigator presents an incident report. It should contain at least the following:

- Circumstances and consequences (fact summary)
- Data gathered during the investigation
- Conclusions
- Action plan: Necessary preventive and corrective measures.

The report should be distributed to all interested personnel, including other similar plants of the same company. Information about the causes and the immediate introduction of preventive measures and corrective actions permit the incorporation of knowledge derived from the error into the organization's log [14, 15]. If the information coming from the investigation is not invested in improving the system, it will have been useless.

8.1.4 Operating procedures

Most accidents are due to incorrect or imprudent actions. Operating procedures promote the gathering of experience accumulated about the optimum operation of the plant, as much regarding safety as output or quality. It is necessary (obligatory in the USA according to the law on process safety management of the OSHA [1]) to have written operational procedures, applied and up to date. It is not easy to generate a good operations manual, and its development and maintenance require time and well-trained personnel with a direct knowledge of the process.

Some of the advantages in editing an operations manual are:
- The systematic consideration of the best way to complete various operations.
- The elimination of different work practices between shifts.
- Its inestimable value as a training tool for new operators.

Normally the operations manual is written during the design phase of the plant by members of the project team. Apart from the process engineer and a production engineer, the participation of skilled operators during its development favours a practical approach and reduces modifications during revision and use of the manual.

Composition and organization

A good operating procedure should be written for the person who will use it in his job. The people for whom it is written, their cultural level, the atmosphere in which they will consult it, the time allowed for doing so, etc., are factors to be borne in mind at all times [16].

The procedures are, or should be, tools for consulting every day. If an adequate language is not used, or the necessary information not given, they will fall into disuse. Too much or too little information is equally negative. The form is as important as the depth when making them manageable in tense or hurried situations when it could be necessary to consult them:

- Using direct sentences, no more than 15 words (exceptionally 20) helps clarity. Vocabulary should be adequate for the intended reader.
- Wherever possible, expressing information by graphic means, using photos, schemes, or flow charts helps and accelerates understanding.
- The area which can be undertaken simultaneously without losing the over all view are two opposite pages. For example one side can be used for graphics and the other for texts.
- To make handling easier, the procedures manual must be well structured and contain a good index. The information should be quick to find. An adequate division in sections could be the one shown in Figure 8.4.

As to its content, the OSHA regulations [1, 17] set the following sections as minimum requirements:

- Process description
- Equipment description
- Start-up, shut-down and waiting procedures
- Normal operation procedures
- Normal operation limits
- Records
- Hazard descriptions
- Operational procedures outside of normal conditions
- Descriptions of alarm and pressure relief systems
- Emergency procedures
- Hazardous work procedures
- Personal safety equipment

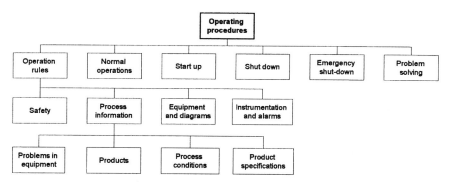

Figure 8.4   Structure of an operating procedure manual.

- Communication procedures
- Maintenance programme
- Process and equipment schemes
- Control loops.

The extent and amount of detail necessary in each section depend on the type of plant and the risks involved when straying from the established procedures. In some cases giving the objective value of a parameter and a brief indication of how to reach it will be enough, while in others it will be necessary to indicate in detail the sequence of opening and closing of valves and even, for example, the temperature ramps to follow. The HAZOP studies help to determine the precise quantity of information.

Written procedures and training are complementary. Training creates the capacity to assess the circumstances and make decisions when situations do not coincide with the procedures. This therefore tries to ensure that an accident cannot be provoked by a wrong decision, but it cannot contemplate every possible situation. The balance between the two elements is difficult to establish, but it should be borne in mind when the procedures are written.

The contents of some of the previous topics will be dealt with more thoroughly when we talk about permits to work in the maintenance section and in Chapter 9.

Example 8.1   Instructions not followed (Kletz [13])

In the instructions for carrying out a reaction it said: 'Add the second reagent at 60°C during a period of sixty to one hundred and twenty minutes'. The operators thought that the equipment available did not allow this and were carrying out the reaction for several months, adding the reagent at room temperature and heating it inside the reactor, until there was an explosion due to a runaway reaction.

1. *If the instructions are not correct or doubts exist, they should be revised and, if necessary corrected. You should never take an action that you think is equivalent.*
2. *Why did the immediate supervisor not warn that the instructions were not being followed? There was probably no monitoring of reports and operation logs.*

Example 8.2   False alarm (Kletz [13])

The operation instructions of a heater clearly stated 'If the low level alarm goes off, stop the heater immediately'. In this situation the alarm went off. A worker stopped the heater causing a general plant shut-down. It turned out to be a false alarm and caused significant economic losses. The feeling of guilt, probably fomented by a supervisor's comments or reprimands, caused the alarm to be ignored the next time it went off and the heater suffered serious damage.

Maintenance procedure updating

Production processes, as with installations, undergo continuous changes. Operating procedures should change with them. An important tool for the continual improvement of procedures is the Job Safety Analysis (JSA). These should be carried out periodically for the more dangerous operations and consist of a detailed, step by step, study of the method used, analysing implied risks and the steps that should be taken to minimize them. In this analysis, the workers who carry out the job should participate directly, so that it serves them as training and motivation. The conclusions are added to the operations manual. To maintain a record of job safety analysis results carried out, the format in Figure 8.5 could be used.

To ensure the person to whom it is directed knows of the changes introduced, those parts of the document that have changed should be marked in a special way (bold black, underlined, etc.). The document can be distributed with a form which is signed, in acknowledgement of receipt by the operators of each shift, as evidence that they have been informed of the change.

A document control system is necessary to make sure that the updated versions are available and the obsolete ones are withdrawn and destroyed to avoid errors. Therefore the documents should be properly approved by the person responsible for their issue and should also have either a review or version index. In no way should modifications be made directly on the copy without authorization. In safety audits documentation control should also be reviewed.

JOB SAFETY ANALYSIS	Job			Date
	Operator		Supervisor	Done by
Department		Section		Revised by
Personal protection equipment required				Approved by
JOB SEQUENCE	HAZARDS AND POSSIBLE ACCIDENTS		SAFE JOB PROCEDURE TO FOLLOW	

Figure 8.5 Typical form for a Job Safety Analysis (JSA). All the stages of the job are described in the left-hand column. In the centre the hazards and possible accidents are identified. Precautions required to minimize the risks are written in the right-hand column.

Special attention should be given to temporary orders. Very often values of variables or rules are modified temporarily either to adapt to a different raw material, for a breakdown or equipment maintenance or for any other cause. If these rules are ambiguous or the duration of the modification of the normal operating procedures is not clearly specified, they could be a cause of accidents. Normally it is not practical to modify the operations manual in these cases and 'plant books' are used where temporary instructions are written. After reading them, the shift manager should sign them as proof that he has been informed of any changes. If the circumstances that cause the modification happen frequently, an alternative working method would have to be included in the operation manual.

Example 8.3   Not very clear instructions (Kletz [13])

The plant manager's written instructions in the book for shifts said 'to clean the reactor shake with 150 litres of nitric acid for 2 hours, heating it to 75°C'. Normally the reactor was cleaned by filling it first with water and later adding a lower quantity of acid. The night shift did not fill the reactor with water, but added the acid directly through the isopropanol feed line in which there were still remains of this liquid which had not been carried away, as normally occurred, by the flow of water. The nitric acid reacted violently to the isopropanol, opening up the rupture disc and emitting nitrous gases.

If the department manager's intention was only to change the quantity of acid and the temperature he should have been more specific. He also did not state clearly whether it was a permanent change to the working rules or only a trial:

'Carry out the reactor washing **TONIGHT** *following the normal procedure, but increase the quantity of nitric acid to 150 l and the temperature to 75°C'.*

If his intention was only to emphasize the quantities to be used, because that shift was not following the procedures, he should have said:

'When washing the reactor follow the written procedure XX-23-C exactly.'

Example 8.4   Instructions not maintained up to date

Because of an accident it was decided to change the regeneration procedure of a catalytic reactor. The new rules were placed on the instrument panel and were emphasized in the shift book. After some time an operator who returned to his old job from one in a different plant, saw that the notice had disappeared, and supposed that the procedure had once again been modified. He read the operations manual, which had not been modified, and carried out the operation the wrong way, causing another accident.

> *If the operations manual is not kept up to date and alternative ways are used to communicate work rules, the system will soon degenerate and the manual will become useless. Shift books and notices are useful during the first few days, to call attention to the changes, but the master book should be the operations manual.*

8.2   Risk reduction in maintenance

In a chemical plant, all the installed equipment suffers deterioration in time due to working conditions, which can give rise to corrosion, wearing out of rotating parts, material fatigue, damage and deformation to internal parts or dirtiness. Before deterioration compromises the safety of the plant, it is necessary to repair or replace it as discussed in Chapter 6.

In many cases the need for maintenance is contrary to the objectives of production and sales. It is necessary that the management gives maintenance the necessary importance in their safety policy, and delegates the responsibility and authority necessary to act, even against pressure for sales, in cases where safety is compromised.

Safety with relation to maintenance can be grouped into three sections:

* *How and in what conditions a job is carried out: Permit-to-work systems.* It is necessary to ensure that sufficient precautions are taken to minimize the risks present in each specific job, such as fires, burns caused by heat or chemicals, electrocution, asphyxiation, falls, etc. The system should guarantee that the work has been carried out correctly and that the equipment is left in working condition.
* *The extension of the maintenance carried out: Maintenance programme.* Lack of maintenance or insufficient maintenance allows potentially dangerous situations to develop.
* *Control of modifications introduced in the plant.* Uncontrolled modifications can alter the safety conditions of the plant if they are not previously subjected to careful and detailed reviews.

The maintenance expenses of a chemical plant are high and the work carried out could involve high risks. The question is immediate: Would it not be possible to reduce the necessity of maintenance? The answer is yes. The nuclear industry has managed to reduce its maintenance needs. The great difficulties involved and the enormous risks a mistake implies obliges keeping it to a minimum. The solution is to apply extremely strict technical standards to the equipment to be used.

Why have these standards not been extended to the chemical industry? Probably the main cause is that they would greatly increase the cost of plants. Normally when a new plant is being designed the principal objective is to minimize the capital invested. However, the cheapest plant is not always the most economic in the long term. During the design stage it is important to bear in mind the 'maintainability' of the plant. The improvement in equipment reliability increases the initial cost of the plant, but decreases maintenance expenses, which constitute an important part of fixed costs, with a typical value of about 6% of sales for chemical industries.

8.2.1   Permits to work

To control risks during execution of maintenance it is essential to have a procedure which regulates the granting of permits to work. The objective of this procedure is to control the work to be carried out, defining :

- The activities to be carried out
- Responsibilities
- Protection and precautionary measures necessary to avoid accidents while carrying out the job
- Tests and inspections to check that the job has been done properly.

Different types of permits exist, according to the nature and risk of the job to be carried out. The most common are:

- Normal permit. For routine work not subject to special permits.
- Permit for work with fire or explosion risk.
- Vessel entrance permit.

Special requirements usually exist for work at heights, high voltage, excavations, mobile cranes, toxic products and pipe openings.

In the same way as the other activities of the management safety system the permit-to-work procedure must be subject to frequent audits and checks to make sure that the established rules are being followed.

Very often maintenance is subcontracted to specialized companies. It is vital that they are bound by the same safety procedures as one's own workers and that they receive the necessary training to ensure that their attitude and capacity are the same as the plant's workers.

Activities to be carried out

The permit should clearly define, without any possibility of error, the work to be carried out. If during the job something unexpected happens that results in carrying out new tasks not included in the original permit to work, a new permit should be applied for. A typical case is that of a check of the bearings of a pump which does not need to be opened up. If abnormalities are found and the decision is made to open up the pump, it could be that it is full of flammable, toxic or corrosive liquid, making it necessary to ask for a new permit. A similar accident provoked a fire in which three people died [13].

If circumstances exist which prevent the original work being done, it should not be exchanged to one that the worker considers equivalent without first obtaining a new permit to work.

> **Example 8.5   (Kletz [13])**
>
> *To isolate a piece of equipment it was decided that a blind should be placed between two flanges. The mechanic found it difficult to separate them to introduce the blind, so he decided to place it between a different pair of flanges which were a bit further along in the same pipeline. These flanges were on the other side of the valve which blocked the line, causing him to receive a jet of corrosive liquid in his face when loosening it. The mechanic did not carry out the job entrusted to him, but a different one, for which the installation was not safe.*

Responsibilities

Normally the person who issues the permit is the person responsible for production in that plant or affected plants, and the person who receives it is the person responsible for maintenance. In this way, production has to make sure that the conditions in the plant are the necessary ones to carry out the job. Maintenance is responsible for the work being carried out in the established manner.

It would be a truism to say that the worker has to read the permit to work. However, it could be that they only read the description of the job to be done. Most tasks are routine and the contents of permits are very similar. However, when conditions are different to the normal, confidence and ignorance of the risks to which one is exposed can cause accidents. Usually the worker is asked to sign the permit as proof that it has been read, but the crux of the matter is whether the workers understand why permits exist, and the risks they are exposed to, with examples of accidents that have happened. It is advisable that the person who has issued the permit to work accompanies the worker to the equipment to explain the principal risks.

> **Example 8.6   (Kletz [13])**
>
> *A mechanic received a permit to change a valve in a corrosive product line. When he loosened the flange, a jet of product splashed him, causing serious burns because he was not wearing protective gloves or mask.*
>
> *The problem would appear to be a clear case of mechanic negligence, because on all permits, 'PROTECTIVE GLOVES AND MASKS MUST BE USED' was printed by default. In the majority of cases it was not necessary to use them so the workers systematically ignored this order. In this case the worker was not to blame, but rather the inadequate format of the permit to work.*

Permit contents

Figures 8.6 and 8.7 show forms that are used in the Dow Ibérica (Tarragona) factory for standard jobs, jobs with a risk of fire or explosion (FER) and for vessel entrance.

The contents of a job permit should include at least the following points:
- Identification of the department and equipment on which the work is to be done.
- Task description.
- A checklist, to make sure that all the possible risks have been considered.
- Protection measures derived from the risks.
- The necessary personal protection equipment.
- Validity period
- Space for observations and signatures.

Hot work, i.e. that which can produce a source of ignition, as in the case of welding, blow lamps, cutters, pneumatic hammers, etc. have a high level of risk when carried out in closed containers or in areas classified 1 or 2. In these cases extreme precautions must be used to ensure there is no presence of any mixture within its flammable range.

Special attention must be paid to the correct isolation of the equipment. Electrical systems, valves, blind flanges or any other element which should be placed in a certain position during the job should be blocked or labelled (normally with a red label), clearly indicating its state. A valve cannot be relied upon as an isolation element because it may leak. The use of double block and bleed, or even better, blinding plates between flanges is recommended.

Work inside closed containers is an operation which has given rise to a number of accidents. The principal risks to which one is exposed are the presence of fumes, toxic or inflammable liquids, dangerous solid residue, oxygen deficiency, intake of process fluid or fumes, movement of internal equipment (e.g. agitators). All of these should be considered and eliminated by taking the necessary blocking precautions, cleaning, purging, ventilation, and use of autonomous breathing equipment.

Final inspection and closure of permit.

Once the job is finished, an authorised supervisor should check that it has been carried out correctly, that the plant is left in the same condition in which it was found and then sign the permit, returning it to the person who issued it.

If after permit expiry the work has not been finished, a new permit should be issued to allow the work to be completed.

Example 8.7   (Lees [18])

A mechanic finished his shift without finishing the job he was doing and he left, with the intention of finishing it the following day. However, a colleague on the night shift finished it. When the following day the first mechanic tried to finish it, the plant was no longer in a safe condition for the job and there was an accident.

IF THERE IS AN EMERGENCY: STOP WORKING AND FOLLOW THE EMERGENCY PLAN

	Cracker	Derivatives
Emergency calls	3222	4222
Medical services	3492	4363

DOW

DOW CHEMICAL IBÉRICA S.A.

Tarragona   **PERMIT TO WORK Nº** 19593   Date:

	SUBCONTRACTOR (or exec. Dept.)		

Order Nº / Job	SUBCONTRACTOR (Or exec. Dept. if applic.)	Nº of operators	Receiving Dpt.	Resp./S.C.
Area / Working place	Description of the work:			
Tank / Line / Equipment				

Type of work (Risk analysis)

Line/Equipment opening	5 ☐ Electrical	☐ Welding equipment
Fire or explosion hazard	6 ☐ Critical lift	☐ Torches
Confined space entry	7 ☐ Modification	☐ Gas cylinders
	8 ☐ High pressure flushing	☐ Grinder
1 ☐ Heights	9 ☐ Asbestos	☐ Portable pump
2 ☐ Floor openings/ grating	10 ☐ Radioactivity	☐ Vehicle
3 ☐ Excavation	11 ☐ Vehicle entrance	☐ Compressor
4 ☐ De-energization	12 ☐	

Type of work (Risk analysis):
- ☐ High pressure truck
- ☐ Elect./Pneum. tools
- ☐ Ladder
- ☐ Scaffolding
- ☐ Crane
- ☐ Excavator

Personal protection (Helmet, glasses, shoes - Mandatory)

HEAD	BODY	EXTREMITIES	OTHER / SP. INSTR.
☐ Mask (Org.)	☐ Body belt	☐ Regular gloves	☐ Semi-aut. equipment
☐ Mask (Inorg.)	☐ Body harness	☐ Rubber gloves	☐ Dosimeter
☐ Face shield	☐ Nomex cloths	☐ Anti-thermal gloves	☐ Other
☐ Goggles	☐ Anti-acid suit	☐ Anti-cut gloves	☐
☐ Mask (Dust)	☐ Disposable suit	☐ Rubber boots	☐
☐ Ear protection	☐ Water suit	☐	☐

CHECKS

(For all works)

	Y N/A		Y N/A		Y N/A
1 Define the area	☐ ☐	3 Isolate	☐ ☐	6 Purge / Inert	☐ ☐
2 Written procedure		4 De-energized equipment	☐ ☐	7 Place tags	☐ ☐
(J.S.A.)	☐ ☐	5 Empty	☐ ☐	8 Clean	☐ ☐

(Measurements)

Time	
Explosivity	
Toxics	
Oxygen	
Signatures	

(Line/equipment opening)

	Y N/A		Y N/A
1 Take precautions, spills		8 Isolate products	
2 Hose with water / steam		Blind plates	
3 Mark line / equipment		Physical disconnection	
4 Assign watchman		Double block and bleed	
5 Check pipe supporting		Block valve / Red tag / Lock	
6 Check safety shower /eye wash baths		9 Check low points	
7 Isolate energy sources (pump)		10 Check toxic gases:	

BEFORE INITIATING FIRE OR EXPLOSION HAZARD (FEH), CALL TO MEASURE FLAMMABILITY

(Line/equipment opening)

	Y N/A		Y N/A
1 Limit the area with ignifugous blanket	■ ■	6 Place water curtains	■ ■
2 Assign watchman	■ ■	7 Check leaks /spills nearby	■ ■
3 Get fire extinguisher P-12/50	■ ■	8 Remove drums from the area	■ ■
4 Hose with water	■ ■	9 Cutting with saw	■ ■
5 Block drains (foam, ignifugous blankets)	■ ■	10 Continuous flammability testing	■ ■

CHECK LIST BEFORE ENTERING CONFINED SPACES

(RECEIVING DEPARTMENT)

Figure 8.6   Permit to work form (courtesy of Dow Chemical Ibérica, Tarragona).

PRECAUTIONS

1 WORKS IN HEIGHTS AND ROOFS
- [ ] Ladder
- [ ] Scaffolding
- [ ] Use of basket or mobile platform
- [ ] Supply means to attach the safety belt
- [ ] Check wind velocity
- [ ] Walkways for the roofs
- [ ] Fence off area below

2 FLOOR OPENING AND GRATING
- [ ] Mark with red tape
- [ ] Protect openings / access with fixed bars
- [ ] Use safety belt
- [ ] Protecting nets
- [ ] Watchman required

3 EXCAVATION
- [ ] Signature electrical workshop _____
- [ ] More than 1.2 m. Vessel entry.
- [ ] Attach drawing
- [ ] Check underground
- [ ] Check interferences
- [ ] Only manual excavation
- [ ] Signals, fences, covers
- [ ] Test boring
- [ ] Shove / slope
- [ ] No piling

DE-ENERGIZATION OF EQUIPMENT
- [ ] Equipment shut-down
- [ ] Open control/manoeuvre circuit
- [ ] Open power circuit
- [ ] Drawout motor starter
- [ ] Remove fuses (L.V.)
- [ ] Drawout motor starter (H. V.)
- [ ] Block with lock
- [ ] Does not energize when pressing
- [ ] Use red tags

5 ELECTRIC WORKS
- [ ] J.S.A. / Procedure
- [ ] Affects other equipment
- [ ] Work on deenergized equipment
- [ ] Install insulating screens
- [ ] Check one line diagram
- [ ] Break control and power
- [ ] Block with lock
- [ ] T.C. circuits for diagnosis
- [ ] Check it does not energize
- [ ] Earth the conductors

SUBCONTRACTOR (or exec. Dpt.) / RECEIVER

6 CRITICAL LIFT — [ ] Signed checklist
7 PLANT MODIFICATIONS — [ ] See authorisation of modification control
8 HIGH PRESSURE FLUSHING — [ ] Signed checklist
9 ASBESTOS — [ ] Approved work procedure
10 RADIOACTIVITY — [ ] Signed checklist
11 VEHICLE ENTRANCE TO THE PLANT — [ ] Flammability control
- [ ]
- [ ]

SIGNATURES (Readable)

I UNDERSTAND AND ACCEPT THE SAFETY MEASURES HERE DESCRIBED

Signature subcontractor supervisor
or Executing Department _____   Signature worker _____

I authorise this work   The conditions have been checked

Signature authorised supervisor _____   Signature area operator _____

Validity period: From _____ to _____

Reported to other Plants / Departments _____

Signature _____

WORK FINISHED [ ]   WORK NOT FINISHED [ ]

Signature subcontractor supervisor
or Executing Department _____   Authorised notified signature (receiver) _____

Renewed period: From _____ to _____

Signature subcontractor supervisor
or Executing Department _____   Signature worker _____

Signature authorised supervisor _____   Signature area operator _____

Figure 8.6 Continued.

CHECK LIST BEFORE ENTERING CONFINED SPACES

PERMIT TO WORK N° _____ Date: _____
Work description: _____

	YES	N/A
1 Are all the lines to and from the equipment with blind flanges or disconnected and with its end open to atmosphere? (According to the line list)	☐	☐
2 Are all the electrical equipment deenergized? (Visible cut)	☐	☐
3 Have the red tags and locks been placed?	☐	☐
4 What materials are normally present in the vessel? (Identify) _____	☐	☐
5 Has the equipment been cleaned?	☐	☐
6 Gas checking. Max. 15 min before entry		
- Flammable mixture % of L.E.L. Done by	☐	☐
- Oxygen contents %. Done by	☐	☐
- Toxic materials % T.L.V. Done by	☐	☐
- Repeat the check every	☐	☐
7 Is a continuous analysis of flammable mixture or oxygen required?	☐	☐
8 Is the equipment now free of flammable, toxic or corrosive materials?	☐	☐
9 Is additional ventilation equipment required?	☐	☐
10 Has the rescue procedure been reviewed?	☐	☐
11 Is autonomous breathing equipment required?	☐	☐
12 Lighting voltage 24 V A.C / D.C. Tools voltage 24 V A.C./D.C.	☐	
13 Does the external watchman need a radio connected to the control room?	☐	☐
14 Special personal protection equipment		
15 If used, has the double block and bleed system been checked?	☐	☐
16 If existing, have the radioactive sources been revised?	☐	☐
17 Have instruction been given, and understood, by the external watchman?	☐	☐

LIST OF PERSONNEL ENTERING THE CONFINED SPACE
NAME   SIGNATURE

_____   _____
_____   _____
_____   _____
_____   _____
                EXTERNAL WATCHMAN
_____   _____

Figure 8.7   Check list before entering confined spaces (courtesy of Dow Chemical Ibérica, Tarragona).

8.2.2   *Maintenance programmes*

Maintenance can be classified in two categories: planned and unplanned, according to whether it is subject to a periodic revision plan or whether it is required for equipment failure. Corrective or preventive classification is complementary, unplanned maintenance is always corrective, but both may occur within preventive maintenance. A preventive maintenance programme should exist which would help to prevent equipment critical to safety from failing unexpectedly. It is not usually practical to place all factory equipment under preventive maintenance, or to carry it out on all equipment with the same frequency. Special programmes usually exist for pressure vessels, safety valves and critical instrumentation. On a critical equipment list the importance, frequency and type of revision for each one should be established, bearing in mind the type of equipment, the risk involved in this equipment breaking down, and the minimum established by legislation in force for some machines (there are regulations for safety valves and pressure vessels in most countries [19]).

For equipment which does not work continuously, but only when there is demand, periodic revisions must be established with enough frequency to obtain low values of the fractional down time as discussed in Chapter 6.

Maintenance management computer packages of database type exist, storing information and providing lists of equipment which should be serviced in a specified period of time, together with breakdown records for each machine.

There also exist jobs which cannot be carried out without stopping the whole plant. Normally general maintenance shut-downs are organized with very variable frequencies from one type of plant to another. While an ethylene production plant via naphtha cracking usually stops every four years for several weeks, in many plants an annual general stoppage of several days is usual, or more frequent stoppages in autonomous sections.

As already discussed, plant maintenance should be borne in mind at the design stage, including in the project team, whenever possible, a maintenance representative on a part-time basis. Aspects such as the reliability of equipment to be used, construction materials and accessibility for carrying out servicing and repairs should be taken into account at all times. Table 8.1 contains a checklist of maintainability to be considered during the design and selection of equipment.

Inspections and tests of the equipment related directly to fire-fighting and the emergency plan, emergency sirens, foam or sprinkler systems, gas detectors etc. should be included in the maintenance programme.

8.2.3   *Control of modifications*

A safe plant could cease to be so if uncontrolled modifications are carried out. The Flixborough accident is a typical case (see Appendix). Confusion should be avoided between plant modifications, which require even stricter control, and maintenance jobs. Control is especially important in temporary or urgent

Table 8.1 Check list of equipment maintainability. It should not be considered that these are exclusively the points to be checked, and in each specific case all the circumstances which can influence the maintainability of the equipment should be taken into account

---

- Is continuous working of the machine necessary? Is spare equipment necessary?
- Is it going to be subject to action from corrosive compounds continuously or occasionally?
- Is there sufficient space for the access of equipment (cranes, tools, etc.) and the necessary staff to do the job and the removal of dismantled parts (heads, impellers, shafts, tube bundles, etc.)
- Will it be necessary to enter the equipment? For what reason?
- Will it be necessary to carry out jobs with a risk of fire or explosion on or inside it?
- Do adequate blocking systems exist in all vessel pipeline entrances and exits, in accordance with the potential risk that each one presents? (Except for air or water, one valve alone is not considered sufficiently reliable.) Does this blockage affect other equipment which should be kept running?
- Could there still be toxic or dangerous products in the container? What are the potential hazards? What measures should be taken to eliminate them?
- Are special lines needed for venting or emptying of the container to a safe place?
- Would it be necessary to enter with personal protection equipment or respiration systems? Which ones? Are the entrances large enough for this?
- What internal elements exist? (Agitators, coils, baffles, trays, etc.) What risks do they present?
- Will a ladder or other auxiliary measures be needed to gain interior access? Has the manner of supporting it been considered? Have rescue measures from the exterior been anticipated?
- Is interior ventilation necessary while working? Have adequate measures been anticipated?

---

modifications, in which safety could be pushed into second place, displaced by the urgency or because of the 'I want it by yesterday!' syndrome.

When modification demands capital investment, with consequent project and revisions, the risk is less, because normal controls are applied to capital projects. The problems usually arise when modifications are not recognized as such and the possible problems that could occur slip by until an accident happens. In a management safety system it is necessary that a procedure exists for approval of modifications.

The first necessity is to clearly establish what constitutes a modification. An adequate definition could be: 'Any permanent or temporary change in the equipment or processes (operation conditions, procedures or products) which could affect the safety of the plant.'

The responsible authority for the design of the modification and its approval should be established in the procedure. The modification should be subject to the same risk evaluation controls as in a project, submitting it to a HAZOP analysis by a trained team.

Henderson and Kletz [20] propose nine points which should be checked when carrying out a modification:

1. Check that the required number or size of safety valves does not change (otherwise, include the re-design in the study).
2. Check that there are no modifications to the electrical classification of areas (or the necessary changes in the electrical equipment be specified).

3. Check that the control system, alarms or emergency stop systems and trips are not affected.
4. Check that there are no other effects which reduce the level of safety.
5. Check that the established engineering standards have been followed.
6. Check that materials and grades of adequate quality have been used.
7. Check that existing equipment will not be subject to working conditions outside its design parameters without checking its capability.
8. Working conditions should be modified in accordance with necessity.
9. The necessary training and information should be given to the production and maintenance teams.

The appropriate changes should be made immediately to all the documentation affected by the modification: operation procedures, piping and instrumentation diagrams, plant lay-outs, process flow-sheets, electrical distribution drawings, etc.

8.3   Human resources management

Up to this point a large part of the book has been dedicated to equipment, processes and installations. The human factor plays a critical part in the functioning of a chemical plant. The sophisticated control systems used nowadays do nothing more than reflect contributions made in the design stage. Some decisions that an operator previously had to take in a few seconds at an instrument panel, are now taken at a keyboard and in front of a monitor, and they are translated into a programming language.

As in any other type of industry, in the chemical industry the people are the most valuable resource in the organization, even though they make mistakes. Rather than persecute the person guilty of a mistake, the knowledge gained should be used to advantage and as an opportunity for improvement. This in no way means permitting or fostering errors, but learning the lessons obtained to prevent a future occurrence.

To succeed in this, efficient management of human resources is necessary through, among other instruments, motivation, training and participation of the workers. People's creativity can only be wholly profited from when they are given the opportunity to assist in the improvement of the system. However, it is not enough to call for participation, but also means and tools must be put into their hands which allows them to gain results effectively. The management principles of total quality are equally applicable to safety management (we will not go into much detail here as this topic has been widely discussed in many references, including Deming [21] and Juran and Gryna [22]).

The majority of the 14 points Deming made are applicable to safety management with the same validity as quality management:

1. Create constancy with the intention of improvement.
2. Adopt the new philosophy, presented by these points.

3. Stop depending exclusively on inspection, move the emphasis to the inherent characteristics of the process.
5. Improve the processes constantly and continually.
6. Introduce training at work.
7. Adopt and implant leadership in the management.
8. Eliminate fear of participation and change.
9. Demolish the barriers between different departments, encourage team work.
10. Eliminate slogans, exhortations and targets for the workers without accompanying them with adequate means.
12. Eliminate the barriers which deprive people of pride in their work.
13. Introduce a vigorous education and self-improvement programme for everybody.
14. Put everyone in the company to work to obtain the transformation.

8.3.1   Human error

Mark Twain said, 'Mankind is a creature who was made at the end of the week... when God was tired.' Mankind is the only element of a plant that cannot be redesigned or modified. One must play with what one has. The only exception to the rule is that people can be selected with an adequate education and can be trained to improve their response and attitude, as will be seen further on.

To know and bear in mind the probability of a human error is as important as having information about the reliability of equipment and instrumentation. Techniques can be applied, as described in Chapter 2, such as fault tree analysis (FTA) or failure modes and effects analysis (FMEA). In Chapter 6 a study of human error and the procedures used to obtain estimates of frequency were discussed.

If the safety management system of the company [13] influences the failure frequency of equipment (through preventive maintenance, tests of safety systems, etc.) it is obvious that this influence should be much greater in the case of human errors. The principal factors which affect the probability of committing errors are:
- Type of activity (routine, special)
- Time available for completion
- Education, training
- Environmental pressure (normal, emergency)
- Work atmosphere.

Most of these can be improved during the design and operation of the plant acting upon aspects dealt with in this and other chapters, such as maintenance, operations procedures, training, risk analysis during the design, etc.

It is said that up to 90% of accidents are caused by human error. The only reservation being that while normally the responsibility for an error falls back on the worker who has physically committed the error, a detailed analysis of the causes makes clear that very often the human error is originally due to the designer, the manager or the department head, who have not been able to anticipate the error and take measures to make it less likely.

8.3.2 Education and training

Safety training and education should not only include the workers but should also extend to the foremen, engineers and, in general, all the company's personnel, including the contractors who work for it. More than just the transmission of knowledge and skills training, a fundamental objective is to convince all the organization of the importance of safety and to motivate to participate in its improvement.

Training and education needs

The training and education needs are not the same for every worker. They depend in each case on the work to be carried out and the responsibility that each person has in the organization. The procedure for training and education should be orientated towards the detection and satisfaction of these needs, and the company should provide the required human and material resources. In Spain, the General Regulation of Safety and Hygiene at Work specifically states as an obligation of the executive the 'promotion of the most complete education in the matter of Safety and Hygiene at work of the management, specialists, intermediate managers and workers' (Art 7.8).

Normally the training requirements for each position are summarized in a matrix as shown in a simplified form in Figure 8.8, where the different functions or responsibilities versus the modules of education or training are given.

Supervisors and engineers

The individuals who have special responsibilities in the safety management system should receive a specific training. Lees proposes a series of subjects which are

POSITION OR RESPONSIBILITY	TRAINING AND EDUCATION MODULES									
	Basic safety	Standards and rules	Operating manual	Reactivity	Safety and hygiene	Hazardous jobs	Permits to work	Emergency planning	Control of modifications	
Production manager	▓							▓	▓	
Plant manager	▓	▓						▓	▓	
Department head								▓	▓	
Plant engineer	▓	▓						▓	▓	
Plant operator			▓	▓	▓		▓			
Maintenance worker				▓	▓	▓	▓			

Figure 8.8 Simplified example of a training needs matrix (not all the positions and training modules are included).

Table 8.2 Possible subjects for safety education for managers and specialists (adapted from Lees [18])

- Safety responsibility – company policy and organization.
- Requirements, rules and legislation of safety and hygiene.
- Safety and loss prevention principles.
- Process and chemical risks.
- Maintenance procedures and permit-to-work system.
- Systems of fire prevention and protection.
- Emergency plans.
- Training procedures.
- Communication and information.
- Pressure vessels, emergency relief systems.
- Control systems, emergency shut-downs.
- Accidents: sources of information, real cases, practical case studies.
- Sources of information on safety and contact persons.

presented in Table 8.2. Depending on the exact responsibility, it is convenient to emphasize different aspects. For a process engineer inherent safety, risk and consequences analysis, rules and regulations, design of pressure relief systems, etc. whilst for a production engineer it might be necessary to insist more on training, communication and accident case studies.

In any case, and due to the great differences between diverse jobs, given the technical content of the material, the necessary training should be determined after a detailed analysis of the responsibilities of the job and of each individual's career plan.

Operators

Chemical industry workers need a much higher level of education and training than the average for other industries. The complexity of continuous chemical processes, together with the risks associated with the handling of toxic, reactive or flammable products, creates a need for highly specialized knowledge and training.

Besides the initial education, which should be carefully evaluated, specific education and training on the installation and the specific job in which the worker is going to be involved is required. Just as in the previous case, it is necessary to start with an evaluation of the training needs bearing in mind the job requirements. Lees gives a list of training ideas to be considered, shown in Table 8.3.

The most frequent situations in which training is needed are:

1. *For a new employee after joining the company*. The training plan for new employees should be clearly established, and an employee should not assume tasks without supervision if he has not completed his training and his performance has not been positively evaluated. At this moment, on 'virgin territory', good skills are adopted more easily as one's own, however if this first lesson has not been passed successfully, then later on it is much more difficult to correct bad habits.

Table 8.3 Possible safety education ideas for workers (adapted from Lees [18])

- Safety responsibility – company policy and organization.
- Requirements, rules and legislation of safety and hygiene.
- Process and chemical risks.
- Fire-fighting. Precautions. Actions in case of fire.
- Toxic products. Actions in case of escapes.
- Local emergency plan.
- Personal protection equipment. Directions for use.
- First Aid
- Handling and lifting of objects.
- Safety. Restricted areas.
- Communication and reporting on incidents and accidents.
- Permit-to-work system.
- Housekeeping.
- Hygiene. Medical aspects.
- Accidents: information sources, real cases, practical case studies.

2. *When a person's job or responsibilities are changed.* Normally in these cases it is enough to supplement the education already received in order to cover the new job's requirements. When it is a supervisors job it should never be forgotten to include training on leadership and organizational skills, which play a key role in safety.
3. *When changes have been introduced in the plant, the process or operating procedures.* It is not enough to introduce the written changes in procedures. The way to ensure they are understood is to discuss, step by step, the modifications and the reasons which brought them about.
4. *Periodically.* It is useful to refresh knowledge which can be easily forgotten. It also contributes to the creation of an atmosphere favourable for safety. Normally, the most critical procedures, in which it is necessary to avoid any error, are those which are less practised, as, fortunately, emergency conditions are usually not frequent. It is necessary to keep fresh this knowledge which is more prone to be forgotten, or dangerously simplified.

Training media

Numerous options exist when selecting the way to develop a determined education or training module. Some of them are mentioned, with their main characteristics, below.

1. *Speeches or conferences.* This is the option with the least participation, it is good for transmitting theoretical information or knowledge. It can be completed with discussions and debates.
2. *Individual education with manuals.* It can be especially useful for the studying of plant operation procedures, following a manual with self-assessment through questions and answers.
3. *Videos.* These are a useful tool to begin a discussion. Normally the authors propose the dynamics to follow after the projection. Numerous institutions such as AIChE, IChemE, Chemical Industries Association (United Kingdom), NFPA, API, etc., have ample catalogues on safety and hygiene. The majority are in English.

4. *Interactive video.* Is a thriving technology, based on the simulation of situations by computer on a subject referring to a practical case. The employee can choose different options and see the consequences in each situation.
5. *Case study discussions.* This was one of the most used techniques in the past. Although not as realistic as the previous mentioned, it leaves a large area for discussion and participation of all the attendants. The work in groups, involving a variety of departments and skills, such as operators, technicians, process engineers, etc. gives very positive results through the interchange of experiences and points of view.
6. *Laboratory practice.* This is used more for manual techniques such as welding or electronics. It is also useful to demonstrate the dangers of chemical products.
7. *On the job education.* This is one of the most widely used. It is normally combined with instructions and manuals. An experienced worker accompanies the pupil and shows the operations to be carried out. In a second phase, the pupil carries out the job under the veteran's supervision, until he is judged capable of working independently.
8. *Simulation training.* Through computer systems the response of all or a part of the plant to different operator actions can be simulated. It is convenient that the consoles and control panels are the same as or very similar to those installed in the plant.

8.3.3 Communication

Apart from the specific education programmes for each circumstance, it is necessary to establish an atmosphere of interchange of information and continual improvement in the organization which permits sharing experiences and obtaining maximum benefit from them.

Especially, as the number of company employees increases, it is more difficult to make sure that all the staff have the information they need. In the majority of cases it is necessary to establish mechanisms to ensure the exchange of information between departments, plants, or even between production centres of the same company.

In addition to the circulation of reports, investigation of incidents, statistics and other information of interest through pre-established routes, one of the more common ways is the holding of periodic safety meetings (once or twice per month). It is convenient that the person responsible for the plant is present, in addition to the assistance of the safety manager. Some of the points around which the meetings could be structured are:

*   Discussion of the incidents in the plant or in other similar ones.
*   Presentation and discussion of personal experiences related to safety.
*   Review of an operating procedure, modifying it later according to the conclusions.
*   Studying practical cases of safety.
*   Short talks on safety and hygiene.

These meetings should also be extended to areas of staff related to production, such as engineering and maintenance or could be included in periodic departmental meetings. Temporary teams could also be established, consisting of people from a department or, perhaps, inter-disciplinary, for the purpose of carrying out specific tasks, such as reviewing job safety analysis or operating procedures. Participation should be voluntary although it is desirable that all the staff are invited.

8.4 Questions and problems

8.1 Carry out a job safety analysis for unloading a cistern of sulphuric acid into a storage tank.

8.2 Study the start-up process of a natural gas burner and propose a safe start-up sequence.

8.3 Many accidents are due to errors from pushing the incorrect button. Describe the factors involved and a way to reduce the frequency.

8.4 Another common cause of accidents is errors of opening or closing of valves. Discuss the possible causes and solutions.

8.5 Which factors would have to be borne in mind in the installation of a heat exchanger to ensure its maintainability?

8.6 Kletz [13] describes the following case: A permit was issued to connect a hose for nitrogen to flange A (Figure 8.9), to check that there were no leaks in the equipment and afterwards to disconnect it (the flange was not identified). When the test was finished, the work supervisor ordered a different mechanic, who was new, to disconnect the hose. This mechanic did not fully understand the instructions and disconnected flange B. The supervisor signed the permit without verifying the job and returned it to the person responsible for production, who started the equipment, producing an escape of toxic gas. Describe the mistakes that took place and propose possible solutions.

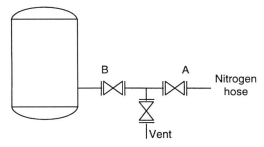

Figure 8.9 Diagram of the installation for Problem 8.6.

8.7   Houston [23] describes the following accident: one of the heads of a vessel used for storage of an organic chemical with a high melting point burst, injuring three workers who were close by. The filling line was heated by steam, but in spite of that it used to block up frequently. Normally the pipe was blown with air at 7 bars before and after each liquid transfer operation in order to eliminate blockages.

On this occasion, the worker observed that the safety valve of the container did not open and supposed that the air entrance line was blocked. It was really the safety valve which was blocked, and when the pressure reached 2 bars the head broke. The remains of the product, at 120°C, which were still in the container, reached three men who were working five metres away and who were seriously injured. Carry out an investigation of the accident analysing the causes which contributed to its occurrence and propose safe working practices.

8.5   References

1. Lo Pinto, L. (1993) Complying with OSHA's new safety law. *Chem. Eng.*, January.
2. Kuryia, M. L. and Yohay, S. C. (1992) New safety rules add to plant managers' worries.*Chem. Eng.*, June.
3. British Standards Institution. *BS 7750:1992, Environmental Management Systems*.
4. International Organization for Standardization. *ISO 9002:1994, Model for Quality Assurance in Production, Installation and Servicing*.
5. American Petroleum Institute (1990) *API Recommended Practice 750. Management of Process Hazards*, 1st edn, API, Washington D.C.
6. CCPS (Center for Chemical Process Safety) (1989) *Guidelines for Technical Management of Chemical Process Safety*, American Institute of Chemical Engineers, New York.
7. CCPS (Center for Chemical Process Safety) (1992) *Plant Guidelines for Technical Management of Chemical Process Safety*, American Institute of Chemical Engineers, New York.
8. Jenssen, T. K. Systems for good management practices in quantified risk analysis. *Process Safety Prog.*, **12**(3), 137.
9. Chemical Industries Association (1992) *Responsible Care Management Systems*, CIA, UK.
10. Armenante, P. M. (1991) *Contingency Planning for Industrial Emergencies*, Van Nostrand Reinhold, New York.
11. King, R. (1990) *Safety in the Process Industries*. Butterworth-Heinemann, London.
12. CCPS (Center for Chemical Process Safety) (1989) *Guidelines for Auditing Process Safety Management Systems*, American Institute of Chemical Engineers, New York.
13. Kletz, T. (1991) *An Engineer's View of Human Error*, 2nd edn, The Institution of Chemical Engineers, Rugby.
14. Kletz, T. A. (1993) *Lessons from disaster. How organizations have no memory and accidents recur,* The Institution of Chemical Engineers, Rugby.
15. Kletz, T. A. (1993) Organizations have no memory when it comes to safety. *Hydrocarbon Processing*, June.
16. Sutton, I. S. (1991) Write a better operating manual. *Hydrocarbon Processing*, December.
17. Ozog, H. and Stickles, R. P. (1992) What to do about process safety audits. *Chem. Eng.*, September.
18. Lees, F. P. (1980) *Loss Prevention in the Process Industries*, Butterworth-Heinemann, London.
19. Spanish Department of Industry and Energy, Publications Centre (1989) *Reglamento de aparatos a presión e instrucciones técnicas complementarias* (Pressure vessels code and related technical instructions), Madrid.
20. Henderson, J. M. and Kletz, T. (1976) Must plant modifications lead to accidents?, in *Process Industry Hazards - Accidental Release Assessment, Containment and Control*, Institution of Chemical Engineers, London.

21. Deming, W. E. (1986) *Out of the Crisis*, MIT, Cambridge.
22. Juran, J. M. and Gryna, F. (1993) *Manual de Control de Calidad*, Díaz de Santos, Madrid.
23. Houston, D. E. L. (1971) New approaches to the safety problem, in *Major Loss Prevention in the Process Industries*, Institution of Chemical Engineers, London.
24. Greenberg, H. R. and Cramer, J. J. (eds) (1991) *Risk Assessment and Risk Management for the Chemical Process Industry*, Van Nostrand Reinhold, New York.
25. Kenney, W. F. (1993) *Process Risk Management Systems*, VCH, New York.
26. Le Vine R., Arthur D. Little Inc./CCPS (Center for Chemical Process Safety) (1988) *Guidelines for Safe Storage and Handling of High Toxic Hazards Materials*, American Institute of Chemical Engineers, New York.
27. Sanders, R. A. (1993) *Management of Change in Chemical Plants*, Butterworth-Heinemann, London.
28. Dow Chemical Ibérica (1990) *Curso sobre seguridad en la industria química*. Polytechnic University of Valencia.
29. Englund, S. M. (1991) Design and operate plants for inherent safety (2 parts). *Chem. Eng. Prog.*, March–May.
30. CCPS (Center for Chemical Process Safety) (1989) *Guidelines for Safe Automation of Chemical Processes*, American Institute of Chemical Engineers, New York.

9 Emergency Planning

Then be alert, for you never know the day or the hour.

Matthew, 25,1

9.1 Introduction

In spite of applying all the techniques related in previous chapters, and as already discussed in this book, it is not possible to reduce to zero the risk of an accident happening. Whatever level of risk may be considered as acceptable, a finite probability exists that a mistake with potentially serious consequences for people, the environment or installations will occur. For the consequences to be kept to a minimum, it is necessary to develop an internal emergency plan (for the employees of the company) and an external emergency plan (for the surrounding communities), which enables the identification of the risks, the prediction of the most probable consequences, the incorporation of safety measures and the protection of the integrity of the people who could possibly be affected in the case of a major accident.

In some accidents action is difficult because of the rapidity with which the events unfold. In the Flixborough case it is estimated [1] that from the moment the cyclohexane cloud appeared, until an ignition point was found, took only about fifty seconds. However, in the Bhopal and Seveso accidents the defects in the emergency plans, or their non-existence, were the principal causes of the very serious consequences for the population. In both of these two cases many lives could have been saved if the necessary communication, co-ordination and evacuation mechanisms had been established.

In the case of Seveso, the opening of the reactor rupture disc, with the emission of 1–2 kg of dioxin, at 12.37 on Saturday 10th July 1976, was followed by the immediate reaction of the personnel of the plant, within its limits. They tried to warn the authorities of the possible effects of the emission, but this was impossible as it was a weekend. Throughout the following days, communication between the authorities and the company was deficient, cases of dead animals were reported and the vegetation dried up. Measures were not taken until four days later, when the consequences of the escape were manifested in a boy. The next day a state of emergency was proclaimed and an area of 5 km^2 was declared contaminated. The first group of citizens was not evacuated until 27th July. Later it was found that an area five times larger than was first thought was affected.

After the Seveso accident and largely because of it, the majority of the developed countries established compulsory legislation, regulating declarations of risk by the industry, the development of emergency plans, internal as well as external, and the creation of co-ordinating organizations for emergency cases. In the EU this rule is covered by directive 96/82/EC, which has superseded its predecessor, commonly known as the 'Seveso Directive'.

United States legislation also requires the development of emergency plans, both internal and external, in laws such as 'Hazardous Waste Operations and Emergency Response' (OSHA, 29 CFR 1910.120), the 'Process Safety Management of Hazardous Chemicals' (OSHA, CFR 1910.119) and Title III on 'Emergency Planning and Community Right-to-know' of the 'Super Fund Amendments and Reauthorization Act' (1988), known as SARA. More information on USA legislation can be found in the bibliography [2–4] and in Chapter 10.

To begin this chapter it is necessary to start with a definition which delimits the concept of emergency or major accident. Lees [5] defines the latter as 'that situation capable of causing serious damage or loss of lives or properties, with considerable risk of extending inside or outside the plant and which could require the use of external resources'. The Seveso 2 Directive defines it as 'an occurrence such as a major emission, fire or explosion resulting from uncontrolled developments in the course of the operation of any establishment covered by this Directive and leading to serious danger to human health and/or the environment, immediate or delayed, inside or outside the establishment, and involving one or more dangerous substances'.

9.2   The EU Seveso 2 Directive

The EU 'Seveso' Directive was adapted to the legislation of European Union countries with significant differences. Holland is the country that has acted most in the restriction of dangerous industrial activities, requiring the determination of the distances at which different levels of risk occur, and limiting the construction inside those areas which surpass amounts considered to be dangerous. In the United Kingdom similar recommendations have been made, but not as restrictive for similar levels of risk.

In January 1997, the revised Seveso Directive was published (Council Directive 96/82/EC of 9 December 1996 on the control of major-accident hazards involving dangerous substances) [7]. Under the scope of this directive, which supersedes the previous Seveso Directive, are all establishments (excluded from its scope are military establishments, hazards created by ionizing radiation, materials during transport, pipelines, mining and waste land-fill sites) in which are present, whether in storage or in processing, a series of dangerous substances, in quantities superior to those indicated in annex 1. Two limiting quantities are given. Depending on which one is exceeded, the requirements change.

Compared with the previous directive, important changes can be seen. The directive includes areas not covered in the previous legislation, such as the safety management system and information to the public on the risks and the actions to be taken in case an accident occurs.

For those establishments which exceed the lower limit, articles 6 (Notification) and 7 (Major accident prevention policy) apply:

- The competent authority has to be notified of information on the company and the products handled: for new establishments, existing establishments or changes in conditions.
- The operator must draw up a document setting his major accident prevention policy according to the principles contained in the Directive.

For those establishments which exceed the higher limit, articles 9 (Safety report), 11 (Emergency plan) and 13 (Information on safety measures) apply. For these sites, article 6 applies, but not article 7.

1. *Safety report*: The operator has to produce a safety report demonstrating that:
   - a major-accident prevention policy and a safety management system have been put into effect.
   - hazards have been identified and the necessary measures to prevent accidents and to limit their consequences have been taken.
   - safety and reliability have been incorporated into the design and operation of the establishment.
   - internal emergency plans have been drawn up and information for the external emergency plans has been made available.

   The aim of this report is to provide the competent authorities with sufficient information. For new establishments, the report has to be made available to the authorities before commencing construction or operation in order to decide upon the approval of the new facilities. This safety report will have to be updated periodically.

2. *Emergency plan*: The directive requires the operator:
   - to prepare an internal emergency plan;
   - to supply to the competent authorities the necessary information for the development of the external emergency plan.

   The competent authority is responsible to draw up, test, revise and update the external emergency plan.

   The Directive also specifies the objectives of the emergency plans and the information they have to contain.

3. *Information on safety measures*: '*Member states shall assure that information on safety measures and on the requisite behaviour in the event of an accident is supplied, without their having to request it, to persons liable to be affected by a major accident originating in an establishment covered by Article 9*'.

   The minimum information to be communicated is defined, and it has to be updated and made permanently available to the public.

The member states are also assigned responsibilities. They have to ensure that the operator is required to prove compliance with this directive. Also they have to ensure that the competent authority identifies possible domino effects with the information received and takes appropriate actions to consider the overall hazard in the area and to ensure co-operation of the operators in the area in the preparation of the external emergency plan.

9.3   Internal emergency plan

This is directed at the actions within the company's premises in the case of an emergency. It comprises all the actions to be carried out inside the installation, including communication with the external emergency plan co-ordinators.

Planning prior to an emergency allows necessary action to be taken more quickly, without the high risk normally associated with rapid decisions. Without a previous study of the possible causes of an emergency, it is highly probable that mistakes will be made in the actions taken owing to confused or contradictory information and to the pressure of the moment.

The main objectives to be fulfilled by an internal emergency plan are:

- Determine the type of accidents which could lead to an emergency situation. Provide the most complete description possible of what is happening.
- Determine the necessary solutions for the different possible emergencies and so be prepared for them. Determine action priorities.
- Ensure that an organization and adequate communication channels have been established so that decisions can be made in an orderly way during the emergency, including the necessary human and material resources, whether internal or external.
- Establish the necessary mechanisms to maintain the plan up-to-date.
- Establish the necessary training and simulations.

9.3.1   The preparation of an emergency plan

The development of the objectives mentioned previously in specific action plans, compiled in a simple structured document, together with the allocation of human and material resources, is the final product of an essentially interdisciplinary task. In the preparation of an emergency plan, representatives of all the affected departments should participate, among others:

- Plant management
- Production
- Safety
- Environment
- Engineering
- Maintenance
- Personnel

- Medical services - Industrial hygiene
- Legal services

Normally it is not necessary that they are all present at every meeting of the team. Some tasks pertaining to very specific areas can be prepared by subgroups, before revision by the full team.

The time and resources necessary for the preparation of the plan depend on the initial situation of the company's safety management system. In general a significant effort is required, and it is vital that the team has the necessary material resources, as well as time, to develop its work.

9.3.2   Development and editing stages

It is necessary that the team that is to develop the emergency plan organizes its work in a structured way, because no relevant possibility should remain without, at least, being considered. The approach to the problem should be meticulous, going through all the stages step by step from the detection of an accident to the conclusion of the emergency situation. In the development of the emergency plan, three stages are usually established, which are described in more detail in the following sections.

Gathering and analysing information

In this first stage two objectives exist: identification and evaluation, on the one hand of the existing risks and on the other, of the human and material resources necessary to tackle an emergency situation.

The principal types of risks to be considered in the chemical industry are a toxic emission, a fire or an explosion, although in each particular case specific risks could arise, although much less frequently. The probability of it happening and the seriousness of the consequences could be estimated for each event applying the methods of risk analysis and evaluation already described in previous chapters. In this phase each and every possibility should be examined exhaustively, in a quantitative way whenever possible.

On the basis of the current situation, and considering possible emergencies and their potential consequences, a preliminary estimation of the human and material resources that would be necessary to confront them can be carried out. Amongst the first are included those who have to develop functions such as the emergency director, co-ordination, communications, fire fighting, process, ... Amongst the second is the emergency control centre, sirens, radios, telephone lines, transport, resources for fire-fighting and many others.

To culminate this stage successfully it is necessary to have all the pertinent information and time to process it, which, in general, represents a very important part of that invested in the emergency plan, unless previous useful studies exist.

Determination of activities in a case of emergency: plan synthesis.

Once the details of possible emergencies to be considered are known and analysed, the activities to be carried out in each case, and the necessary resources, should be planned for. During this stage the final contents of the emergency plan are developed including:

- Assignment of functions and responsibilities.
- Levels of emergency and communication procedures.
- Actions to be taken in each case.
- Communication to the authorities responsible for the external emergency plan and request for external help.

Preparation of the emergency plan

The editing and final preparation of the plan are of great importance, especially due to the circumstances in which it is going to be used. An emergency plan is not a text or consultation book, but is referred to in an extreme situation, under pressure.

An adequate division of chapters helps to find the information that is needed quickly at any time (an alternative is proposed in the following section). The order and structure of the plan should follow the logic of the person who is going to use it. That is why it is convenient to make as simple and graphical as possible the taking of decisions, using flow charts, organization charts and tables.

Computerized artificial intelligence systems are increasingly being introduced in numerous areas of science and technology. Their usefulness to guide actions in the case of an emergency is indubitable. The use of expert systems which collect the knowledge summarized in the two previous stages and permits its application by following a logic tree is an alternative to paper, especially in taking decisions. The principal reasons that support the use of these systems in emergency situations are [3]:

- The reduced frequency with which these situations happen and the scarcity of experts in their treatment.
- The uncertainty of the available information and the technical complexity of consequence analysis.
- The high environmental pressure and the need to make quick decisions.
- The need for the actions to be correct, with regard to the company, the local administration, the press and the public opinion.

The principal applications of an expert system in an emergency plan start when an emergency is detected, they continue with an estimation of the possible consequences and finish with a determination of actions to be carried out, including the consideration of any communications necessary in accordance with the legislation in force.

There is a wide variety of computer programs for emergency planning that are available commercially. The usefulness of these programs is two-fold: used as simulators with meteorological and topographical information corresponding to local characteristics, they help to identify the possible whereabouts of emergencies and evaluate their seriousness. Used during an emergency they can make predictions in real time, as the emergency unfolds, using information from meteorological stations connected on line to the computer. As, for example, the possibility of carrying out corrections to the trajectory of a toxic cloud, when meteorological conditions change, in real time, and make the corresponding decisions.

9.3.3   Elements of an emergency plan

The principal elements which should be contemplated in all emergency plans are the following:

- Organization and resources.
- Procedures for the evaluation of the seriousness of accidents.
- Reporting and communication during the emergency.
- Co-ordination and management during the emergency.
- Intervention during an emergency.
- Simulations and up to date maintenance of the plan.

A more detailed description of each one of these follows.

Resources

In order for emergency response to be effective it is necessary to consider the human and material resources necessary. The determination of these resources is one of the first activities of the team who develop the plan. In this section we shall concentrate on material resources, while human resources will be dealt with later under the heading of organization and co-ordination.

The material resources necessary can be divided into two main groups: (a) equipment and installations for co-ordination and communications, and (b) equipment for reduction of the consequences.

The best way of co-ordinating an emergency (especially with external parties) and the communications, is through an emergency co-ordination centre (ECC). This centre should be provided with the necessary systems to fulfil this function. Normally the following are needed:

1. A room equipped as a meeting and communications centre containing:
    - Audio-visual equipment such as blackboards, maps and area plans. It is especially useful to have a system for superimposing information on a map of the area, such as magnetic strips, transparent sheets for drawing vapour cloud profiles or areas to be evacuated, computers for emergency simulations, etc.
    - All the documentation of the emergency plan, internal and external, regulations and current legislation, MSDS (material safety data sheets [6]) of the products handled by the plant, etc.
    - A list of addresses and telephone numbers of the plant personnel who need to be contacted.
    - Telephone numbers of the authorities, institutions, fire brigade, hospitals and other industries in the area. It is very useful that these numbers are pre-recorded.

2. Telephones, both internal and external, including external lines not susceptible to blockage by calls from reporters or the public during the emergency. It is convenient to have available a recording device for external calls, both sent and received.
3. Radios. Normally this is the best method of assuring communication with the teams working on the emergency. It is appropriate that they should be controlled from the co-ordination centre. If they are battery operated they should be periodically checked to ensure they are ready for use.
4. Facsimile. Could be useful to communicate the situation to the external emergency controller by means of maps or computer printout.
5. Alarms or sirens. They are the most widely used system of communicating a state of emergency, using different sounds.

Apart from the measures permanently installed in the process, with regard to the reduction of the consequences it is important to consider the necessity of:

- Equipment giving information of the meteorological situation, principally wind speed and direction indicators.
- Fire-fighting equipment: smoke detectors (in buildings, electrical substations, etc.), extinguishers, fire hydrants, motor pumps, fire engines (water or foam), sprinklers or deluge systems.
- Equipment to control escapes of liquids or gases, such as gas detectors, water curtains, polyurethane foams of rapid solidification or other measures for constructing containment barriers.
- Medical assistance and transport services, doctors, first aid workers, medical supplies, ambulances, etc.
- Consequence evaluation software, including simulation of cloud dispersion, together with area maps, for predicting the areas which could be affected.
- Personal protection equipment and the necessary supplies for fighting escapes and fires, such as clothing, boots, gloves, facial protection, impermeable clothing, autonomous respirators, torches and others.

Very often it is not necessary to have on site all of the above-mentioned resources, as the collaboration of external entities such as the fire brigade, the Red Cross, hospitals and others can be relied on for help. In these cases the extent of collaboration and the amount of help available to ensure that this collaboration is realistic, should be established in a written agreement. There should always exist a reasonable minimum within the company as the back up forces could be at a considerable distance, and a rapid reaction is the key to success. In this sense it is necessary that the back-up forces participate in emergency simulations. The establishment of 'mutual help pacts' between companies in the same industrial complex facilitates the supply of necessary materials during an emergency.

Procedures for evaluating the seriousness of accidents

As soon as an accident is detected through any of the available means, the first emergency plan mechanism begins: the evaluation of the seriousness of the accident and the decision whether or not to declare an emergency, initiating the actions and communications necessary. A typical flow chart of this procedure is shown in Figure 9.1.

Immediately after detection, the magnitude of the possible consequences of the situation should be assessed and the responses to be initiated decided upon. A widely accepted practice is the definition of various levels (normally numerical) of emergency. These levels help different evaluators to judge these situations in the same way.

The list of levels should be accompanied by guides or rules which permit selecting in each case, with the minimum inaccuracy, what the corresponding level is. Three emergency levels are very often used [2]:

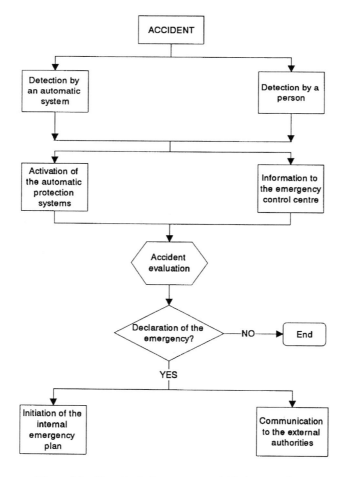

Figure 9.1   Flow chart of an emergency initiation procedure.

1. Alert.
   This is associated with those accidents or incidents which can be controlled with the internal resources available.
2. Plant Emergency.
   This also includes events which will probably have no effect on areas external to the company, but because of their seriousness, they may need external help to be put under control.
3. General Emergency.
   In this case the accident could affect areas external to the company, also requiring outside help to reduce the consequences, internal as much as external. This implies the activation of the external emergency plan.

Rapidity in the detection and initial communication of an emergency is a critical factor in minimizing the consequences. Normally it is convenient to establish a fast means of communicating the message of a possible emergency, either by use of a button or signal (as is done usually for fires) or by use of a special telephone number known to all the workers (hot line). This alarm should go directly to a place that is permanently attended, usually to the emergency co-ordination centre.

At all times, including night shifts and at weekends, it should be clearly defined who is the person responsible for deciding when an emergency should be declared. In a typical chemical plant, working 24 hours per day in 8 hour shifts, the probability of an accident happening outside the working hours of office staff is more than three times greater than when they are there, and this is the reason why it should be the shift manager who normally makes this decision, communicating it immediately to the plant engineer on duty. This person should receive specific training to enable him to assume the responsibility.

Communication procedures and notification of an emergency

Internally, the commonest way of communicating an emergency is by sirens. The level of the emergency is communicated by different sounds. It is necessary that all personnel know the significance of each signal and how to act when on hearing each one of them. Three sounds are usually used, corresponding to three messages:

- Alert.
- Emergency.
- End of emergency: all clear.

In each case it should be clearly defined what should be the behaviour of the personnel, especially of those not directly implicated in the emergency activities, such as office personnel and contractors. According to the level of alert the personnel could congregate at a meeting point or be evacuated to a safe area.

The legislation [7, 8] requires that the competent authorities be informed in the event of a major accident. The external emergency plans determine in detail the type of accidents which should be notified to them. In general it is convenient to inform to the authorities of any type of accident, even though it is small and has no probable consequence outside the plant. This favours the establishment of a positive collaboration and in case the events do not go as planned, it permits the more rapid intervention of external help.

The information to be given usually includes:

- Clear identification of the company where the accident has occurred (name and address)
- Name and responsibility of the person who notifies the emergency. Person and contact telephone.
- Description of the situation: chemical product or products involved, principal risks (explosion, toxicity, etc.), quantity emitted, in the case of continuing escape, flow and estimated duration.
- Internal measures taken to reduce the consequences.
- Critical atmospheric conditions, such as wind direction and speed.
- Possible consequences and measures to be taken.
- Classification of the emergency.

In all cases this information should be communicated by an authorized spokesman.

Organization, co-ordination and control throughout the emergency

It is vital that responsibilities are clearly defined during an emergency. At the beginning it is normal that all of the decisions should be made by one person alone. Later, when the emergency situation has been notified, emergency teams should be constituted immediately in accordance with the organization charts established in the plan. In Figure 9.2 a typical organization chart which could

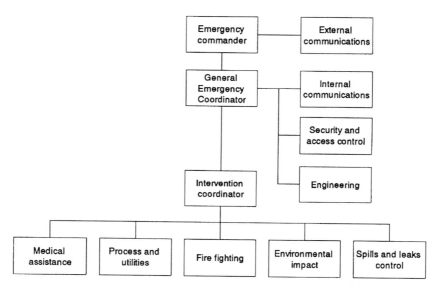

Figure 9.2   Typical organization chart for an internal emergency.

serve as a guide to explain different functions is shown. Depending on the size of the company, more than one of these could fall on the same person. Likewise functions which have not been mentioned can be defined when considered convenient.

At the initial moment all of the co-ordination responsibilities are assumed by the same person, normally the shift superintendent, until the required personnel are present. At this time, it is normal that the shift superintendent becomes the Intervention Coordinator. The principal responsibilities of each function are:

1. Emergency Commander. He is the person with maximum responsibility during the emergency. He co-ordinates the intervention of all resources, both internal and external, to minimize the harm to employees, public and property. He controls communications and exchange of information with the authorities. He determines at each moment the level of alarm, taking decisions at the highest level, such as evacuation, requests for external help, or notifying the correct authorities. This person is usually the Manufacturing Director.

2. General Emergency Coordinator or Operations Coordinator. Usually the production manager as he must know the factory and the process perfectly. He receives from the Emergency Commander the direct responsibility for the activities aimed at reducing the consequences of the accident. He co-ordinates the forces that work in the field, and determines the need for support personnel. He centralises internal communications from the Emergency Control Centre.

3. Intervention Coordinator or Field Operations Coordinator. Reporting directly to the General Emergency Coordinator he supervises and co-ordinates all of the operations and resources involved at the scene of the accident, assessing its seriousness and channelling communications with the emergency co-ordination centre.

4. Engineering. Gives technical support to the General Emergency Coordinator. Its main function is to determine the possible consequences of the accident and recommend the actions to be taken to isolate potentially dangerous equipment or to evaluate the effectiveness of possible alternative actions.

5. Security and access control. Their principal mission is to control access thus avoiding the entrance of unauthorized personnel, and to lead the people whose presence is required or authorized to the co-ordination and communications centre. They should also account for personnel present in the emergency area, searching for them if they have not been located.

6. Medical assistance. Provides urgent emergency medical help to injured people during the accident and transfers them to medical centres. Also advises the emergency coordinator on the risks of exposure to the chemical products present in the accident. (A summary of this information is found in the material safety data sheets, MSDS [6] or in records and equivalent databases.)

7. Process and utilities. They are in charge of maintaining the safety of that part of the factory not directly affected by the emergency, generating the services required for the emergency operations, isolating threatened equipment or stopping the process when necessary. They also isolate the affected equipment and stop leaks to minimize the consequences of the accident (in the case of toxic leaks experts or special personal protection equipment may be needed)

8. Fire fighters. Their mission is to control and extinguish any fires which break out. They also usually take charge of controlling the leakage of flammable liquids. Their training is critical to the success of their task, one of the most risky and technically most complex.
9. Environmental impact. Here the responsibility is to minimize the environmental impact of the accident, determining the level of the risk, recommending the preventive measures to be taken and remedying the consequences through necessary operations during and after the accident.

Activities during the emergency

The plan should contain strategies for any situation that might arise during the emergency, in such a way that it can guide effectively the co-ordinators. There are two types of activity: preventive and corrective. Obviously, the first is preferable to the second, but a choice cannot always be made and therefore both types should be considered.

Preventive actions are directed at minimizing the consequences of an accident. They depend on the specific situation of the emergency and on the type of products handled:

- The prevention of catastrophic ruptures of equipment and tanks, by cooling them down with jets of water, insulation, protection barriers, etc.
- Extinction or control of outbreaks of fire.
- Rescue or evacuation of those who are possibly affected.
- Control and alleviation of the situation: containment of escapes, absorption of spills, spraying and absorption of toxic clouds, establishment of physical barriers, control of the process conditions, emergency plant shut-down, etc.
- Protection of areas under risk of being exposed to damage by 'domino effect'.

Co-ordination and organization are fundamental to establishing action priorities. The specific action for each circumstance should be defined in the emergency plan. In Example 9.1 the actions for a specific case are described.

Corrective action, such as cleaning, decontamination and restoration of the area, usually comes after the emergency ends. This type of action is specially important after liquid spills, as in the case of oil slicks or escapes of toxic products of low degradability as in the Seveso accident.

Example 9.1   Action when faced with a chlorine gas leakage (Armenante [2])

Chlorine is a highly toxic gas, even lethal, with an IDLH value of 25 ppm and a TLV of 1 ppm; and therefore, any leakage, however small, is considered extremely dangerous. The easiest way of detecting a chlorine gas leak consists of spraying a liquid ammonia solution, which upon reacting to chlorine, forms a white cloud of ammonium chloride. All personnel who are going to work in proximity to a leak should wear adequate personal protection equipment and autonomous respirators.

In any case, the area of danger should be immediately isolated and evacuated, this could have a radius from 50 m (in the case of leaks in small containers) up to several kilometres (in the case of large tanks). The work to attenuate or reduce the leakage should start as soon as possible. The return of personnel should not be permitted until the concentration of chlorine has been checked and is tolerable in the whole area, especially in points protected from the wind or low lying, where chlorine gas which has a high density could have accumulated, or in the presence of liquid residue.

Depending upon the origin of the leak, action could be the closing of a valve, tightening or closing a flange, or placing an adequate plug.

In general an attempt should be made to minimize the evaporation of the liquid chlorine, using proteic foams to reduce the evaporation surface. Water should never be added, as this would form hydrochloric acid, which, moreover, due to its corrosive power could augment the magnitude of the leak. Only if it is really necessary to cool a tank threatened by the leakage should use of water be considered.

In the design stage, the usual option is to install tanks within dikes which contain the leak and reduce the area of evaporation. Also, gravel floors are usually avoided, as they increase the heat transfer from the ground to the liquid mass.

Training, simulations and up to date maintenance of the plan

As already manifested in earlier sections, it is vital that all personnel understand perfectly at least the part of the emergency plan which affects them. It is normal for emergency situations to be infrequent, so experience is of little use as the probability of forgetting is high. The only way of ensure that all personnel act correctly under circumstances of doubt, even of panic, is through simulations and training.

The principal objectives of a training and simulation plan are:

- To check the knowledge of personnel in that part of the emergency plan which affects them, assuring a good response capacity in the case of an emergency.
- To check the adequacy of the plan in close to reality cases, detecting weak or improvable points.
- To check the state of the necessary resources for confronting the emergency (alarms, vehicles, extinguishers, etc.).
- To check the response capacity of external entities involved and the level of co-ordination reached.
- To provide the necessary training for those employees who have special responsibilities in the plan.

To assure good functioning of the alarm systems it is necessary to do frequent tests, normally weekly, which assure its functioning. In the case of sirens, which are usually used to communicate a state of emergency to the personnel, it also achieves an improvement in the understanding of the significance of the different sounds. Communication systems should also be tested weekly.

All personnel who could intervene in the case of an emergency should be involved in simulations. Some activities which are usually the object of simulations, with or without prior warning are:

- External communications systems
- Alarms and evacuations
- Fires, spills or toxic leaks
- Accidents with requests for external help (fire brigade, hospitals, etc.).

It is convenient to designate an observation team which knows the emergency plan perfectly, to gather the maximum information during the simulation, through videos or other alternative means. All of the information gathered should be analysed afterwards before issuing a report with an evaluation of the exercise including the changes to be introduced in the emergency plan, material which it is necessary to obtain or the necessary additional training.

9.4   External emergency plan

This is directed at actions external to the plant. It includes communication and co-ordination between companies within the area and the authorities and evaluation of the consequences, evacuation decisions, co-ordination between civil protection, fire brigade and the army, etc. Its objectives are to:

- Identify the emergencies that could occur and ensure preparation for confronting them through previous planning.
- Ensure that the decision-making throughout an emergency is done in an orderly way.
- Ensure that the necessary human resources, equipment and services are made available to mitigate the consequences of the emergency and co-ordinate their actions.

The elaboration of an external emergency plan, as with the co-ordination of the different intervention forces, is normally the competence of the local administration [7, 8], although internal emergency plans of the companies involved should be considered and their participation taken into account.

9.4.1   Communication and co-ordination of the emergency

The procedures for communication of the emergency to those responsible for the external emergency plan should be shown clearly in the internal emergency plan. The external emergency plan should define which types of accident should be reported, in what way and what information should be given.

Special attention should be paid to the external telephone communication lines, which are frequently blocked in the case of an accident because of calls from the inhabitants of the area. It is necessary to have dedicated emergency communications lines which do not appear in telephone directories.

In Example 9.2 the external emergency plan organization and co-ordination structure of an important concentration of chemical companies is described. In reference [9] the external emergency plan for a fertilizer plant emergency plan is also described.

Example 9.2   The PLASEQTA

The chemical safety plan of Tarragona (PLASEQTA) is an example of an external emergency plan, which seeks the following objectives:

- *The identification of possible major accidents which could occur in the industrial complexes in the Tarragona area (La Pobla de Mafumet, Salou - Vilaseca and Flix) and evaluation of the consequences.*
- *Defining the necessary human and material resources to minimize the consequences of the accidents under consideration.*
- *The co-ordination and collaboration of the authorities and organizations involved.*

The plan includes all of the companies which store or process products included within the scope of the Seveso Directive. A co-ordination committee exists, of which these three administrations are members:

- *Local administration, represented by the Mayor of Tarragona.*
- *Regional Administration, represented by the local governmental delegate of the Generalidad of Catalonia.*
- *Spanish State administration, represented by the General Sub-director of Civil Protection, Planning and Operations.*

The organization chart is shown in Figure 9.3. Through this organization the four active emergency action groups are co-ordinated.

A 'mutual help pact' also exists which establishes that the companies will share the necessary resources made available during the emergency. The text of the plan and the information necessary during an emergency are structured in four volumes.

1. *Emergency Plan*
2. *Design scenarios to evaluate risks.*
3. *Consequence analysis and action patterns for the cases under consideration.*
4. *The Computer System Manual, which combines a series of programmes and databases and allows handling the information which is available and analysis of consequences for situations which are different to those found in the plan. Through this system, products handled by each company are known, along with their specific risks, permitting possible consequence analysis and giving pertinent recommendations to the action groups.*

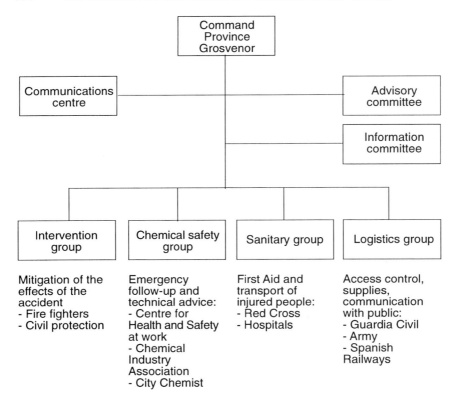

Figure 9.3   Structure of Tarragona's Chemical Safety Plan (PLASEQTA).

9.4.2   Community right-to-know

Growing public concern with chemical risks demands the availability of reliable and understandable information on the development of the situation, not only when the danger directly affects the population, but also in circumstances in which they could have been involved but were not. Special attention should be paid to the media, as lack of information could give rise to headlines based on hasty interpretations made by technically incompetent people, who arrive at the wrong conclusions, cause alarm and irreparable damage to the image of the company, [10].

In general, the best way of confronting the situation is by taking the initiative and offering information to the media, thus maintaining maximum control and endeavouring to present it to the public with the maximum truth and in the quantity necessary. The company's attitude and that of its spokesmen at this moment in time are critical for its future image. The relation of trust or mistrust, transparency or concealment of information, which is established between the media, the company and the authorities will immediately be transmitted to the public.

Information should be given to the media as frequently as necessary, and as events unfold, whether through press conferences or press releases. It is especially important not to transmit doubtful or unconfirmed information which may need rectifying at a later date.

Rarely are the problems of information, which could come about during an emergency situation, purely a consequence of the tension of the moment. Frequently they are rooted in minor incidents or confrontations between the company and the community in the past. This is the reason why the attitude of collaboration and information during an emergency is built day by day through continuous contact and daily collaboration between the authorities, the media and the company. In the section 'Community awareness and emergency response' of the voluntary programme Responsible Care [11], five points are defined for improving relations and communications with the community.

1. Continuously evaluate the questions and preoccupations of the community about the company.
2. Develop an education programme for the community, media, support institutions, public representatives, etc. on the risks that the company represents for the community and the emergency plan.
3. Talk constantly to the citizens and answer their questions and worries on safety, hygiene, environment and any other subject of interest.
4. Promote an open door policy which admits people who are interested in familiarising themselves with the installations, way of working, the products which are handled and the effort made to protect the safety of the people and the environment.
5. Periodic evaluation of the results of these efforts.

Example 9.3

Kenney [4] describes a case which he himself saw. The people responsible for a company which had suffered an oil slick in the sea gave a press conference in which they tried to give answers from the company's point of view without speculating or incriminating anybody, protecting all the legal aspects. In the tension of their presentation they failed to recognise the Mayor of the city, who tried to ask them a question for more than 5 minutes. They had spent too long preparing their presentation with their lawyers, and all that they achieved was an accusation of incompetence by the media, and the feeling transmitted to the public was that somebody should go to prison.

In the 'Exxon Valdez' case something similar occurred. The company reacted slowly, trying to delay for as long as possible until approval of the clean up plan, causing a brief period of good weather to be lost (because of

> bad meteorological conditions at the time) which could have significantly reduced the area which suffered the consequences of the leak. The company was condemned for negligence, slow reaction and arrogance in their management, spending the same amount of money (or more) it would have cost to react adequately and in time: $2000 million dollars without counting the damage to their public image.
>
> Kenney compares these cases with the plant manager of a company which suffered a serious accident with several deaths and which, because of transparency and close public relations, achieved the support of public opinion. In this case all of the recommendations of the legal advisers were ignored and he went into the streets to explain what had happened, meeting with the families affected, checking damage personally and placing all of the company's resources at the public's disposal. Nobody asked for the imprisonment of anyone. The company's image suffered no harm thanks to him.
>
> In this type of situation the search for culprits is inevitable. Any posture carries a certain risk and the people responsible must exercise their leadership instead of sitting and waiting for events to unfold.

9.5   Transport of hazardous materials

The transport of hazardous materials has traditionally received separate treatment [12–14]. Although it has similarities with the risks of chemical processing, principally by handling the same substances, the circumstances of this handling gives rise to specific risks. The most important of these being, although maximum precautions are taken, that an accident could occur in numerous places, affecting populated areas, rivers etc.

Although strictly speaking the product is the responsibility of the company which transports it and not of the manufacturer, an inevitable responsibility exists. For this reason, when there have been accidents in the transportation of dangerous merchandise, companies not involved in the accident have given their support, moving their resources a long way from their production centres.

The principal elements to be considered in planning for transport emergencies are:

- Product information and labelling
- Procedures to be followed in case of an accident
- Incident control network and emergency action teams
- Information to the public.

One of the systems most used at present is the 'TREMCARDS' (Transport Emergency Cards) prepared by the European Chemical Industry Council (CEFIC). These cards allow drivers to carry with them instructions of what to do in case of an accident or breakdown.

It is equally convenient that the units which would intervene in the case of an accident have the necessary information on how to act in each specific case. The mode of action is very specific depending on the product being transported, the type of container and the type of accident, and the Tremcards do not contain the necessary information as they are meant for the drivers. Normally this information is presented by means of cards for each product or group of similar products with the same level of danger and type of approach.

To obtain maximum efficiency it is necessary that these cards contain:

- The main characteristics of the product or group of products to which the card refers.
- Hazards of the product.
- The personal protection equipment necessary.
- Appropriate extinguishing agents.
- Prevention measures to be adopted.
- Action in the case of an accident, in accordance with its characteristics: fire, spill, etc.
- First Aid.

A collection of cards with these characteristics have been edited in Spain by the Interior and Transport Ministries with the participation in their elaboration of the Spanish Federation of Chemical Industries.

During the past few years accidents in sea transportation have been especially serious as much for their impact on the environment as on public opinion. The large quantities of product transported and the severe harm to the population and to marine life, because they could happen in ports or near the coast, are the principal factors which differentiate transportation by sea and by land. Cases such as those of 'Urquiola', 'Amoco Cádiz', 'Exxon Valdez', 'Aegean Sea'... have started discussions on the adequacy of safety measures demanded from ships, the rapidity and efficacy of measures applied to contain oil slicks and the correction of the consequences. The principal types of accidents which could happen are: spills of toxic or flammable liquids, emissions of liquefied gases with evaporation and formation of vapour clouds, as well as fires or explosions on the ship.

Example 9.4   The 'Aegean Sea' Accident [15]

The actions carried out after the accident of the oil tanker 'Aegean Sea' (Figure 9.4) are an example of emergency co-ordination with a great social repercussion. After the ship ran aground it split in two and fire broke out after one of the crude oil tanks exploded, during very adverse meteorological conditions, with strong winds and reduced visibility. A co-ordination centre for operations was established immediately and the most important activities were started, including:

- Crew rescue.
- Evacuation of the surrounding areas and the closure of the port.

Figure 9.4 A view from the Aegean Sea accident (Example 9.4). (Courtesy of El Pais newspaper, Spain. Photo: Bernardo Perez.)

- Extinguishing the fire.
- Removal of the remaining cargo
- Salvage of the tanker, cleaning of the beaches and protection of sensitive areas of the coast.

A technical Commission met daily to follow through the plan of operations. All of the competent organizations in this matter (Minister, government of Galicia, Town Halls, Civil protection) took part in these. The resources mobilised included, amongst others, freighters and towboats, rescue helicopters, skimmers and floating hoses and private companies for the cleaning of beaches and rocks and the treating of residuum.

Faced with the protests from ecological organizations on the environmental impact, the president of the company who had chartered the tanker (Repsol) directed a letter to the members of Greenpeace explaining the company's attitude and the actions carried out to repair the damage. It was answered by this organization which manifested its disagreement to their arguments [16].

9.6 Questions and problems

9.1   Faced with a fire affecting a storage tank of flammable liquid, three options can be taken: (a) Attack the fire, (b) Control the fire, allowing the fuel to burn, (c) Withdraw. All of them could be adequate for different situations. Discuss possible cases and when each strategy is most appropriate, giving details of a concrete course of action in each case.

9.2   Describe the actions to be carried out in the case of a spill of a hazardous liquid, covering the largest number of possible alternatives.

9.3   Elaborate an emergency plan (general) for a loading station of flammable hydrocarbons in cisterns, from atmospheric tanks built inside containment walls. The station is built into a refinery, although it can also be handled by lorry drivers. Discuss the most important factors which could affect the approach to the problem.

9.4   Consider a situation in which in a factory an escape of a toxic substance in vapour phase occurs. Draw a map of the chemical plant, access routes, near by population, the situation of emergency services, hospitals, etc. It can be carried out on a hypothetical case or on a real industrial facility. Consider the actions to be taken in different situations in the external emergency plan. Discuss, with a reasonable probability, which is the worst case.

9.7   References

1. King, R. (1990) *Safety in the Process Industries*, Butterworth-Heinemann, London.
2. Armenante, P. M. (1991) *Contingency Planning for Industrial Emergencies*, Van Nostrand Reinhold, New York

3. Greenberg, H. R. and Cramer, J. J. (eds) (1991) *Risk Assessment and Risk Management for the Chemical Process Industry*, Van Nostrand Reinhold, New York.
4. Kenney, W. F. (1993) *Process Risk Management Systems*, VCH, New York.
5. Lees, F. P. (1980) *Loss Prevention in the Process Industries*, Butterworth-Heinemann, London.
6. Ministerio de Trabajo y Seguridad Social. Instituto Nacional de Seguridad e Higiene en el Trabajo (1992) *International Material Safety Data Sheets (MSDS)*, INSHT, Madrid.
7. Council Directive 96/82/EC of 9 December 1996 on the control of major-accident hazards involving dangerous substances.
8. Royal Decree 952/1990, of June 29th, to modify and complete the Royal decree 886/1988, of July 15th. *Spanish Boletín Oficial del Estado* (BOE) no. 174, Saturday July 21st 1990.
9. Ramabrahmam, R. V. and Mallikarnujan, M. M. (1995) Model off-site emergency plan. Case study: toxic gas release from a fertilizer unit. *J. Loss Prev. Process Ind.*, **8**(6), 343–8.
10. Pérez de Tudela, C. (1994) *La Información en las Catástrofes*, Editorial Mapfre, Madrid.
11. Chemical Industries Association (1992) *Responsible Care Management Systems*, CIA, UK.
12. de las Alas Pumariño, E. (1986) Algunas cuestiones sobre la situación del transporte de mercancías peligrosas. *Ingeniería Química*, September.
13. Royal Decree 1723/1984, of June 20th. Reglamento nacional para el transporte por carretera de mercancías peligrosas (Spanish rules for road transport of hazardous products).
14. Royal Decree 811/1982, of March 5th. Reglamento de transportes por ferrocarril de mercancías peligrosas (Spanish rules for transport by railway of hazardous products).
15. Respuesta Cooordinada. Protección Civil, no. 18. November-December 1992.
16. López de Uralde, J. *Repsol responde a nuestras protestas*, Boletín Informativo Trimestral de Greenpeace nº 28, III/93.

10 Legislation, standards and design codes

Deep-rooted virtue is justice, according to the wise ancients, may it last forever in the
will of just men, and give and share with each one their equal right. And however men
may die when justice within never weakens but lives always in the hearts of living men
who are straight and good. And the script says that the just man errs seven times each
day, for he still cannot do what he should, for nature's weakness which lives in him,
and even so he must make an effort to do good and to fulfil the commandments of
justice.

Part III, Title I, Law I, Alfonso X The Wise

The growth during the past few years in the number of laws directed at regulating
environmental and safety aspects in industry is well known to everyone. According
to Cralley [1], the number of federal laws in the USA which contain aspects relevant
to industrial process safety increased more than three-fold in the period between
1960 and 1980. There is no updated information, but it is evident that after 1980
the approval of regulations has accelerated more than ever.

The objective of this chapter is to offer a summary of regulations, both
compulsory (legislation from various countries) as well as voluntary (good practice,
recommendations, design codes, etc.). In no way do we pretend that this is an
exhaustive list, given the great number of sources of this type of information, the
volume of documentation already existing and the continuous changes and updating that make it practically impossible. Even if this could be achieved, within
a few months it would be of little use. Only those aspects which from our point of
view are most important and which form the minimum baggage that a specialist
in the chemical industry should possess are included.

In any case we believe that by adding to the bibliography a list of institutions
that continuously generate information it is made easier for the reader to up-date
this information.

10.1   EC legislation

* **Directive regarding the application of measures to promote the improvement
  of the health and safety of employees at work.**
  89/391/EC of 12th June 1989
* **Decision of the Council regarding the creation of an advisory committee for
  safety, hygiene and health protection in the work place.**
  74/325/CEE, of 27th June 1974.

* **Directive on the control of major-accident hazards involving dangerous substances.**
  96/82/EC of 9 December 1996
* **Directive on the protection of workers from hazards related to exposure to chemical, physical and biological agents at work.**
  80/1.107/CEE, of 27th November 1980
  Modified and completed later by directives on specific products.
* **Directive on the minimum safety and health protection measures regarding the use of individual protection equipment by employees at work.**
  89/656/CEE, of 30th November 1989.
* **Directive regarding the indicative limiting values for the protection of workers against the hazards involved when exposed to physical, chemical and biological agents during work.**
  91/332/CEE, of 29th May 1991.
  Develops Directive 80/1107/CEE.
* **European agreement on the international transport of hazardous chemicals by road (ADR)**
  Geneva, 30th September 1957.
* **Directive regarding integrated pollution prevention and control (IPPC)**
  96/61/EC of 24th September
* **Directive on the ambient air quality assessment and management**
  96/62/EC of 27th September

10.2   Spanish legislation

* **Industrial Legislation**
  Law 21/1992 of 16th July.
  Chapter 1 on industrial safety.
  Article 9. Objects of safety.
  Article 10. Hazard prevention and limitation.
* **Royal Decree on the prevention of major accidents in some industrial activities**.
  Royal Decree 886/1988 of 15th July.
  Some annexes and dispositions modified later by the Royal Decree 952/1990.
  Creation of the Technical Committee of Chemical Hazards by Order of 21st March 1989.
  (Will be phased out when the new directive is implemented).
* **Law on the prevention of work hazards.**
  Law 31/1995 of 8th November.
* **Royal Decree 473/1988, of 30th March.**
  Includes the dispositions for the application of the Directive 76/767/CEE, of 27th July 1976, on pressure vessels, which provides that Spain cannot refuse or limit the use of pressure vessels type EU which comply with the said Directive or the specific Directives which follow it.
* **Decrees that develop the Law 38/1972 for the protection of the atmospheric environment.**
  Decree 833/1975, of 6th February.

Modified by Royal Decree 1.613/1985, of 1st August. Containing the regulations on emissions of pollutants to the atmosphere.

* **Regulation on public water supplies (which develops the Law 29/1985).**
  Royal Decree 849/1986, of 11th April.
  Containing the regulations on waste emissions to rivers, lakes and underground watersheds.
* **Emissions of hazardous substances into the sea (which develops the coastal Law 22/1988).**
  Royal Decree 258/1989 of 10th March.

10.3   USA legislation

Even though the legislation which is cited in this section is not in force in the EU, it was thought interesting to detail it to mark tendencies which in the majority of cases have been included in the European legislation, or which may be in the near future.

* **Hazardous waste operations and emergency response.**
  Principally concerned with emergency actions to combat toxic substance emissions and the actions to clean up and eliminate toxic products.
  OSHA (Occupational Safety and Health Administration) 29 CFR 1910.120.
* **Process safety management of hazardous chemicals.**
  Regulating the minimum requirements for safety management systems in plants that handle quantities above certain limits of some dangerous chemical products.
* **Extremely Hazardous Substances list and threshold planning quantities.**
  List of extremely toxic substances and quantities for which emergency plans must be established.
  Environmental Protection Agency (EPA), 40 CFR Part 300, 1987.
* **Emergency planning and release notification requirements.**
  Defines the requirements which should be fulfilled with regard to emergency planning and notification to the authorities in the case of emissions.
  Environmental Protection Agency (EPA), 40 CFR Part 355, 1987.
* **Toxic chemical release reporting; community right-to-know.**
  Defines the regulations for reporting toxic product emissions and informing the community.
  Environmental Protection Agency (EPA), 40 CFR Part 372, 1987.
* **Title III, on 'Emergency Planning and Community Right-to-know' of the 'Superfund Amendments and Reauthorization Act' (1988), commonly known as SARA III.**
  Continues the programme 'Superfund', established in the 'Comprehensive Environmental Response, Compensation and Liability Act' (CERCLA), of 1980. Establishes the obligation of industries to cooperate with local communities to develop external emergency plans through local emergency planning committees. Under this rule, the industries should inform the authorities of the quantity and location of dangerous substances being used if they exceed 500 lbs.

10.4   Design codes and regulations

On a world-wide level, the largest standards organization is the ISO (International Standards Organization), however, in every country, there are standards organizations, either of state origin or formed by manufacturers' associations.
Some of the most important are:

- Germany   DIN.
                VDI/VDE (Verein Deutscher Ingenieure).
- Great Britain BSI (British Standards Institution)
- USA   ANSI (American National Standards Institute)
                ASTM (American Society for Testing and Materials)
                API (American Petroleum Institute).
                NFPA (National Fire Protection Association)
- Spain   AENOR (Spanish Association of Standardization and
                Certification). Publishes the UNE standards (Una Norma
                Española: A Spanish Standard). It is the Spanish representative in
                the ISO, the CEI (International Electrotechnical Commission,
                IEC) and other international organizations for standardization,
                and transposes their standards into UNE standards.

Normally, the national standards body in each country is a member of ISO and acts as the representative and distributor of the majority of the national standardization organizations. Usually the easiest way to obtain standards from other countries is through them. For information or purchase of other institutions' publications (not members of ISO) it is best to contact them directly. (The addresses of the most important are found further along in the book).

In general, the standards are recommendations, unless there is additional legal demand for their fulfilment, and therefore voluntary. However, a lot of them have gained extensive recognition, even at international level. Here is a selection of some of the more important. Others have been mentioned in corresponding chapters.

* **American Petroleum Institute (API)**
  API Recommended Practice 520. Recommended Practice for Design and Installation of Pressure Relief Systems in Refineries.
  API Recommended Practice 521. Guide for Pressure Relieving and Depressurizing Systems.
  API Standard 620. Recommended Rules for Design and Construction of Large Welded, Low-Pressure Storage Tanks.
  API Standard 650. Welded Steel Tanks for Oil Storage.
  API Recommended Practice 750. Management of Process Hazards.
  API Recommended Practice 2000. Venting Atmospheric and Low Pressure Storage Tanks.
  API Recommended Practice 2001. Fire Protection in Refineries.

* **American Society of Mechanical Engineers**
  ASME Boiler and Pressure Vessel Code.
  ASME Code for Pressure Piping.

* **National Fire Protection Association**

NFPA 30.	Flammable and Combustible Liquids Code.
NFPA 36.	Solvent Extraction Plants.
NFPA 49.	Hazardous Chemical Data.
NFPA 58.	Storage and Handling of Liquefied Petroleum Gases.
NFPA 59.	Storage and Handling of Liquefied Petroleum Gases at Utility Gas Plants.
NFPA 59A.	Production, Storage and Handling of Liquefied Natural Gas.
NFPA 68.	Deflagration Venting.
NFPA 69.	Explosion Prevention Systems.
NFPA 70.	National Electric Code.
NFPA 231.	General Storage.
NFPA 325M.	Fire Hazard Properties of Flammable Liquids, Gases and Volatile Solids.
NFPA 491M.	Hazardous Chemical Reactions.
NFPA 704.	Identification of the Fire Hazards of Materials.
NFPA 901.	Uniform Coding for Fire Protection.

10.5   Institutions

American Conference of Governmental Industrial Hygienists (ACGIH)
6500 Glenway Avenue, Building D-7, Cincinnati, Ohio 45211-4438, USA.

American Industrial Hygiene Association (AIHA)
2700 Prosperity Avenue, Suite 250, Fairfax, VA 22031, USA.

American Institute of Chemical Engineers (AIChE)
(Including the CCPS, Center for Chemical Process Safety)
345 East 47th Street, New York, NY 10017, USA.

American Petroleum Institute (API)
1220 L Street Northwest, Washington, DC 20005, USA.

American Society of Mechanical Engineers (ASME)
1950 Stemmans # 5037C, Dallas, TX 75207, USA.

Spanish Association for Standardization and Certification (AENOR, Asociación Española de Normalización y Certificación)
Fernández de la Hoz, 52, 28010 Madrid, Spain.

Chemical Industries Association (CIA)
King's Buildings, Smith Square, London SW1P 3JJ, United Kingdom.

Environmental Protection Agency (EPA)
Washington, DC 20460, USA.
Office of Technical Information and Publications
Research Triangle Park, North Carolina 27711, USA.

European Chemical Industry Council (CEFIC)
Av. E. Van Nieuwehuyse, 4, B - 1160 Brussels, Belgium.

Institution of Chemical Engineers (IChemE)
Davis Building. 165-171 Railway Terrace.
Rugby, Warwickshire CV21 3HQ, UK.

International Organization for Standardization
1, Rue de Varembé,
CH-1211 Genève 20, Switzerland.

Spanish Department of Industry and Energy
Paseo de la Castellana, 160, 28046 Madrid, Spain.
Publications Centre
Dr. Fleming, 7 , 2°. 28036 Madrid, Spain.

National Fire Protection Association (NFPA)
One Batterymarch Park, Quincy, MA 02269-9101, USA.

National Institute for Occupational Safety and Health (NIOSH)
4676 Columbia Parkway, Cincinnati, OH 45226, USA.

Occupational Safety and Health Administration
1726 M Street NW, Washington DC 20210, USA.

TNO
P.O. Box 45, 2280 AA Rijswijk, The Netherlands.

10.6   Internet sites

Legislation databases

European Union

European Documentation Centre	WWW.UV.ES/CDE	- Free indexes
European Union Document Delivery Service (EUDORA)	WWW.EUDOR.COM	- Free document search. - Document delivery (Pay).

United States of America

The U.S. House of Representatives Internet Law Library	LAW.HOUSE.GOV	- Free on line document search and display
Thomas: Legislative Information on the Internet	THOMAS.LOC.GOV	- Free on line document search and display

United Kingdom

Government Direct	WWW.OPEN.GOV. UK/GDIRECT	- Free on line document search

Spain

Boletín Oficial del Estado	WWW.BOE.ES	- Free indexes. - On line documents (Pay).

Institutions

Institution of Chemical Engineers (IChemE)	ICHEME.CHEMENG.DE.AC.UK
American Institute of Chemical Engineers (AIChE)	WWW.AICHE.ORG
Occupational Health and Safety Administration (OSHA)	WWW.OSHA.GOV
Environmental Protection Agency (EPA)	WWW.EPA.GOV
Canadian Centre for Occupational Health and Safety (CCOHS)	WWW.CCOHS.ORG
National Institute of Science and Technology (NIST)	WWW.NIST.GOV
American Chemical Society (ACS)	WWW.ACS.ORG

10.7   Reference

1.   Cralley, L. J. and Cralley, L. V. (1984) *Industrial Hygiene Aspects of Plant Operations*, vol. 2, Wiley, New York, p. 13.

Appendix: Some cases of industrial accidents

A1 Explosions caused by runaway reactions: the accident in the Union Carbide facilities at Seadrift, Texas, 1991

On 12th March 1991, at 1.18 a.m., an explosion took place in a small boiler of the ethylene oxide redistillation still (ORS) in the facilities of the Union Carbide company, at Seadrift, Texas. The explosion and the subsequent fire caused the death of a worker and damage of great importance to the plant. An analysis of the accident which is shown below follows the work of Viera and colleagues [1], of Union Carbide, who carried out an extensive investigation of the causes which provoked the accident

A1.1 Installation characteristics

The ORS in question was a unit of 36.5 m in height and 2.6 m in diameter, constructed of steel with a thickness of 3/8" to 1/2". The design pressure or maximum allowable working pressure (MAWP) was 90 psig (6.1 gauge atmospheres). The normal operating temperature was 60°C. The column was supplied with refined ethylene oxide, eliminating as head product traces of light components (typically parts per million of formaldehyde) and as bottom product some heavy components (parts per million of acetaldehyde and traces of water) . The redistilled ethylene oxide was obtained as a side stream and was sent to other units to be used in the manufacture of different derivatives.

The reboiler was a thermo siphon, with more than 800, 3 metre high 1 inch diameter vertical tubes. In the shell side steam was condensed at 75 psig (5.1 gauge atmospheres), with its flow regulated by a control valve. The condensation temperature of the steam at this pressure is about 160°C. The reboiler was designed for a high liquid/vapour ratio, although there was no direct measure of the flow of recycled liquid.

There was an automatic trip for high pressure in the still, high temperature in the base of the column or in the steam, or low liquid level in the bottom of the ORS. The trip provoked stoppage of the steam feed to the reboiler and the injection of high pressure nitrogen in its shell with the aim of blocking the heat transfer.

A1.2 Chronology

Some days before the accident a programmed shut-down was carried out in the ethylene oxide unit to carry out maintenance work and to repair a piece of equipment not related to the accident which happened later. No maintenance work was carried out on the ORS unit.

The unit was started again on the afternoon of the 11th. The feed of ethylene oxide was resumed at approximately 7.30 p.m., and the steam to the reboiler at 8.30 p.m. A few minutes later there was an increase in the pressure, activating the automatic trip, closing the steam feed to the reboiler. Two fresh attempts were made to start the unit before the cause of the increase in pressure was found and corrected.

The definitive start occurred at midnight, and apparently the unit worked normally for about an hour. At 1.18 a.m. there was an explosion. The number 2 unit was functioning normally in parallel with the first, and was not involved in the explosion.

A1.3   Causes of the accident

The accident started with a hot point in the upper part of the reboiler tubes. A previously unknown reaction, catalysed by iron oxide, increased the temperature of this hot point until it surpassed the 400–500°C necessary for the decomposition of the ethylene oxide. The self-decomposition does not require the presence of oxygen, and generates gaseous products such as carbon monoxide and methane.

The reaction front of the self-decomposition reached the base of the column and continued moving up and accelerating. The release of heat and the increase in the number of moles that accompanied the process, pressurized the ORS up to four times its design pressure causing the explosion. The whole sequence of events was completed in about a second, and no conventional system of pressure relief could have avoided the explosion.

According to Viera and colleagues [1], a series of coincidental circumstances provoked the accident:

- The recirculation in the ORS boiler had been reduced.
- The top part of the boiler pipes dried out.
- The ethylene oxide vapour heated until it reached temperatures close to that of the heating vapour.
- A local pocket of ethylene oxide vapour near the top part of one of the boiler tubes occurred.
- A strong exothermic reaction developed, catalysed by iron oxides which generated local temperatures greater than 500°C.
- Self-decomposition of the ethylene oxide retained in the top part of the pipes, took place.

A1.4   Design and safety considerations.

Liquid–vapour flow in the pipes

The thermosiphon type of boiler with vertical tubes is used widely in the chemical industry. In a system of this type different boiling patterns are found as the fluid goes up the tubes: heat transfer without boiling, by convection from the wall; initiation of nucleated boiling; bubble boiling; slug flow and annular flow. These systems are described in conventional texts [2] of heat transfer.

If the system design is inappropriate, or if the operational conditions make it possible (for example, with low liquid recirculation), in the upper parts of the tubes the liquid film on the heat exchange surface could be eliminated completely, drying this, with the consequent increase in temperature, approaching that of the heating steam. The maintenance of a high liquid–vapour ratio (L/V) in the two-phase mixture which leaves the upper part of the evaporator is the best way of ensuring that the interior of the pipe stays wet and, therefore, at a low temperature. The L/V ratio depends on the physical properties of the liquid and vapour and their two-phase equilibrium, as well as such design variables as the length and diameter of the pipes, the level of liquid at the base of the column and the head losses in the accessories which determine the recirculation flow. The L/V ratio also depends on the difference between the internal and external temperature of the tubes. In the case we are dealing with, the external temperature could be considered constant and equal to the steam condensation temperature. On the other hand, the temperature of the internal wall could vary considerably: after the initial heating up, until reaching the boiling point, and then the overheating, which gives way to the forming of bubbles, the internal side stays close to the boiling point, although this changes with height, as the hydrostatic pressure varies. In accordance with what has been indicated, if the inner face dries out, the temperature is no longer the boiling point which makes sudden increases of temperature on the inner wall possible, approaching that of the outer wall.

It is important to realise that the L/V ratio is not normally measured and that tube wall drying is difficult to detect. A reboiler could be operating 'normally', that is to say, providing the required flow of vapour and, however, the upper parts of the pipes could be totally or partially dry. In this respect, it is important to maintain a sufficient level of liquid at the base of the column, as this gives way to superior values of the L/V ratio, improving the safety of the operation, even though it is at the cost of lower heat flow in the evaporator.

Another factor to bear in mind is the presence on the shell side of inert gases not condensable at the operating conditions. These gases, as such as air and CO_2, can be introduced at start-up, or accumulate as impurities contained in the steam, and its effect is that of diminishing the heat flow. Another circumstance capable of diminishing the heat transfer is the accumulation of condensate. Whether inert or condensed gases are accumulated, the effect from our point of view is that of reducing the L/V ratio at the tube outlet for a determined level of liquid at the base of the column [1]. Lastly, Viera and colleagues also postulate a mechanism of intrinsically unstable operation through which a change of small magnitude in the velocity of the vapour in the tubes could give way to a process which notably accelerates the drying of the pipe.

Chemical reactions

In the system studied there are two competing reactions of the ethylene oxide: polymerization and disproportionation. The polymerization to ethylene polyoxide could be represented as:

$$n\ C_2H_4O \rightarrow -(CH_2CH_2O)_n-$$

The above reaction is highly exothermic, and has a lower activation energy than the reactions of disproportionation but occurs more slowly.

The disproportionation consists in a chain of oxidation and reductions that produce ethylene and carbon dioxide, as well as hydrogen and/or water:

$$4\ C_2H_4O \rightarrow 3\ C_2H_4 + 2\ CO_2 + 2\ H_2$$
$$5\ C_2H_4O \rightarrow 4\ C_2H_4 + 2\ CO_2 + H_2O + H_2$$
$$6\ C_2H_4O \rightarrow 5\ C_2H_4 + 2\ CO_2 + 2\ H_2O$$

These reactions liberate a quantity of heat similar to that of polymerization, but at high temperatures are much faster, and can form localized hot points.

An examination of the remains found in the ORS boiler tubes showed in its upper part the existence of a layer of polymer containing iron oxide. The iron content varied between 7 and 63%, with lesser quantities of other metals. The polymer was not only capable of providing a physical support to disperse the iron, but turned out to be a bad heat conductor and an efficient means to retain liquid ethylene oxide, providing in this way a pool of reactant for local reactions. Laboratory tests carried out after the accident showed that the iron oxide carried in the polymer was a catalyst for the previous reactions (polymerization takes place essentially at 155°C, but the heat flow is relatively low and can be dissipated; at 200°C the disproportionation takes place with appreciable velocity provoking a rapid increase in temperature), as well as for the self-decomposition in the ethylene oxide vapour phase, which basically gives carbon monoxide and methane. This last reaction, strongly exothermic, takes place without the need of oxygen, but with temperatures in the order of 525°C (in the absence of a catalyst) although authors cite lower temperatures under certain conditions, which would situate the initiation of the decomposition at around 400°C [3, 4]. As implied by the laboratory tests carried out after the accident, the temperature necessary to start the reaction could be reached locally in the dry areas of the tubes, helped by the polymerization and disproportionation reactions previously mentioned.

A1.5   Conclusions

It is very difficult or impossible to avoid small quantities of ethylene oxide polymer in operations which involve liquid ethylene oxide. Moreover, it has been found that the polymer is capable of fixing the metals and metallic ions present in this liquid, therefore the catalyst could always be present. The key to a safe operation consists in maintaining an adequate contact of liquid in all heat transfer surfaces, eliminating the possibility of the vapour reaching temperatures higher than the boiling point of the liquid.

After the accident and with the aim of assuring an adequate L/V ratio, Union Carbide have modified their operations in such a way that the ORS units maintain a level of liquid in the base of the column at least up to the upper part of the tubes, with an automatic trip when the liquid falls below this level. Moreover ways of avoiding the accumulation of condensate and inert gases in the shell will be provided and the temperatures of the heating medium used will be maintained at the lowest possible level.

A.2 Accidents originating with operations of loading and unloading of containers: the accident in the EMPAK facilities, Deer Park, Texas, 1988

On August 28th, 1988, at about 1.00 a.m. an explosion occurred in an ethylene oxide cistern while it was situated in a storage area in the EMPAK company facilities in Deer Park, Texas. Although no lives were lost, the witnesses to the accident were able to see the formation of a fire ball. Ten cisterns close by were damaged, as were several buildings, including one situated at more than 240 m from the accident, that suffered the impact of one of the cistern heads. Pieces of the cistern were found at distances of more than 750 m from the place of the accident. The following description of the accident is taken from the work of Vanderwater [3].

A2.1 Installation characteristics

The internal dimensions of the cistern were 15 m in length and 2.9 m in diameter. The dome of the container had the form of a disc and had three valves: two of them led to drip legs for loading and unloading of liquid, and the third was a nitrogen inlet. There was, logically, a pressure relief valve, and also a level measuring system, although this could only indicate filling levels of 50% or more.

A2.2 Chronology

The cistern in question had been returned to Shell Chemical by a client who found that the ethylene oxide was out of specifications due to the colour. The cistern was refilled and when sampled it was found that the ethylene oxide was still outside specifications because of the colour, so it was decided that the cistern should be cleaned. The ethylene oxide was unloaded and the cistern sent to EMPAK. Routine procedures were used for the unloading, admitting nitrogen to the cistern as the ethylene oxide was displaced. The procedure continued until the flow meter and loss of suction of the pump indicated that all the ethylene oxide had been discharged.

The cleaning began at EMPAK's installations at 7.30 a.m. on the 27th of August. The initial manometer pressure in the cistern was of about 69 kPa. A 1" pipe was connected to the purge valve of the cistern with the aim of sending gases to a container with a caustic absorber and from there to an incinerator. To displace the gases another 1.5" pipe was connected through which water was introduced into the cistern, until the tank was full. Then the vent line was disconnected at which time the worker noticed an unusual smell, so it was decided to stop the procedure and to contact Shell; the tank valves were closed, the pipes were disconnected and the cistern was removed to the storage area at about 11 a.m. Nothing else happened until the explosion took place, about thirteen hours later.

A2.3   Causes of the accident

The cistern contained about 13 000 kg of ethylene oxide when unloading began. After the accident, a review of the speed of the cistern unloading and from the times involved suggest that the flow interruption during the unloading of the ethylene oxide happened before all of the liquid had been unloaded. Approximately 1/6 of the tank's volume was full of liquid ethylene oxide then, the rest corresponding to brine introduced to remove the gases. The causes of the premature interruption of the liquid unloading are unknown, as examination of the remains of the cistern and specifically of the tank dome and the drip legs did not reveal defects which could have been the direct cause of this type of failure.

The circumstances of the accident clearly point to an explosion caused by a chemical reaction in which the remaining ethylene oxide had been involved. Ethylene oxide is a very reactive material capable of multiple reactions among which are:

1. Reaction with water thus forming glycols, a reaction catalysed by acids, bases and some salts.
2. Reaction with water in the presence of chloride ions to form chlorhydrin.
3. Polymerization to form polyethylene oxide, catalysed by acids, bases and some salts.
4. Decomposition, at higher temperatures, essentially producing methane and carbon monoxide.
5. Combustion in the presence of air or oxygen, to carbon dioxide and water.

It seems reasonable that as ethylene oxide and water were present in the cistern a reaction between these occurred, and in fact from the analysis of the remains of liquid found at the scene of the accident, it is deducible that the first two reactions took place. However, estimations of the increase in temperature which could have taken place because of these reactions in the conditions of the accident do not reach sufficient value as to provoke an explosion [3]. Therefore other possibilities were considered, among them that a stratification of the liquid ethylene oxide and the brine occurred, caused by the difference in density, reducing the mixture between both layers.

To corroborate with this hypothesis experiments were carried out on a scale model of the installation made of transparent material, using coloured liquids of similar densities and viscosity and recording the system evolution on video. In all cases an important proportion of the initial liquid remained unmixed with the brine when the filling of the tank was completed, as shown in Figure A1. The existence of stratification in the tank opened new perspectives to explain the causes of the accident: in the interfacial mixing area below the layer of liquid ethylene oxide and the water, the reaction between the ethylene oxide and the water takes place. The heat generated heats the tank walls until they reach a temperature capable of initiating the decomposition of the ethylene oxide which is found stratified in the upper part of the tank. The decomposition generates sufficient overpressure to provoke the rupture of the container.

Calculations were made of the evolution of the reaction between the water and the ethylene oxide in different scenarios. The reactions proceed slowly at the beginning but accelerate considerably as the temperature increases. In the case considered, other accelerating factors would be the presence of chlorides and a pH between 8 and 9. On the other hand, as the tank is full of liquid there is almost no vapour formation, therefore latent heat cannot be eliminated in this way. The estimations carried out, assuming adiabatic conditions and an initial temperature of 27°C showed that the time available for the reaction was sufficient to provoke the required increase in temperature. Other circumstantial evidence came from the examination of the remains of the cistern where the longitudinal rupture lines corresponded approximately to the position where the interface between the layers of stratified liquid was expected.

A2.4   Safety considerations and conclusions

Ethylene oxide is a dangerous material, which has given rise to numerous accidents resulting in fires or explosions. Its aqueous solutions are flammable even in low concentrations [4], with flash points of 31°C at 1% concentration weight, and −2°C at 5%. The flammability limits in air go from 2.6 to 100%, and the explosion can generate very important overpressures, in the order of 10 or 20 times the initial pressure, depending on whether there is only vapour phase or liquid and vapour phases present. In the absence of oxygen its decomposition could begin

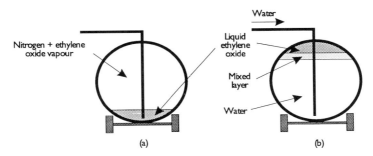

Figure A1   (a) Initial and (b) final situations in the loading operation which gave rise to the Deer Park accident.

from about 400°C, essentially providing CO and CH_4, with smaller quantities of C_2H_6, C_2H_4, H_2, C and CH_3CHO. The presence of water is an additional hazard factor, as it can cause an exothermic reaction with significant velocity in mild conditions. It is also possible that the polymerization of the ethylene oxide takes place in mild conditions, especially if in the medium there exist traces of polymerization initiators (e.g. amines). In this case a runaway reaction can be expected, which is able to cause the rupture of the container. In this respect it is worthwhile pointing out that steel containers are only adequate for ethylene oxide if precautions are taken to avoid the formation of oxides, which can act as catalysts to polymerization.

A fundamental recommendation, following the analysis of the accident, is the need to verify that the quantity of ethylene oxide (or any other material in similar operations) to be discharged is consistent with the quantity previously estimated, and to confirm this information by weighing the cistern before and after the unloading. However, the simple difference in weights is not sufficient to guarantee the complete elimination of the liquid, so it is convenient to purge the vapour space with nitrogen after the unloading, promoting in this way the evaporation of the residual liquid. In cistern cleaning operations the possibility of stratification of the liquid with the cleaning water has to be borne in mind, even if both are completely miscible. Finally the team in charge of the cleaning should carry out an independent test of the tank's emptiness before beginning operations, not taking for granted that the unloading has been adequately completed.

A3   Destruction of process and storage containers by vacuum: various cases

The dangers of creating vacuum in process containers or storage tanks have already been explained in Chapter 7. The possibilities of creating a vacuum are often less obvious than those of overpressure, and moreover, the vacuum necessary is of small magnitude (atmospheric storage tanks are designed for a vacuum in the order of 0.006 – 0.007 gauge atmospheres [5]), which makes the collapse of tanks and containers a relatively frequent phenomenon within the panorama of relevant industrial accidents. With the objective of illustrating some of the circumstances which can give way to effects of this type, three cases of equipment destruction by vacuum creation are detailed next.

A3.1   Case 1:

Installation characteristics

Sanders [6] describes the collapse of a stripping column during the start-up operations of a plant in the Caribbean. The column had a total height of 25 m, with two sections of different diameters: the inferior, of 2.7 m diameter, up to a height of 4.6 m, and the upper part of 1.7 m in diameter. The container was designed for operation at about 5 psig (0.34 gauge atmospheres) and to resist up to 25 psig (1.7 gauge atmospheres).

The gas outlet in the stripping column constituted the feed of a re-absorption column, where the pressure/vacuum vent was located. In this way, the only vent system for both columns (stripping and re-absorption) was in the re-absorption column.

Chronology of the accident

The column had been recently installed and was in the preparation phase to enter service. As part of the trials, it was decided to proceed with a simulation of the operation of the column by circulating water in the system.

A few hours after starting the system, the workers observed that water was coming out of the re-absorber vent. This meant that both the stripping and the re-absorber columns were full of water, due, as was determined later, to a leak through a water valve in the upper part of the column. It is important to note that this prevented the possibility of venting the stripping column through the line which joined both columns, because this line, through which in normal operating circumstances gas should circulate, was full of water.

The plant operators decided to empty out the water by opening the valves in the suction line of the liquid pumps in the base of both columns. Obviously, doing this without any possibility of venting in one of the columns tends to create a partial vacuum in that column. The situation got worse when the pump at the base of the stripping column was started and increased the partial vacuum. Ten minutes after the pump was started a witness observed that the column started to deviate from the vertical, bending at the point where the change in diameter was located. The column inclined 45° collapsing rapidly at this moment. Fortunately, the direction of the fall avoided impact of the column with other important equipment and there were no victims. Neither were there fires nor significant emissions of dangerous products, as the column had not entered into operation and it was full of water.

Safety considerations and conclusions

After the incident vacuum breakers were installed in all the process vessels (except in those designed to withstand total vacuum). Moreover the vent line of the stripping column was changed.

A3.2   Case 2:

Installation characteristics

Also in Sanders' work [6] the case of an oil refinery in the United States, which installed a new unit to increase production of petroleum coke through a process of delayed coking, is described. The unit had four coker drums operating in couples. In these units the material, previously heated, undergoes thermal decomposition, the gas by-products leaving the drum by the head while the coke remains in it.

When the drum is full, the feed is diverted to another waiting drum. This is why these units work in couples, as indicated in Figure A2. The coke drums outlined in Figure A2 were 32 m high and 8.2 m in diameter in the cylinder. They were designed for an internal pressure of 55 psig (3.7 gauge atmospheres) but not for vacuum, in spite of the considerable thickness of the wall, which reached 21 mm in the lower part. The pressure indicators installed in the unit were graduated at 0 to 60 psig, and did not indicate, therefore, pressures below atmospheric.

Chronology of the accident

Before starting the operation of the new unit a test with 50 psig steam was carried out, with the objective of verifying the existence of leaks and of removing the air from the inside of the drums. To vent the steam to the atmosphere, an 8" line was temporarily installed on the original line of 24 inches. However, as indicated in the figure, the design of the conduction created a U-shaped section where condensed steam could accumulate. The steam flow in unit B was started, venting through the 8" line. Two days later the steam flow to Unit B was cut off and steam to Unit A was started. Two days later Drum B collapsed, due to implosion, destroying the structure and leaving no possibility of repair. There were no victims or other consequences.

Safety considerations and conclusions

The accident happened because steam from Drum A condensed in the vent line of Drum B. As this steam cooled a partial vacuum was created which could not be relieved through the vent line, thus provoking the collapse of the unit. After the accident the design of the vent line was modified to eliminate the possibility of liquid accumulating in it. Also, the installation of a low pressure alarm to alert the operator in the case of a partial vacuum developing in the interior of the coker drums was recommended.

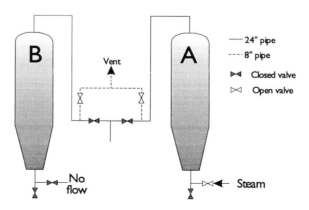

Figure A2   Coke drums A and B during pre-start-up.

A3.3   Case 3:

Installation characteristics

In this case the affected container was a stainless steel atmospheric tank, 4 m high, with flat bottom and elliptical superior head which received process water, irregularly produced in a washing operation, acting as a buffer and heater before returning it to another process point. At about 300 mm from the upper head there was an overflow outlet connected to a vertical pipe which nearly reached the floor of the plant. The inlet pipe entered the tank through the top head. There was a re-circulation with a pump and a heat exchanger.

Chronology of the accident

While the tank was being filled (apparently with a higher flow than normal), an operator observed that the level (measured by a differential pressure sensor) had risen to values higher than normal. When verifying on site the condition of the tank he observed that it was overflowing through the vertical pipe. The inlet valve was closed immediately, after which a sudden fall in the level was recorded, due to the vacuum created. As the tank had been overflowing at a rate equal to that of intake and, therefore, the overflow pipe was completely full of liquid, the act of closing the inlet water valve left the tank isolated and subject to the vacuum caused by the water column (–0.4 gauge bars) an unbearable circumstance for the tank and more than enough to cause the damage which is shown in Figures A3 and A4.

Safety considerations and conclusions

The original design did not take into consideration the necessity of protection against vacuum, as the tank was vented to the atmosphere through the overflow. If in the design stage a HAZOP had been carried out, the application of the guide word 'more' to the water inlet line would have detected the possibility that the tank would be completely filled, creating a column of water in the overflow line. This would have meant the identification of the risk allowing the designer to take the necessary preventive measures.

After the accident the design criteria to be applied to containers which could be subject to vacuum were reconsidered, including the necessity of installing vacuum relief valves in these cases.

A4   Unconfined vapour cloud explosions. The case of the Nypro plant at Flixborough (United Kingdom) [7, 8]

A4.1   Installation characteristics

The explosion happened in the reaction section of the caprolactam production plant, one of the raw materials in the production of nylon 66. The process presented the innovation of carrying out the first stage, the production of cyclohexanone directly from the oxidation of cyclohexane instead of using the classical way

Figure A3 Detail of the tank described in Case 3.

Figure A4 Detail of the tank described in Case 3.

based on the hydrogenation of phenol. The process consisted of injecting air into the liquid cyclohexane in the presence of a catalyst in a battery of six agitated reactors in series, where the flow from one reactor to the next was by gravity. The reaction is exothermic and took place at a gauge pressure of 8.8 kg/cm² (kg/cm²g) and 155°C, and reached a conversion of approximately 6%. The temperature was maintained by eliminating heat through the evaporation of part of the circulating cyclohexane in each reactor, together with the nitrogen from the air and some non-reacted oxygen. An explosive atmosphere in the reactors was avoided by injecting nitrogen coming from liquid nitrogen tanks. There was a safety system which cut off the air injection and purged the reactors with nitrogen when too high a concentration of oxygen was detected or too low a level in the liquid nitrogen tanks. This system could be blocked manually if the nitrogen purging time was set at zero.

The feed was a mixture of fresh cyclohexane and recycled product from the separation section, where the unreacted cyclohexane was separated by distillation from the products of the reaction (cyclohexane and cyclohexanol), which were converted in another section to caprolactam.

Two months before the accident happened a leak was detected in reactor 5, which later became a fissure almost 2 m long. The seriousness of the situation led to the decision to remove reactor 5 and substitute it with a bypass pipe different in size and design (Figure A5) than the expansion joints connecting the other reactors (designed to absorb the expansions and contractions produced in the shut-downs and start-ups of the plant). In January 1974 the agitator of reactor 4 had been removed because of a failure. In which ever way it had been working before, for several months, without agitators owing to electrical restrictions, and there having been no problems, it was not only considered unnecessary to replace it, but a study was started to remove all of them.

A4.2   Chronology of the accident

A leak located the day before provoked a stoppage for maintenance, after which, on the morning of Saturday 1st June 1974, the plant start-up was initiated. When the temperature in the reactors was still low (110°C in reactor 1) a pressure of 8.5 kg/cm²g was detected, higher than normal for these conditions. When normal

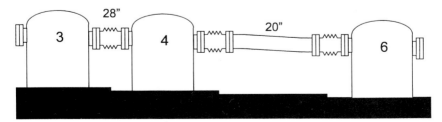

Figure A5   Oxidation reactors in the Nypro facilities showing the bypass that replaced reactor 5.

temperatures were reached, the pressure was about 9.2 kg/cm²g. In these circumstances it would have been normal to reduce the pressure by venting part of the gas of the reactors to the flare. However, this manoeuvre meant the loss of an important quantity of nitrogen, of which there was hardly enough for the start-up. It was not possible to obtain more nitrogen until midnight, so it was decided to save as much nitrogen as possible by trying not to vent.

In the evening there was an escape of about 40 Tm of cyclohexane due to the rupture of the temporary connection between reactors 4 and 6. Almost immediately the vapour cloud found a source of ignition and at 16:53 an unconfined vapour cloud explosion occurred which completely destroyed the plant, causing the collapse of the control room and the death of all 18 occupants. The total number of deaths was 28, with 36 seriously injured and an unknown number of slightly injured people, estimated at several hundred. About 1800 houses and 167 commercial establishments suffered damage of different consideration. It is calculated that the TNT equivalent of the explosion was between 15 and 45 Tm, the largest explosion registered in the UK in times of peace.

A4.3   Analysis of the causes of the accident

The causes of the rupture of the temporary connection are not completely clear, and there are three main hypotheses:

Hypothesis of the rupture of the 20" pipe [7]

This is the hypothesis put forward by the official investigation committee in their report. The reasons for the increase in pressure which caused the rupture were not sufficiently established, although some studies showed that, although to approach a 100% probability of rupture a pressure of about 10.5 kg/cm²g was necessary, at a pressure of 9.2 kg/cm²g there already existed a significant probability, although substantially lower. (For a 50% probability the required pressure was 9.8 kg/cm²g).

The rupture was produced in one of the welded oblique connections, as a consequence of the shear stress they had to withstand, caused by the configuration of the pipe. A formal project was not done and neither was a modification plan drawn up, except for a scheme done with chalk on the floor of the factory. The plant engineer had left the company several months before and had not been replaced, so that the mechanical calculations for the connection they were going to install were also not done, and it was considered as a straight length of pipe. This connection did not fulfil the applicable design standards.

Some of the alternatives suggested [8] as causes of the increase in pressure are: ingress of nitrogen under pressure due to an instrument failure, temperature increase caused by a control failure in a boiler in the separation section, explosion of peroxides formed in the process, leak in a tube of the same boiler (causing an increase in temperature and entrance of water into the system) and an explosion in the reactors, due to an excessive concentration of oxygen.

Hypothesis of the 8" pipe [7]

This hypothesis assumes that the cause of the rupture in the 20" line was a fire which was started by a leak in an 8" valve in a line near the reactors and which originated an explosion in one of the fans located close by. The official investigation committee dedicated nearly all of their time to discard this possibility.

Hypothesis of the overheated water

This hypothesis was hardly touched by the investigation committee and has been defended by King [8]. It proposes that the cause was the presence of water in the reactors, either due to a leak in an heat exchanger, or as the remains of the washing out of the cyclohexane with water during the plant shut-down the day before. The lack of an agitator in reactor 4 allowed the decanting of the water, forming a third liquid phase, which boiled suddenly causing a rapid pressure increase of more than 1 kg/cm^2, causing the rupture of the bypass.

Underlining the direct causes of the accident lies a group of serious defects in the safety management system of the company, which are worth while analysing:

Organization

In the Nypro plant not only was there no safety management system for aspects as elementary as control of modifications, but the plant also lacked sufficient qualified personnel to supervise its operations. Specifically, there was no mechanical engineer, as the previous one had left the plant and had not been replaced. The role of the person responsible for safety was not defined.

Production taking priority over safety

The changes which caused the accident were introduced with urgency because of the need to keep the plant running because of sales pressure. If the causes of the fissure in reactor 5 had been seriously analysed (it was attributed to corrosion by nitrates caused over a period of several weeks in which the said reactor worked with the fire sprinkler system running to prevent its being damaged by a leaking valve located above it), the plant would have been stopped to review the condition of all of the reactors. A later study showed that the expansion joints had been designed to support four times less pressure than that which they were subjected to and therefore they were transmitting this pressure to the reactors with a similar effect to that of a nail on a tyre. This circumstance was common to all the reactors so that all of them should have been tested to check the magnitude of the damage and take necessary measures.

The added circumstance of the plant start-up without knowing if there was a sufficient quantity of nitrogen to operate outside the inflammability range indicates the low priority given to safety by the Nypro management.

Design and control of modifications

The urgent introduction of two modifications, the sub-standard bypass and the removal of the reactor 4 agitator, to maintain production, were the direct causes of the accident. There was no system to control modifications, or qualified personnel to review the designs. There already existed important failures in the original design of the plant, for example, the already mentioned mistake in the design of the expansion joints, which connected the reactors. The large amount of flammable material handled in the plant also had a major influence on the magnitude of the accident.

A5   Toxic emissions caused by runaway reactions: the accidents of Seveso (Italy) and Bhopal (India)

These two accidents are among the most serious and widely known in the history of the industry, because of the serious consequences they caused to the population of large areas adjacent to the complexes where they occurred. The European Union, with the release of the so-called 'Seveso Directive' (1982), and the United States of America with SARA III, legislation which has already been dealt with in previous chapters, have tried to reduce the possibility of similar catastrophes from happening.

A5.1   The dioxin emission at Seveso (1976) [7, 8]

Installation characteristics

The plant of Icmesa Chemical Company in Seveso, a town with a population of about 17 000 inhabitants near Milan, was dedicated to the production of weed killers and insecticides, a process in which trichlorophenol (TCP) is used as an intermediate product. The production had increased significantly in the past years, as some plants in other countries had closed due to safety and hygiene problems with the products involved.

The TCP was produced in an agitated reactor from tetrachlorobenzene and excess caustic soda, to produce at first sodium trichlorophenate. The reaction is carried out in the presence of a solvent and at about 160–200°C. During the strongly exothermic reaction, the heat generated is removed by evaporation of the solvent, which is normally condensed and returned to the reactor. Towards the end of the reaction the temperature is raised to increase the conversion. Once the reaction is considered finished part of the solvent was distilled for re-use, and in the same reactor water and hydrochloric acid were added to obtain the TCP. The working pressure depends on the volatility of the solvent used. In the reactor at Seveso they worked at about 160°C at atmospheric pressure, except during distillation of the solvent (a mixture of ethylene glycol and xylene), which was carried out in a vacuum. The reactor was protected by a rupture disc set to open at a pressure of 3.6 bars gauge, and leading directly to the atmosphere. The heating of the mixture was done using a jacket, heated with medium pressure steam with a maximum temperature of 190–200°C.

As a by-product of the reaction, in normal conditions, 2, 3, 7, 8-tetrachlorodibenzoparadioxin (TCDD), commonly known as dioxin, is produced in quantities of about 25 ppm. This reaction is also exothermic and the quantity of TCDD produced increases with the temperature. Dioxin is insoluble in water, very stable and lethal in doses above 10^{-9} times the body weight. This makes it one of the most toxic products known. It causes damage in the liver, kidney and to the foetus, and it can cause cancer and mutations. Its action during a pregnancy is especially harmful. In slight intoxication it produces chloracne.

The Imecsa plant worked continuously by shifts 5 days per week. In principle each day a new reaction was started at 6 a.m., when a new shift entered, and which was terminated by the night shift. However, due to slight problems, frequently, throughout the week, delays in the start time of the reaction occurred. In these cases the mixture, already reacted, was usually left in the reactor on Friday and during the weekend, without the addition of water and acid. The first shift on Monday had to heat the mixture, which had solidified (the melting point of TCP is 68°C), until the agitator could be started and the batch terminated. To avoid losing time by having to re-heat the reacted mixture, instructions were given to the workers to shut off the steam in these circumstances, but not to open the refrigeration water, so that the reactor would cool more slowly and on the Monday the reaction could be completed more quickly, with the consequent saving of time.

Chronology of the accident

On Friday, 9th July, 1976, a reaction was started in the afternoon. The night shift only had time to begin distillation of the solvent, so this operation was left unfinished, shutting off the steam and stopping the agitator. At 12.37 the following morning an exothermic runaway reaction produced an increase in pressure in the reactor causing the opening of the rupture disc and the emission of a toxic cloud which was estimated to contain a concentration of about 3500 ppm of TCDD, and a total quantity of TCDD present in the cloud of between 0.5 and 2 kg. To reach the operating pressure of the rupture disc a temperature of 400°C would have normally been needed.

The emission of the cloud was followed by an immediate reaction from the plant personnel on the premises. They tried to warn the authorities of the danger of the escape, but this was impossible as it was a weekend and they could not be contacted. During the following days, communication between the authorities and the company was very deficient; dead animals and dried vegetation were detected. The first measures were taken four days later when the consequences of the escape appeared in a boy. The following day a state of emergency was declared and an area of 5 km^2 was declared to be contaminated. The first group of citizens were not evacuated until July 27th. Later it was discovered that the area which had been affected was more than 5 times greater. The total number of people affected was about 2000. The Italian government had to ask for the help of international experts for medical treatment of the intoxication and cleaning of the contaminated area.

Analysis of the causes of the accident

The accident was caused by a runaway exothermic reaction, due to leaving the reactor without refrigeration, with no agitation and with a mixture which was probably still reacting slowly. It was believed before the accident that the initial temperature of the exothermic reaction was 230°C, but in tests carried out later with more sensitive equipment, it was discovered that the reaction already started with a moderate activity at 180°C. Kletz [9] suggests that the existence of a hot zone in the upper part of the reactor wall, just above the level of liquid, in contact with vapour phase and, therefore, with worse heat transfer, could have been the cause of the initiation of the runaway reaction, as the working temperature was about 160°C.

Other theories are postulated by different authors, which in general appear to be less probable and which range from the unintended addition of hydrochloric acid to the reactor on the Saturday morning, to other reactions, such as the exothermic condensation of two molecules of ethylene glycol or the reaction with the oxygen in the air of some component present in the reacting mixture.

In any case there are three main causes at the root of the problem and its consequences.

- *Leaving a reactive and dangerous mixture over a weekend without vigilance or safety measures* is assuming an unnecessary risk, due to in a large part to the shift work system used in the plant. Nearly every weekend the reaction was left interrupted, although distillation and the addition of water and hydrochloric acid were usually completed.
- *Allowing an emergency relief device to discharge directly into the atmosphere.* It seems clear that the rupture disc was not designed for a runaway reaction scenario, as in these circumstances it should have been connected to a treatment system to avoid the emission of toxic substances. Its high set pressure favoured the dispersion of the emission to greater distances and permitted a large rise in temperature which increased the production of TCDD.
- *The lack of an organization to react in the case of an emergency and an external emergency plan* were the cause of the important delay in recognizing the seriousness of the accident and the evacuation of the people affected. There had been previous experience in other production plants [7] of the seriousness which accidents involving TCDD could reach.

The nature of these causes is such that for certain, without any doubt, the accident could have been avoided, or at least the consequences reduced, through an analysis of the risks and a more careful design and operation of the reactor.

A5.2   *The escape of methyl isocyanate (MIC) at Bhopal (1984) [8]*

The Bhopal case has been the biggest industrial disaster in the world, with about 2500 deaths and between 100 000 and 250 000 injured and affected people. Its origin was an escape of 26 MT of MIC, a highly toxic product which was produced in a Union Carbide plant situated in an urban area, surrounded by houses and businesses, in Bhopal, a city with 700 000 inhabitants.

Product and installation characteristics

MIC is an intermediate product commonly used in the production of insecticides. Its vapours are extremely toxic and principally attack the mucus, eyes and lungs. It is also highly reactive, with a tendency to polymerize in the presence of catalysts such as iron or chorides. These reactions are highly exothermic, and although slow below 20°C, their speed increases with temperature, being capable of producing a runaway reaction with the generation of a great quantity of heat, sufficient to vaporize most of the MIC.

The production of MIC, in which other highly toxic products intervene, was carried out in four stages:

- Phosgene production.
  $$CO + Cl_2 \rightarrow COCl_2$$

- Production of methylcarbamyl chloride from phosgene and methylamine by reaction in chloroform:
  $$COCl_2 + CH_3NH_2 \rightarrow CH_3NHCOCl + HCl$$

- Pyrolysis to obtain MIC:
  $$CH_3NHCOCl \rightarrow CH_3NCO + HCl$$

- Separation by distillation of the MIC from the chloroform and the non-converted reactants.

The storage facilities for the MIC, where the accident happened, consisted of two horizontal cylindrical tanks of 604 stainless steel, with a capacity of 57 m³ each and a design pressure of 2.8 gauge bars at 121°C and full vacuum. The tanks were completely buried and isolated from the outside by a layer of concrete with the objective of protecting them from any impact or fire and also thermally isolating them from the surroundings. In addition there was a cooling system through which the MIC was recycled to maintain its temperature below 0°C and minimize polymerization. For unknown reasons the cooling system was dismantled in June 1984. The tanks had a temperature indicator and a high temperature alarm; a pressure indicator and controller set to maintain it within the 0.14 to 1.7 gauge bar range, introducing nitrogen or venting vapour to the flare or to the gas scrubber; and a level indicator with high and low level alarms.

The emergency relief system consisted of a rupture disc and a safety valve in series, with a set pressure of 2.8 gauge bars. The discharge line lead to a gas scrubber 33 m high, where a solution of caustic soda was recirculated as an absorbing agent. The system could neutralize 4 MT of MIC in the first half an hour of operation, reducing its capacity to some 2 MT afterwards, because there was no refrigeration system for the soda solution. There was also the possibility of leading the gases to the plant flare system, which was 30 m high.

The premises of the complex were surrounded by a group of water cannons which permitted creating a curtain of water of about 12–15 m high to absorb possible vapour escapes.

In the years the plant had functioned various serious accidents had taken place, there was even a case of death. There had been no consequences outside of the plant, but the morale of the personnel was low and the local newspapers had spread the information about the danger that the Union Carbide plant meant for the population. In addition, sales of final products in which MIC was used had dropped due to its high toxicity and the consequent risk of its use. There were even rumours that Union Carbide was considering dismantling the plant.

In 1982 an audit carried out by safety experts of the Union Carbide Corporation detected important corrosion problems and warned of the possibilities of a gas escape. Also in September 1984 a copy of the audit to which the MIC plant at Institute (USA) had been subjected was received in Bhopal, describing the problems detected and stating a special preoccupation for the lack of sufficient safety measures guaranteeing adequate action in the case of a runaway reaction in the MIC storage tanks. Some months after the accident in Bhopal there was an escape of MIC at the Union Carbide plant in Institute.

Chronology of the accident

The last batch of MIC produced before the accident, on 22nd October was out of specification (it contained 15% of chloroform, the established limit was 0.5%), but it was not stored in the tank which existed for products out of specifications, instead it was placed in one of the already described final product tanks, number 610.

A day later, as the plant was stopped and apparently to reduce expenses, it was decided to stop the scrubber. Also at this time the flare was taken out of service, as corrosion was detected in numerous points of the header.

At 23.00 on Sunday 2nd December 1984 a pressure above normal in tank 601 was detected. 75 minutes later the pressure was off scale (more than 3.8 gauge bars). The tank insulation was cracking due to the high interior temperature and the safety valve opened, producing an emission of MIC. The operators started the scrubber and at 1.00 gave the alarm. Immediately an attempt was made to attack the escape with the water canons, which did not reach the gases leaving the scrubber. The escape lasted until the safety valve reclosed, at 2.00.

It was determined after the accident that the pressure inside tank 601 reached 12.2 gauge bars at a temperature of 200°C. Despite these conditions, much more severe than those it was designed for, the tank surprisingly held, avoiding an even bigger disaster. After the accident it was proved that the total quantity liberated had been 36 MT, of which 25 MT were MIC and the rest products of polymerization.

Causes of the accident

The immediate causes of the start of the runaway reaction in tank 610 are not very clear, although what is clear is that the lack of refrigeration was the key. There are two principal hypotheses:

The spontaneous reaction starting at a very reduced speed. The good isolation of the tank had made the generated heat accumulate slowly in the product, increasing the speed of the reaction. Once ambient temperature of 15–20°C was reached, essentially by transmission of heat from the floor, the reaction could have accelerated by itself. The main objection to this theory is the 5 months which passed between the disconnection of the refrigeration system and the accident.

The beginning of the reaction by entrance of water, coming from the washing of a pipe, which would have allowed the hydrolysis of a part of the high percentage of chloroform to form hydrochloric acid which acts as a catalyst to the polymerization reaction of MIC. Different theories exist [8] about the evolution of the reaction and the quantity of water needed for its initiation (from 0.5 kg to 500–1000 kg), based on the composition of the solid remains found in tank 610 after the reaction.

This accident gives rise to important questions on the gravity of the problems, at least in the following areas of the company's safety management system:

The procedures for introducing modifications and maintenance. When the accident happened the principal safety systems of the plant were simultaneously out of service: the refrigeration unit, the flare and the scrubber. Apparently, nobody was conscious of the risk that this meant. The dismantling of the refrigeration system was the key to the initiation of the runaway reaction which provoked the accident.

The suitability of the process safety systems. The audits the Bhopal plant had undergone and other similar occurrences indicated that the existing protection systems were not sufficient to lower the risk to an acceptable level. The scrubber unit did not have the necessary capacity to deal with the flow of MIC which could reach it in the case of a runaway reaction and the safety valve was designed for a much lower gas flow than that which could arise in those circumstances, in which, moreover two-phase flow was present.

The siting of the plant and the adequacy of the emergency plan. The plant was built in an area of dense population. The improvisation and lack of resources during the evacuation due to the lack of an emergency plan meant that one of the evacuation roads crossed one of the areas most affected by the toxic cloud, causing the death of many evacuees. The toxic gas alarm sounded an hour after the escape began, when half of the MIC leakage had already taken place, and a few minutes later it was disconnected to avoid alarming the population.

Training and morale of the personnel. The company's situation, with production at a minimum, closing down rumours and frequent serious accidents was the cause of, according to the plant's safety manager, the fact that nobody who could choose another job with better conditions would stay with the company. The Union Carbide specialists recognized the danger that the location of the plant meant, too close to the city, and their morale was very low, because they were conscious of how little they could do if problems arose.

A6   References

A1. Viera, G. A., Simpson, L. L. and Ream, B. C. (1993) Lessons learned from the ethylene oxide explosion at Seadrift, Texas. *Chem. Eng. Prog.*, **89**(8), 66–75.
A2. Costa Novella, E., *et al.* (1986) *Ingeniería Química. Vol. 4. Transmisión del calor.* Alhambra Universidad, Madrid.
A3. Vanderwater, R. G. (1989) Case history of an ethylene tank car explosion. *Chem. Eng. Prog.*, **85**(12), 16–20.
A4. *Ullmann's Encyclopedia of Industrial Chemistry*, Vol. 10 (1987) VCH Publishers, Weinheim.
A5. Kletz, T. A. (1987) *An Engineer's View of Human Error.* Proceedings of the International Symposium on Preventing Major Chemical Accidents (ed. J. L. Woodward), CCPS/AIChE, New York.
A6. Sanders, R. E. (1993) Don't become another victim of vacuum. *Chem. Eng. Prog.*, **89**(9), 54–7.
A7. Lees, F. P. (1980) *Loss Prevention in the Process Industries*, Butterworth-Heinemann, London.
A8. King, R. (1990) *Safety in the Process Industries,* Butterworth-Heinemann, London.
A9. Kletz, T. (1985) *Cheaper, Safer Plants or Wealth and Safety at Work. Notes on Inherently Safer and Simpler Plants*, Institute of Chemical Engineers, Rugby.

Index